W9-BYV-817

Brock/Springer Series in Contemporary Bioscience

Microbial Enzymes in Aquatic Environments

Brock/Springer Series in Contemporary Bioscience

Series Editor: *Thomas D. Brock*
University of Wisconsin–Madison

Tom Fenchel
ECOLOGY OF PROTOZOA:
The Biology of Free-living Phagotrophic Protists

Johanna Döbereiner and Fábio O. Pedrosa
NITROGEN-FIXING BACTERIA IN NONLEGUMINOUS
CROP PLANTS

Tsutomu Hattori
THE VIABLE COUNT: Quantitative and Environmental Aspects

Roman Saliwanchik
PROTECTING BIOTECHNOLOGY INVENTIONS:
A Guide for Scientists

Hans G. Schlegel and Botho Bowien (Editors)
AUTOTROPHIC BACTERIA

Barbara Javor
HYPERSALINE ENVIRONMENTS: Microbiology and Biogeochemistry

Ulrich Sommer (Editor)
PLANKTON ECOLOGY: Succession in Plankton Communities

Stephen R. Rayburn
THE FOUNDATIONS OF LABORATORY SAFETY:
A Guide for the Biomedical Laboratory

Gordon A. McFeters (Editor)
DRINKING WATER MICROBIOLOGY:
Progress and Recent Developments

Mary Helen Briscoe
A RESEARCHER'S GUIDE TO SCIENTIFIC AND
MEDICAL ILLUSTRATIONS

Max M. Tilzer and Colette Serruya (Editors)
LARGE LAKES: Ecological Structure and Function

Jürgen Overbeck and Ryszard J. Chróst (Editors)
AQUATIC MICROBIAL ECOLOGY:
Biochemical and Molecular Approaches

John H. Andrews
COMPARATIVE ECOLOGY OF MICROORGANISMS
AND MACROORGANISMS

Ryszard J. Chróst (Editor)
MICROBIAL ENZYMES IN AQUATIC ENVIRONMENTS

Ryszard J. Chróst
Editor

Microbial Enzymes in Aquatic Environments

With 122 Figures and 46 Tables

Springer-Verlag
New York Berlin Heidelberg London
Paris Tokyo Hong Kong Barcelona

Ryszard J. Chróst
Institute of Microbiology
University of Warsaw
Pl-00-064 Warsaw
Poland

Library of Congress Cataloging-in-Publication Data
Microbial enzymes in aquatic environments/Ryszard J. Chróst, editor.
p. cm — (Brock/Springer series in contemporary bioscience)
Includes bibliographical references.
1. Aquatic microbiology—Congresses. 2. Microbial enzymes—Environmental aspects—Congresses. I. Chróst, Ryszard J.
II. Series.
QR105.M52 1991
576'.192—dc20 90-48880

Printed on acid-free paper.

Production and editorial supervision: Science Tech Publishers.
Printed and bound by Edwards Brothers, Ann Arbor, Michigan.
Printed in the United States of America.

9 8 7 6 5 4 3 2 1

ISBN 0-387-97452-0 Springer-Verlag New York Berlin Heidelberg
ISBN 3-540-97452-0 Springer-Verlag Berlin Heidelberg New York

This volume is dedicated to

Professor Jürgen Overbeck

who, in the early 1960s, was one of the first scientists to examine the ecological importance of microbial enzyme activities in aquatic ecosystems.

Preface

Most organic matter in nature consists of molecules that cannot directly enter cells, because of their polymeric structure, high molecular weights, and large size. The hydrolysis of such polymers is a rate-limiting step in the microbial utilization of organic matter in aquatic environments. Before they can be incorporated into microbial cells, polymeric materials must undergo step-wise degradation by a variety of enzymes, which are located either on the cell surface, or in the periplasmic space (ectoenzymes), or may be particle-bound, or released from the microorganisms into the environment as free, dissolved enzymes (extracellular enzymes). Low-molecular-weight compounds, the products of enzymatic action, can then be taken up by microbial cells to meet energy requirements and to build up biomass. Thus, the enzymatic mobilization and transformation of organic matter is the key process regulating the turnover of organic and inorganic compounds in aquatic environments.

Studies of the enzymatic degradation processes of organic matter are being performed by workers in disciplines as diverse as water and sediment analysis, bacterial and algal aquatic ecophysiology, eutrophication, nutrient cycling, and biogeochemistry, and in both freshwater and marine ecosystems. This diversity has resulted in inadequate communication among the scientists working in this field. Therefore, Professor Jürgen Overbeck, Dr. Uwe Münster, and I felt it was very important to bring workers together for an exchange of ideas and experimental data. Therefore, we organized the First Workshop on Enzymes in Aquatic Environments. This meeting was held from 23 to 27 July 1989, under the auspices of the Max-Planck Gesellschaft zur Förderung der Wissenschaften and the Deutsche Forschungsgemeinschaft, at Ringberg Castle, situated above Lake Tegernsee in the Bavarian Alps (Germany). The aim of the workshop was to discuss problems relevant to the study of microbial enzymes in aquatic ecosystems, such as methods for measuring their activity, regulation of synthesis, distribution patterns in waters and sediments, their role in microbial metabolism, and their ecological significance. The contributions of scientists from 12 different countries yielded much useful information on recent approaches and developments in this field. The workshop also facilitated closer contacts between specialists from different areas of aquatic research, improved our knowledge of environmental processes mediated by microorganisms in aquatic ecosystems, and also contributed to a better understanding among scientists from different parts of the world.

The large number of enzymes considered in this volume indicates the richness of microbial metabolism and its leading role in the decomposition and transformation processes of organic matter in the aquatic environments. The study of microbial enzymes is a relatively new but rapidly developing field of aquatic microbial ecology; it is important from both a practical and an ecological point of view. On one hand, microbial enzymes now are widely used in water biotechnology; on the other hand, their study contributes significantly to our knowledge of the principles of aquatic ecology and microbial processes that affect the whole ecosystem. Aquatic microbial enzymology also represents a border area between different scientific disciplines, such as microbiology, biochemistry, physiology, and ecology, and demonstrates that their integration enhances our understanding of many ecological processes.

The chapters in this volume are based on papers presented at the workshop and represent the state of the art in this field of enzyme studies in both freshwater and marine ecosystems. The first part of the book (Chapters 1 to 6) covers general aspects of extracellular enzymes and ectoenzymes in aquatic environments, such as their storage and distribution, regulation of their synthesis and activity, methodology of enzyme activity measurements in water and sediment, and the enzyme-substrate relationship. The next section (Chapters 7 to 9) deals with the activities of enzymes responsible for degradation of proteinaceous materials. The activity and significance of algal and bacterial phosphohydrolytic enzymes is described in Chapters 10 to 15. The last part of the volume discusses the activity of hydrolytic enzymes (e.g., chitinase, glucosidase, and esterases) in various aquatic habitats.

The purpose of this volume is to provide a range of examples of areas in which "ecological enzymology" are being employed in aquatic research. It is addressed to the students and scientists who are interested in using an enzyme approach to study the ecophysiology of microorganisms, organic matter degradation and transformation processes, and biogeochemical cycles in aquatic ecosystems. In microbial ecology many fascinating and important questions urgently need to be answered and I believe that the enzyme approach presented throughout this volume will be extremely useful for solving them. I hope that this work will draw the attention of microbial ecologists and biochemists to the major role of the enzymatic activity of the microorganisms in aquatic ecosystems. To my knowledge, this is the first book to discuss both cell-bound and extracellular microbial enzymes and their ecological significance in aquatic ecosystems.

I want to express my gratitude to Dr. Uwe Münster for his valuable help in mounting the workshop, and to Karin Schmidt and Cordula Stielau for technical and administrative assistance during the meeting. The financial and technical support of the Max-Planck Institute for Limnology, Department of Microbial Ecology (Plön), during the preparation and editing of this book is also acknowledged. Special thanks go to the contributors for coming in from the field long enough to prepare their manuscripts.

Ryszard J. Chróst

Contents

Contributors

J.W. Ammerman Department of Oceanography, Texas A & M University, College Station, Texas 77843, USA

J. Barfield Department of Biology, University of Alabama, Tuscaloosa, Alabama 35487-03444, USA

T. Berman Kinneret Limnological Laboratory, P.O.B. 345, Tiberias 14102, Israel

G. Billen Université Libré de Bruxelles, Groupe de Microbiologie des Milieux Aquatiques, Campus de la Plaine, CP 221, 1050 Bruxelles, Belgium

P.I. Boon Murray-Darling Freshwater Research Center, P.O. Box 921, Albury, NSW 2640, Australia

R.J. Chróst Institute of Microbiology, University of Warsaw, Nowy Swiat 67, 00-064 Warsaw, Poland

J.B. Cotner, Jr. Great Lakes Environmental Research Laboratory, 2205 Commonwealth Blvd., Ann Arbor, Michigan 48105-1593, USA

D.A. Francko Department of Botany, Miami University, Oxford, Ohio 45056, USA

H.G. Hoppe Institut für Meereskunde, Abt. Marine Mikrobiologie, Düsternbrooker Weg 20, D-2300 Kiel 1, Germany

T.R. Jacobsen Rutgers University, P.O. Box 687, Port Norris, New Jersey 08349, USA

P. Jensen Marinbiologisk Laboratorium Helsingør, Københavns Universitet, Helsingør, Denmark

S.E. Jones School of Biological Sciences, University College of North Wales, Bangor, Gwynedd LL S7, UK

B. Kaplan Kinneret Limnological Laboratory, P.O.B. 345, Tiberias 14102, Israel

C.K. Kim Department of Microbiology, College of Natural Sciences Chungbuk National University, Cheongju, Chungbuk 360-763, South Korea

M. Köster Institut für Meereskunde, Abt. Marine Mikrobiologie, Düsternbrooker Weg 20, D-2300 Kiel 1, Germany

M.A. Lock School of Biological Sciences, University College of North Wales, Bangor, Gwynedd LL S7, UK

J. Marxsen Limnologische Flußstation des Max-Planck Institut für Limnologie, D-6407 Schlitz, Germany

L.A. Meyer-Reil Institut für Meereskunde, Abt. Marine Mikrobiologie, Düsternbrooker Weg 20, D-2300 Kiel 1, Germany

U. Münster Max-Planck Institut für Limnologie, Abt. Mikrobenökologie, Postfach 165, D-2320 Plön, Germany

H. Olsson Swedish Environmental Protection Agency, Freshwater Section, P.O. Box 7050, S-750 07 Uppsala, Sweden

J. Overbeck Max-Planck Institut für Limnologie, Abt. Mikrobenökologie, Postfach 165, D-2320 Plön, Germany

H. Rai Max-Planck Institut für Limnologie, Abt. Ökophysiologie, Postfach 165, D-2320 Plön, Germany

R.A. Smucker P.O. Box 1269, Solomons, Maryland 20688, USA

R.G. Wetzel Department of Biology, The University of Alabama, Tuscaloosa, Alabama 35487-0344, USA

K.P. Witzel Max-Planck Institut für Limnologie, Abt. Mikrobenökologie, Postfach 165, D-2320 Plön, Germany

D. Wynne Kinneret Limnological Laboratory, P.O.B. 345, Tiberias 14102, Israel

1

Early Studies on Ecto- and Extracellular Enzymes in Aquatic Environments

Jürgen Overbeck

The idea that free enzymes could act as "catalysts" in freshwater and marine environments is not new. Vallentyne (1957) in his review on the molecular nature of organic matter in lakes and oceans pointed out "some provocative and controversial evidence" that free enzymes may be important for chemical transformations in both lakes and oceans, and he listed some data from early studies. Fermi (1906) was the first to observe proteolytic enzymes in stagnant waters of pools. Harvey (1925) called attention to the possible presence of catalases and oxidases in sea water. Kreps (1934) assumed that extracellular enzymes originating from bacteria and from marine plants and animals may be present in sea water. In his filtration experiments he reported that bacteria were retained on a Seitz "ultrafilter" and that "the enzymes liberated could pass through and were able to continue in the filtered water their specific catalytic action of nitrate reduction, ammonia oxidation, etc." In 1938 the biochemist Maximilian Steiner reported that "intrabiocönotischer Phosphatkreislauf" (Elster and Einsele, 1937; Ohle, 1952), i.e., repeated incorporation of molecules of phosphorus by epilimnetic phytoplankton during one production period, would not be possible without the active participation of phosphatases. He showed in filtration experiments that phosphatases were excreted by zooplankton, and the evidence for enzyme activity was demonstrated by the cleavage of organic phosphorus compounds.

Thirty years ago, during the fourteenth International Association of Theoretical and Applied Limnology (Societas Internationalis Limnologiae; SIL) Congress in Vienna in 1959, I presented data on free dissolved enzymes in lake water for the first time. At that time I was interested in the phosphate metabolism of phytoplankton. The object of the study was a small artificial pond in the park of Sanssouci near Berlin. The pond was filled once a week with water from the highly eutrophicated Havel river; therefore the content of phosphorus was high, about 600 μg of phosphorus (P) per liter. However, only a very small part of the total P was present as inorganic phosphate (on average 13 μg $P\text{-}PO_4^{3-}\ l^{-1}$ during the 4-year study). From this observation arose the question of how the dense phytoplankton communities could use different phosphorus compounds other than inorganic phosphate.

Scenedesmus quadricauda was the main alga of summer phytoplankton in the experimental pond (10,000–80,000 coenobium per ml). For the study of phosphate metabolism, an axenic culture of *Scenedesmus quadricauda*, strain 276-4E from the Pringsheim algal collection in Göttingen (originally isolated by W. Rodhe from Lake Erken, Sweden), was used (Overbeck, 1961). Measurements of uptake kinetics under sterile conditions showed very clearly, that *Scenedesmus* used only the inorganic phosphate. Various organic phosphorus compounds (Na-glycerophosphate, Na-nucleinate, Ca-phytate, and Na-pyrophosphate) were not taken up by *Scenedesmus* in an axenic culture. Similar results were also obtained with samples of lake water which were sterilized by filtration. However, when a natural water sample without filtration was used, the uptake rates amounted for 70–100 fg P $cell^{-1}$ and agreed well with the results of Rodhe's studies on the physiology of phytoplankton (Rodhe, 1948).

For a better understanding of these results, which indicated the possible participation of free phosphohydrolytic enzymes outside the *Scenedesmus* cells, I began to study the phosphatases in this alga. Homogenized cells from axenic batch cultures of *Scenedesmus* were extracted with 0.25 M saccharose solution (16 hours, 4°C) and the crude enzyme preparation was precipitated with acetone (Overbeck, 1962). This enzyme powder, after freeze drying, was used for experiments. It soon became clear that the alga possessed two active types of phosphatases, an acid and an alkaline phosphatase, that instantaneously hydrolyzed organic P-compounds. The question then arose of how *Scenedesmus* in the pond could utilize the organic phosphorus compounds whereas in axenic cultures and in bacteria-free lake water, this alga was not able to do so.

I then added an *Escherichia coli* suspension to an axenic culture of *Scenedesmus* grown on medium supplied with sodium glycerophosphate. The results were surprising: the organic P-compound was hydrolyzed immediately by bacterial phosphatases and the phosphate was taken up by algae. Apparently phosphatases had been excreted by *E. coli*. To prove this assumption I filtered *E. coli* suspension and measured the phosphatase activity. I found that free phosphatases were present in the culture solution of *E. coli* as well as in the water of the experimental pond. Measurements of free phosphatases in lake water in that time applied the following assay (Overbeck and Babenzien, 1963, 1964): 1 ml 0.5% Na-glycerophosphate, 1 ml buffer (pH 6.4, 8.2, and 9.6), and 1 ml lake water. The samples were incubated for 16 h at 32°C.

The activity of free enzymes in the experimental pond (besides phosphatases, I studied amylase and saccharase) was closely related to the standing crop of phytoplankton, i.e., high abundance of phytoplankton was proportional to increased enzyme activity. High enzyme activity was observed in winter (December-January) because probably the enzyme protein remained stable for a longer time at low water temperatures. Much higher activity of phosphatases was found in the sediments of the pond where an enrichment of the enzyme by precipitation with ammonium sulfate was possible. On the basis of these studies the following model of phosphorus cycling within the epilimnetic water zone, mediated by free enzymes, was proposed (Figure 1.1; Overbeck 1967): Bacteria generally take up inorganic and organic phosphorus using their active transport systems. A variety of bacterial species can store polyphosphates inside their cells. Free, extracellular phosphatases excreted by bacteria hydrolyze dissolved organic phosphorus compounds which cannot be directly incorporated into the algal cell in the absence of cell-bound phosphatases, as in *Scenedesmus*. Otherwise, organic phosphorus compounds, e.g., phosphate esters such as glycero-6-phosphate, and fructose-1,6-diphosphate, have to be hydrolyzed by cell-surface phosphatases, which are widely distributed in various bacteria, algae, and fungi.

A great improvement in the accuracy of the measurement of phosphatase activity was the application of p-nitrophenyl phosphate as substrate instead of the previously used Na-glycerophosphate. W. Reichardt demonstrated in my laboratory that the distribution pattern of phosphatase activity in Plußsee was well correlated with the abundance of phytoplankton and bacteria. Very high activities of lake water phosphatases were detected in the extremely stratified lake at the end of the summer stratification. Also ecological studies on pronounced diurnal activity and on the regulation of intracellular and extracellular

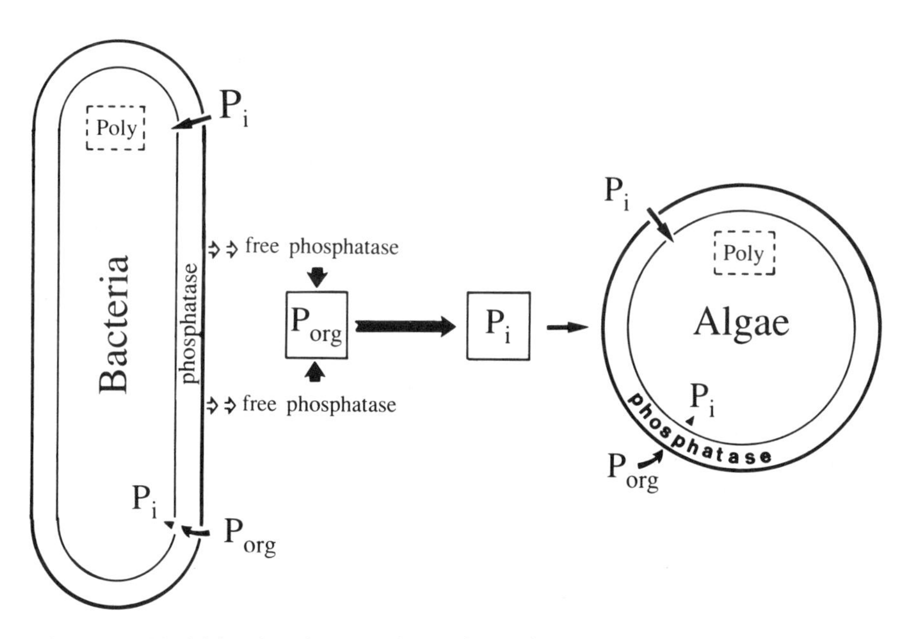

Figure 1.1 Model for phosphorus cycling within epilimnetic phytoplankton and bacteria mediated by free phosphatases (P_i, soluble inorganic P; P_{org}, soluble organic P; Poly, polyphosphate; modified from Overbeck, 1967).

phosphatases in the lake were conducted (Reichardt et al., 1967; Reichardt and Overbeck, 1969; Reichardt, 1969).

These above mentioned studies were the first steps in the exciting field of study of ecto- and extracellular enzymes in aquatic ecosystems. Today, it is a matter of fact that ecosystem functioning cannot be understood without the active participation of enzyme processes. But one important point must be stressed: we should be very careful and cautious in the interpretation of data from measurements in the field. Because of the complexity of metabolic pathways and the possible interference of manifold uncontrolled conditions, it is extremely difficult to understand enzyme regulation only from field data. All data must be complemented by biochemical studies with purified enzymes. I believe that the rapidly expanding field of biochemical and molecular ecology will develop methods in the near future for an adequate coupling of field and laboratory work.

References

Elster, H.J. and W. Einsele. 1937. Beiträge zur Hydrographie des Bodensees (Obersee). *Internationale Revue der Hydrobiologie* 35: 522–585.

Fermi, C. 1906. The presence of enzymes in soil, water and dust. *Zentralblatt für Bakteriologie und Parasitenkunde* 26: 330–334.

Harvey, H.W. 1925. Oxidation in sea water. *Journal of Marine Biology Association U.K.* 13: 953–969.

Kreps, E. 1934. Organic catalysts or enzymes in sea water. pp. 193–202 in *James Johnstone Memorial Volume*, University of Liverpool Press, Liverpool.

Ohle, W. 1952. Die hypolimnische Kohlendioxyd-Akkumulation als produktionsbiologischer Indikator. *Archiv für Hydrobiologie* 46: 153–285.

Overbeck, J. 1961. Die Phosphatasen von *Scenedesmus quadricauda* und ihre ökologische Bedeutung. *Verhandlungen der Internationalen Vereinigung für Theoretische und Angewandte Limnologie* 14: 226–231.

Overbeck, J. 1962. Untersuchungen zum Phosphathaushalt von Grünalgen. II. Die Verwertung von Pyrophosphat und organisch gebundenen Phosphaten und ihre Beziehung zu den Phosphatasen von *Scenedesmus quadricauda* (Turp.) Breb. *Archiv für Hydrobiologie* 58: 281–308.

Overbeck, J. 1967. Biochemisch-oekologische Studien zum Phosphathaushalt von *Azotobacter chroococcum*. *Helgoländer wissenschaftliche Meeresuntersuchungen* 15: 202–209.

Overbeck, J. and H.D. Babenzien. 1963. Nachweis von freien Phosphatasen, Amylase und Saccharase im Wasser eines Teiches. *Naturwissenschaften* 50 571–572.

Overbeck, J. and H.D. Babenzien. 1964. Über den Nachweis von freien Enzymen im Gewässer. *Archiv für Hydrobiologie* 60: 107–114.

Reichardt, W. 1969. Ökophysiologische Untersuchungen zur Katalyse der Phosphatmobilisierung im Gewässer unter besonderer Berücksichtigung alkalischer Phosphomonoesterasen aus Cyanophyceen. *PhD Dissertation, University of Kiel.* 129 pp.

Reichardt, W. and J. Overbeck. 1969. Zur enzymatischen Regulation der Phosphatmonoesterhydrolyse durch Cyanophyceenplankton. *Berichte der Deutschen Botanischen Gesellschaft* 81: 391–396.

Reichardt, W., Overbeck, J. and L. Steubing. 1967. Free dissolved enzymes in lake water. *Nature* 216: 1345–1347.

Rodhe, W. 1948. Environmental requirements of fresh-water plankton algae. *Symbolae Botanicae Uppsalienses* 10: 1–149.

Steiner, M. 1938. Zur Kenntnis des Phosphatkreislaufes in Seen. *Naturwissenschaften* 26: 723–724.

Vallentyne, J.R. 1957. The molecular nature of organic matter in lakes and oceans, with lesser reference to sewage and terrestrial soils. *Journal of Fisheries Research Board Canada* 14: 33–82.

2

Extracellular Enzymatic Interactions: Storage, Redistribution, and Interspecific Communication

Robert G. Wetzel

2.1 Extracellular Enzymes: Their Position Relative to the Cell in the Environment

I begin with the naive question: Where are the enzymes located in relation to the cells that are attempting to acquire a needed exogenous substrate or element? This question is important because many scientists speak rather casually about endo- and exoenzymatic activity in relation to cellular metabolism. Obviously most of the enzymatic activity is intracellular, associated with biosynthetic reactions driven by energy liberated from the hydrolysis of ATP.

An extracellular enzyme is one that hydrolyzes substrates external to cells, and is either bound to the cell membrane or free in the water. Many important enzymes are associated with cell membranes (e.g., carbonic anhydrase, alkaline phosphatase). Presumably these membrane-bound enzymes functionally bridge the finite distance between substrates in the surrounding milieu, their acquisition, and the transport of substrates and/or nutrients to and across the membrane for subsequent intracellular metabolism. Such membrane-bound enzymes are functionally exoenzymes, even though they are chemically bound to the cell surfaces. Important to subsequent discussion in this chapter, these "exposed" membrane-bound enzymes also are susceptible to inhibitory influences similar to the situation among "free" exoenzymes, as delineated below.

First, I emphasize that the separation of free from bound enzymes is quite variable with changes in the phases of growth and growth conditions. For example, in *Bacillus*, both membrane-bound and soluble alkaline phosphatase (APase, orthophosphoric monoester phosphohydrolase, E.C. 3.1.3.1.) forms occur (Hulett, 1986). A peripherally bound APase (salt extractable) is found on the inner leaflet of the cytoplasmic membrane (Figure 2.1). An integrally bound form, requiring detergent extraction, occurs on the outer leaflet. Soluble APase is secreted through the periplasmic membrane to form a truly extracellular enzyme. No discernible physico-chemical differences were found in the enzymes from these different locations. Many investigators have evidence to suggest that some enzymes, particularly phosphatases, are secreted or released by algae into the ambient medium and that they function extracellularly (e.g., Mills and Campbell, 1974; Aaronson and Patni, 1976; Walther and Fries, 1976; Yamane and Maruo, 1978; Cembella et al., 1984, 1985). Bacteriological evidence indicates that Gram-negative bacteria with multi-layered cell walls release little of the periplasmic enzymes, whereas Gram-positive and other bacteria lacking one or more cell-wall layers, tend to release enzymes extracellularly (cf. review in Cembella et al., 1984).

In both soil and aqueous ecosystems, it is commonly assumed that enzymes are released from cells into the environment (Figure 2.2). Affiliation of these compounds with inorganic and organic particulate surfaces can vary greatly with characteristics of both the enzymes and the particles (see discussion below). Numerous workers have specifically sought, in ecological studies, to differentiate between free enzymatic activities and those associated with particles (see Chapter 3). Separation methods vary, of course, with the questions being addressed. Physiological methods use various bond disruptors (salts, detergents) or cell wall disruptors (e.g., lysozyme) and membrane-impermeable enzyme-specific inhibitors to separate surface-bound and intracellular enzymes and their activities. Separation methods, such as centrifugation and filtration, that address ecological questions, such as the enzyme distributions in the en-

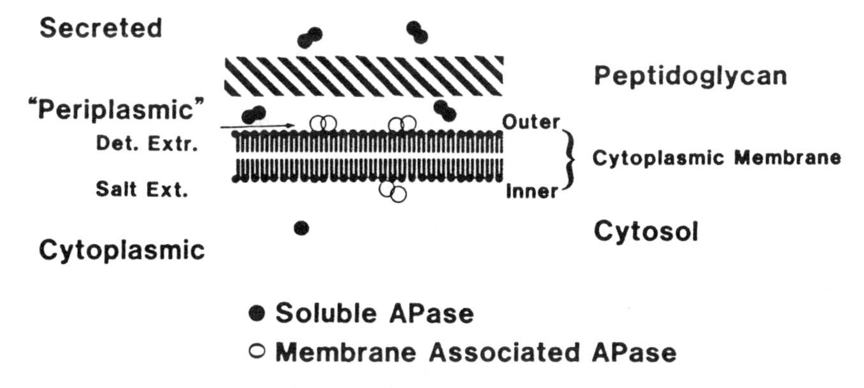

Figure 2.1 Distribution of alkaline phosphatase (APase) in *Bacillus licheniformis*. Single circles = inactive monomers; double circles = active dimers; solid circles = soluble enzyme protein; open circles = membrane-associated APase (modified from Hulett, 1986).

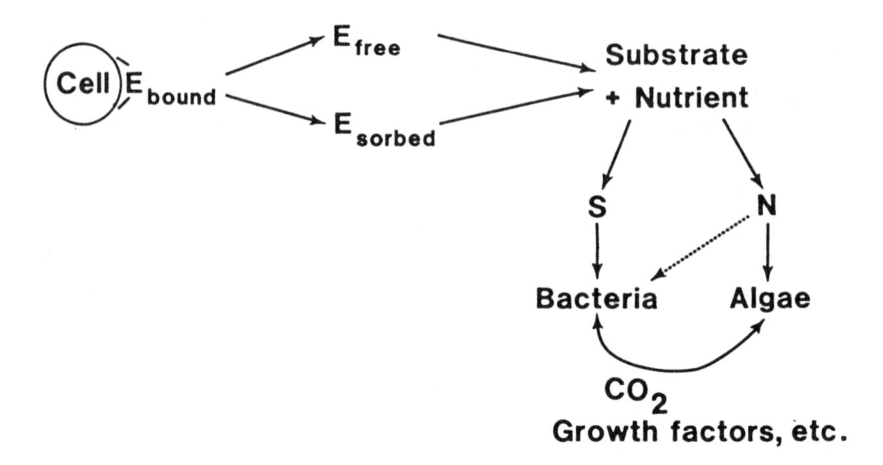

Figure 2.2 Generalized functional pathways of released extracellular enzymes. E = enzyme; S = substrate; N = nutrient.

vironment, allow distinction between enzyme activities of free and particulate phases. Although some workers suggest that certain exoproteolytic activity in filtrates may result from cell fragments due to the lysis of bacteria (Rego et al., 1985; Sharp, 1977), many careful studies have demonstrated very appreciable amounts of total enzyme activities in dissolved phases, particularly among the phosphatases (Fogg, 1977; Aaronson, 1978; Stewart and Wetzel, 1982b).

Among the first to point out the potential occurrence of "free" enzymatic activities in fresh waters were Overbeck and Babenzien (1963, 1964; cf. also Reichardt et al., 1967). Because of the central importance of phosphorus as a commonly regulating nutrient of primary productivity in fresh waters, much attention was given to the phosphatase activities of algae. Wetzel (1981) demonstrated, for example, over numerous annual periods that 30 to 50% of the pelagic alkaline phosphatase was in a dissolved "free" form (less than 0.6-μm-pore-size filtration). These findings were subsequently verified and extended by centrifugation separation methods and additionally showed that much (15 to 73%) of the particulate alkaline phosphatase activity was associated with bacterioplankton (Table 2.1; cf. Stewart and Wetzel, 1982b; also Chróst and Overbeck, 1987). A number of other investigations have also shown that a significant portion of the phosphatase activity in the pelagic zone is "soluble" or "free" (Aaronson and Patni, 1976; Pettersson, 1980; Chróst et al., 1989; Münster et al., 1989). In contrast, relatively little (<10%) of the peptidase, galactosidase, and glucosidase activity was found in dissolved form, perhaps related to differences in enzyme structure, binding characteristics, and recalcitrance to degradation.

2.2 Strategy of External vs. "Free" Enzymes

Because it has been clearly established that a considerable amount of extracellular enzyme activity is free, I would like to address the biochemical and eco-

Table 2.1 Percentage of total alkaline phosphatase activity contributed by algae, non-algal particulates, and "dissolved" enzyme in four lakes in southwestern Michigan

		Percentage of total APase activity		
Sample source	Depth (m)	Algae	Non-algal particulates	"Dissolved" enzyme
Lawrence Lake				
	2	5.3	42.9	51.8
	4	6.1	44.7	49.2
	6	12.3	39.6	48.1
Lefebre Lake				
	2	9.6	46.1	44.4
	4	4.7	53.7	41.6
Little Mill Lake				
	2	25.1	60.5	14.4
	4	13.8	72.5	13.7
Gull Lake				
	4	29.4	31.0	39.6
	9	33.2	14.8	52.0
	15	32.0	6.8	61.2

Modified from Stewart and Wetzel, 1982b. Separations were made by differential centrifugation; "dissolved" APase activities were measured in the supernatant (20,000 x g; 5 min).

logical significance of the existence of both free and bound enzymes. Enzymes in solution are susceptible both to degradation and to chemical alteration. Through these processes, the functional capacities of the enzymes are reduced. From the standpoint of the individual cell or organism, it would appear energetically inefficient to release enzymes from cells and to increase the distance between sites of enzymatic reactivity and cellular utilization. The probability of a cell being able to utilize the hydrolytic products decreases precipitously as the distance of the enzyme from the cell increases (Figure 2.3). Nonetheless, if the hydrolyzed molecule is within circa 500 μm of the cell surface, the probability of assimilation is sufficiently high to warrant enzymatic release.

Surface-bound enzymes exhibit boundary-layer diffusional problems. Diffusion of a substrate to bound enzymes is relatively slow, and the substrate concentration is lower in the microenvironment at the cell boundary surface than in the free macroenvironment (Engasser and Horvath, 1974a; 1974b). Although the concentration gradients can enhance diffusional rates, flux rates are inadequate and the diffusional resistances can cause both substrate depletion and product accumulation in the microenvironment of bound enzymes. Thus, diffusional inhibition could be reduced by releasing enzymes into the immediate medium surrounding the cells (Figure 2.3). Although some losses would result from products diffusing away from the cell, these losses may be compensated for by improved availability of substrates. For example, Ammerman

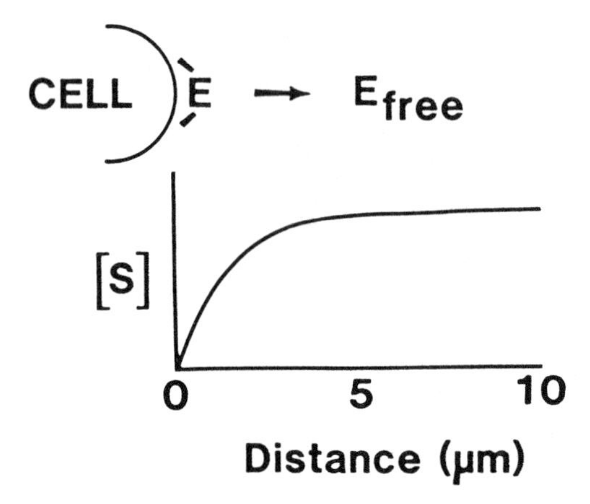

Figure 2.3 Hypothetical potential advantage of the release of extracellular enzymes from cells into the immediate ambient environment of high substrate concentration [S] (based on models of Engasser and Horvath, 1974a; 1974b).

and Azam (1985) found that from 15 to 100% of the hydrolyzed dissolved organic phosphorus was assimilated. Moreover, from the standpoint of the community and ecosystem, it is possible for multiple interactions among many cells and different species to be operational in which, for example, the mobilization of nutrients from organic compounds could be increased by exoenzymatic activities of microfloral metabolism while simultaneously increasing nutrient and carbon availability to algae cohabiting the same water parcel (Figure 2.2). Thus both bacteria and algae could benefit from the interactions.

2.3 Immobilization of Enzyme Activity

The hydrolytic enzymes catalyze the cleavage of covalent bonds such as C-O (esters and glycosides), C-N (proteins and peptides), and O-P (phosphates). Often these same enzymes can also affect synthesis by condensation reactions. Hydrolytic cleavage of C-C bonds is rare and hydrolytic enzymes do not react with aromatic nuclei of phenolic compounds but do react with phenolic hydroxyl and other substituted functional groups that often occur in phenolic compounds.

Conversely polyphenolic compounds, largely of plant origin, form a major component of dissolved organic acids of fresh waters. The dissolved organic acids frequently constitute some 80% of the total dissolved organic matter (ca. 5–10 mg dissolved organic carbon per liter; Wetzel, 1983; 1984). Hence, concentrations of 4–8 mg polyphenolic organic acids per liter or more are common (Perdue and Gjessing, 1990). Polyphenolic organic acids induce precipitation of proteins by binding to one or more sites on the protein surface (Figure 2.4a)

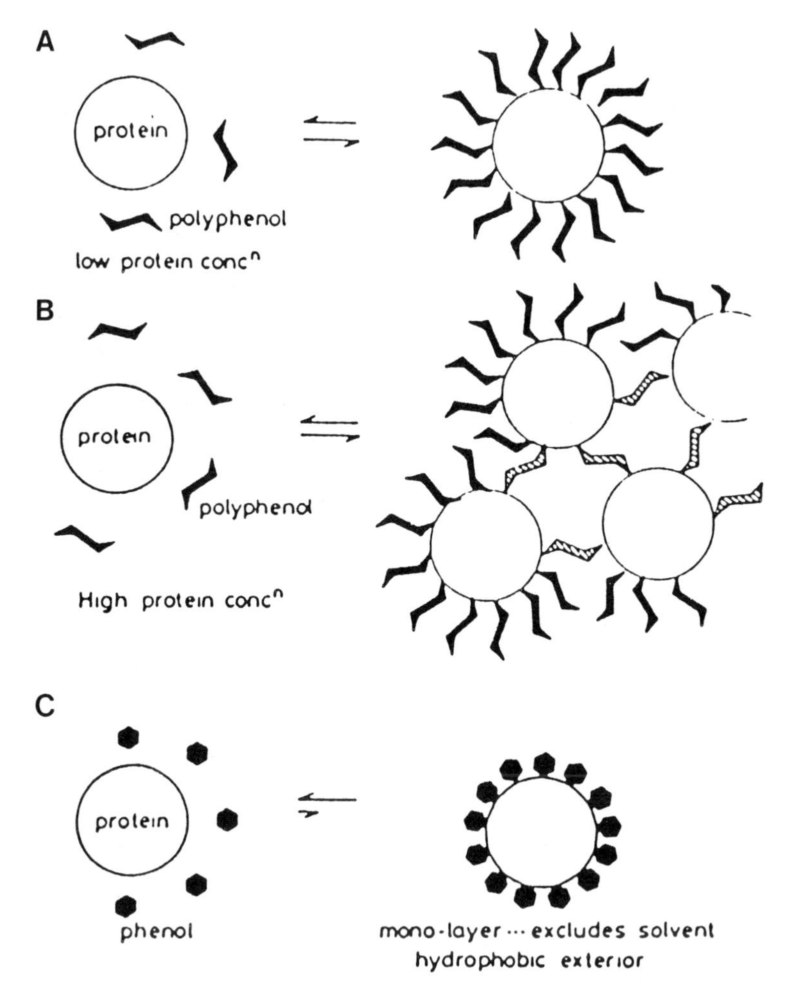

Figure 2.4 Protein complex formation and precipitation by phenols and polyphenols (modified from Haslam, 1988). See text for explanation. Hatched = cross linkages among phenol-protein molecules.

to give a monolayer that is less hydrophilic than the protein itself (Figure 2.4c; Haslam, 1988; Haslam and Lilley, 1988; Spencer et al., 1988).

Aggregation and precipitation can ensue, but clearly enzymatic activity is inhibited (e.g., Ladd, 1985). At higher concentrations of enzymes, cross-linking can occur among phenol-protein molecules (e.g., Haslam, 1988); the result is that less polyphenol is required to precipitate proteins from concentrated enzyme solutions (Figure 2.4b). More aromatic and condensed humic acid molecules are more rigid and can distort bound enzymes to a greater extent than is the case with simpler compounds, such as fulvic acid (e.g., Ladd and Butler, 1975).

The effectiveness of plant polyphenols to complex protein molecules derives from their polydentate ligands. These ligands have many potential hydrogen binding sites provided by numerous phenolic groups and aryl rings on the periphery of the molecule (Beart et al., 1985). Polyphenolic compounds have the proper molecular size and structure to form stable cross-linked structures with a number of different protein molecules (Figure 2.5). The inhibition of enzymes occurs in a classical noncompetitive manner, in which the inhibitor, polyphenol, and substrate bind simultaneously to the enzyme (Figure 2.5).

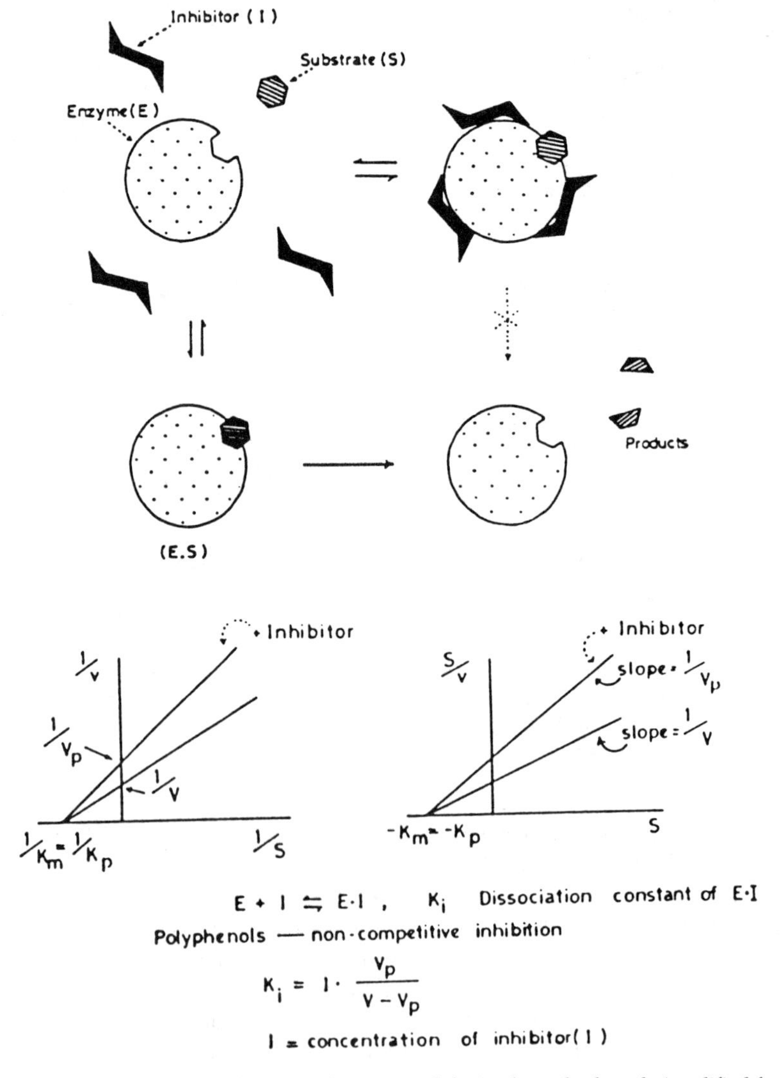

Figure 2.5 Mechanism and kinetics of enzyme inhibition by polyphenols (modified from Haslam and Lilley, 1988).

Let me digress briefly with some examples of the interactions of humic compounds with enzymes in fresh waters. This subject has been evaluated to a much greater extent in terrestrial soil systems. For example, proteolytic enzymes can be reversibly bound to humic compounds in soils by ionic linkages (Ladd and Butler, 1975). The extent of binding by covalent, ionic, or hydrogen bonding to humic acids in soils is unclear. Covalent bonding appears to dominate; the rate of extraction of enzymes by buffers is low, and thus the amount of ionic binding of enzymes to humic compounds must also be low.

Earlier we examined the effects of dissolved humic compounds upon the phosphatase activities of planktonic bacteria and algae of hardwater, phosphate-limited lakes (e.g., Stewart and Wetzel, 1981; 1982a). Strong interactions were determined experimentally between dissolved humic compounds, phosphorus availability, photosynthetic rates, and dissimilation rates of recent photosynthate released from the algae. It was apparent that high-molecular-weight humic compounds of emergent macrophyte origins suppressed photosynthetic carbon fixation and enhanced APase synthesis as presumably greater amounts were inactivated by means of humic-enzyme complexation. Abiotic mechanisms associated with iron and dissolved humic materials can also be involved with the availability of phosphorus under certain conditions (e.g. Francko, 1986; Jones et al., 1988). However, my recent analyses demonstrate an additional mechanism by which the dissolved humic compounds regulate phosphorus availability.

The formation of polyphenolic-enzyme complexes can partially or entirely inactivate the enzyme. Often the reaction is reversible, for example, by means of ultraviolet photolytic degradation of the phenolic compounds or autodigestion by other enzymes. As inorganic phosphorus availability commonly becomes limiting spatially and temporarily in fresh waters, the microflora depend increasingly upon the hydrolysis of dissolved organic phosphorus (DOP) compounds. Loading of polyphenolic and related dissolved organic compounds to an aquatic environment can result in the formation of enzyme complexes and the inhibition of enzyme activities and can become a dominant regulatory mechanism of phosphorus and hence photosynthetic and respiratory metabolism.

Much of the dissolved organic matter in fresh waters, as already noted, emanates from organic compounds of plant origin. The more recalcitrant portions of these compounds have relatively slow turnover rates and persist in the dissolved phase for considerable time periods (average decomposition rates of 0.5 to 1% per day). Phenolic organic acids usually dominate this pool. Particularly important for complex formation mechanisms are those phenolic compounds which include an array of plant substances that possess an aromatic ring with one or more hydroxyl substituents. In regard to this discussion, we are here more concerned with phenolic compounds of structural tissue origin (phenols, phenolic acids, phenylpropanoids, and polymers) because of the high quantitative loadings of these organic materials to fresh waters from wetland and littoral sources. Other phenolic compounds (flavonoid and other pigments, flavonols, flavones) can have important regulatory functions but are quantitatively less significant.

Certain phenolic compounds are more reactive with proteins than others. For example, certain simple phenols and phenolic acids (e.g., *p*-hydroxybenzoic acid) are less reactive with proteins than phenylpropanoids (e.g., the hydroxycinnamic, *p*-coumaric, caffeic, and ferulic acids). In contrast, hydrolyzable tannin polymers (gallo- and ellagitannins), esters of gallic and diphenic acids with glucose, strongly complex with proteins and enzymes. Positively charged groups, such as protonated amino groups of proteins, could bind electrostatically with the negatively charged ionized organic acids. Adsorption of organic acids increases as their charge decreases and their molecular weight and hydrophobic properties increase (Perdue and Gjessing, 1990). At the pH of most natural waters (pH 5–8), a large percentage of carboxylic acid groups will be ionized. Major cations (e.g., Ca^{2+}) will react with organic acids and can reduce complex formation with enzymes and their inactivation.

Several representative purified and generic compounds were assayed for interactions with enzyme activities, and alkaline phosphatase was selected as a model enzyme because of its ubiquitous importance in many fresh waters. Simply evaluating the effects of organic acids upon alkaline phosphatase activity (APase) in buffer demonstrated a marked suppression of activity, using simple low-molecular-weight phenolic acids, phenylpropanoids, or larger polymers (Figures 2.6 and 2.7). The enzyme is more stable in the enzyme-humic acid complex, in part because of stearic effects that interfere with proteolytic activity. When the same assays were performed in lake water or a synthetic medium that simulated the ionic composition of hardwater lakes of the gla-

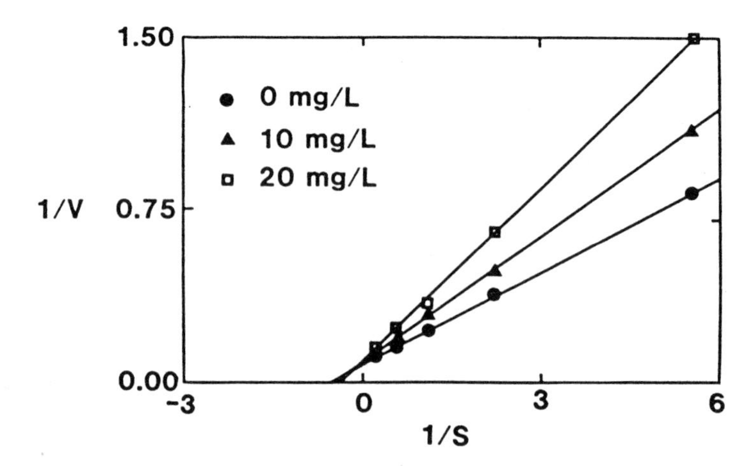

Figure 2.6 Effect on alkaline phosphatase activities of increasing concentrations of a humic acid mixture. Methylumbelliferyl phosphate concentrations were measured at 0.2, 0.5, 1.0, 2.0, and 5.0 x K_m.

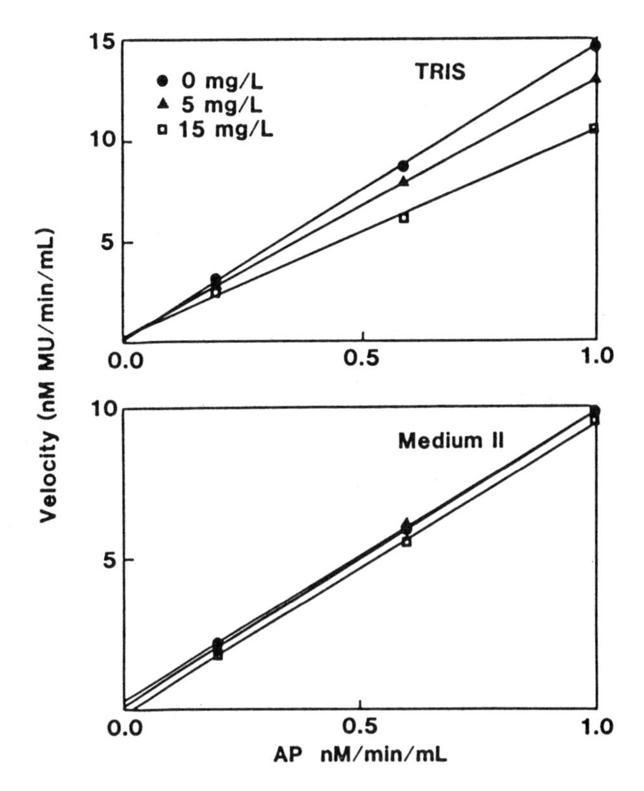

Figure 2.7 Effects of increasing concentrations of a humic acid mixture on the reaction velocities of alkaline phosphatase in Tris (pH 8.0) and in a synthetic hard lakewater medium (Wetzel Medium II, pH 8.0).

ciated midwestern USA, the extent of inhibition was still highly significant but less than in organically buffered water (Tris, pH 8.0) (Figure 2.8). Further assays in which the divalent cation concentrations were varied showed a reduced inhibition of dissolved humic compounds upon enzyme activity with increasing cation concentrations (Figure 2.9). In both soils and fresh waters, humic acids are known to reversibly bind enzymes by a cation exchange mechanism (e.g., Scheffer et al., 1962; Ladd, 1972; Ladd and Butler, 1970; Tipping et al., 1988). At high concentrations, inorganic cations, particularly divalent cations, can displace enzymes from complexes by binding at the humic acid carboxyl group where the enzymes bind.

The rationale underlying these experiments emanates from a complex series of potential interactions in natural waters among humic compounds, cation concentrations, and enzyme activities. The following hypotheses are congruous with manifold observations that have been made along lake gradients (Figure 2.10):

1. Organic acids complex chemically with enzymes. Their enzymatic ac-

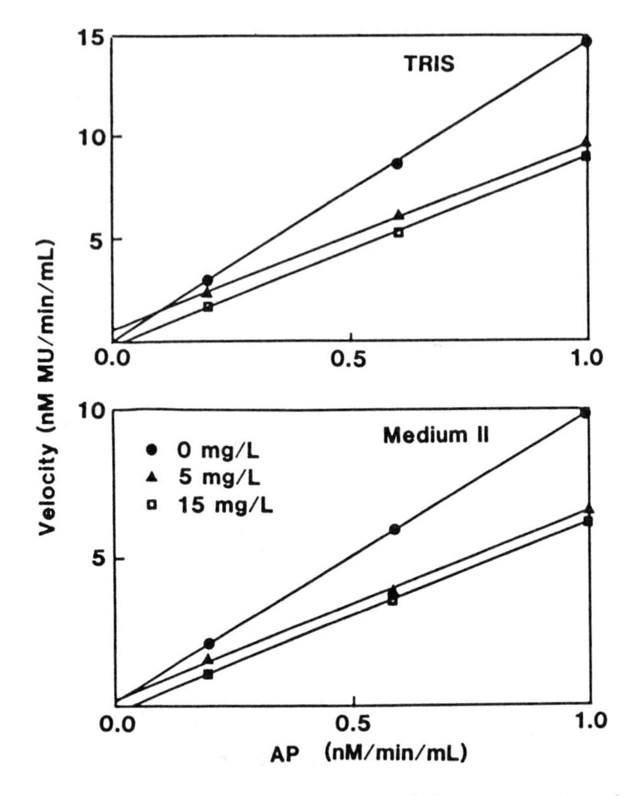

Figure 2.8 Effect of increasing the concentrations of the representative phenolic compound pyrogallol upon the reaction velocities of alkaline phosphatase. (Media as in Figure 2.7)

tivities can be greatly altered or completely inhibited. As a result, metabolic and growth processes, such as photosynthesis, can be markedly suppressed, particularly if the availability of limiting nutrient is involved (e.g., alkaline phosphatase is inhibited). Increased enzyme production may not be able to counter losses or may be energetically too costly.

2. Divalent cations also adsorb to organic acids. High cation concentrations can reduce the reactivity of organic acids with proteins and enzymes. Enzyme activity would be less inhibited in solutions with high divalent cation concentrations.

3. Along a lake gradient of increasing salinity from soft to hard waters, productivity commonly increases. Although in part related to the loading rates of essential nutrients (especially P and N) from the drainage basin and internally from storage sites (e.g., sediments), the causal relationships are not always this direct and simple. As dissolved organic compound loadings increase, productivity in soft waters is much lower

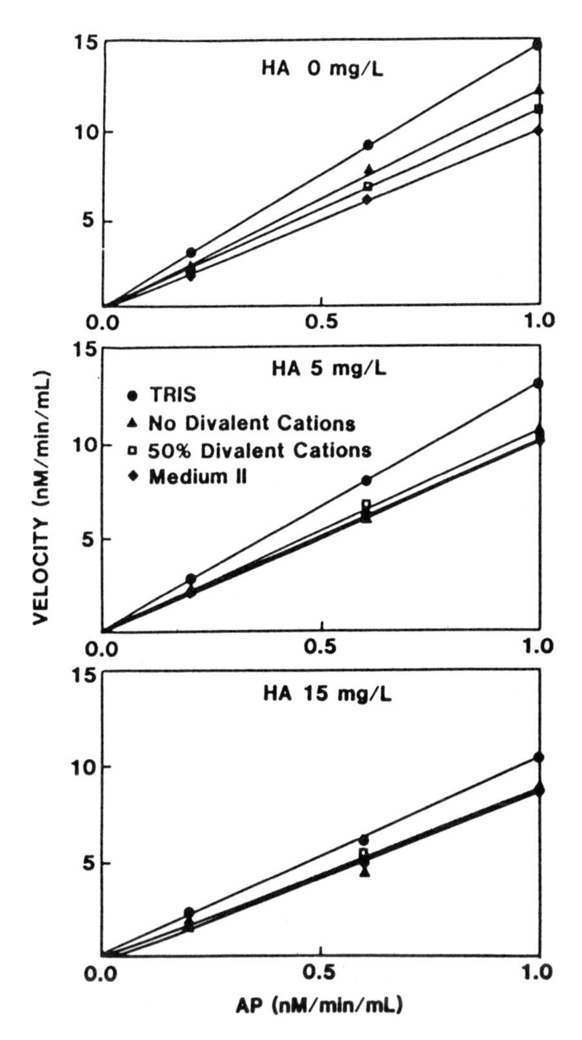

Figure 2.9 Effects of varying divalent cation concentrations on the humic acid inhibition of the reaction velocities of alkaline phosphatase. (Wetzel Medium II with full, half, and no calcium and magnesium concentrations).

than is the situation in hard waters with common concentrations of Ca^{2+} and Mg^{2+} of 40–60 and 15–25 mg l^{-1}, respectively. Part of this improved productivity in organically stained hard waters may be related to a cationic suppression of humic compound interference with enzyme activities, which in turn permits enhanced use of nutrients and/or organic substrates of organic compounds (Figure 2.10). In addition, if $CaCO_3$ is precipitating in hardwater ecosystems, significant quantities of dissolved organic acids can adsorb and/or be coprecipitated with the nucleating carbonates (Otsuki and Wetzel, 1972; 1973).

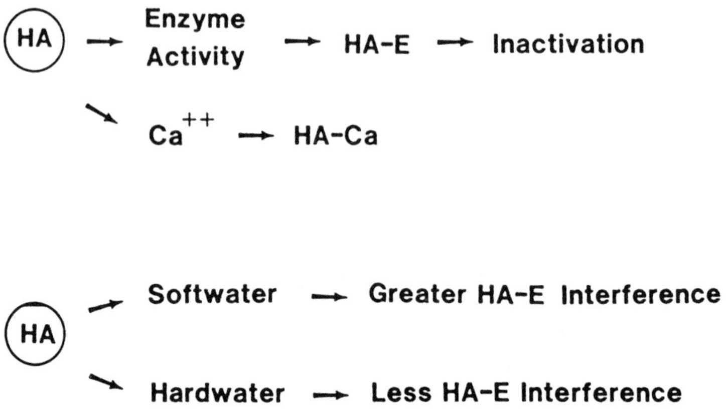

HA → Enzyme Activity → HA-E → Inactivation

HA → Ca^{++} → HA-Ca

HA → Softwater → Greater HA-E Interference

HA → Hardwater → Less HA-E Interference

Figure 2.10 Generalized diagram of humic acid (HA) interactions with enzyme activities and divalent cations, such as calcium, along a gradient of soft to hard waters.

A further example of these binding effects of humic compounds upon enzymatic activities was found among attached microorganisms in littoral zones. In controlled, replicated littoral zones in glasshouse conditions, a number of artificial substrata (0.5-mm-diameter monofilament nylon lines) were suspended among cultures of the mature submersed macrophyte *Scirpus subterminalis*. The macrophytes and epiphytic substrata were allowed to stabilize and then some were exposed to an increase of 2.5 mg l^{-1} (some 20% above controls) of additional humic acids extracted and concentrated on XAD-8 resins from a slurry of senescent emergent macrophyte tissue (*Typha latifolia* leaves). Changes in the chemical composition of the water was followed for over 30 days. After this period of exposure, samples of the periphyton from these treated and untreated systems were exposed experimentally to humic acids and other compounds and the effects upon alkaline phosphatase activities determined. The general results were similar to those chemical experiments of humic-enzyme analyses described earlier: Namely, measured enzyme activities were reduced, presumably resulting from a partial humic-enzyme immobilization, among the periphyton exposed to increasing amounts of humic acids (Figure 2.11). The periphyton that had been exposed earlier to some weeks of increased organic acids from *Typha* leachate had increased alkaline phosphatase activities over those epiphytic communities that had not been exposed to increased amounts of leachate. It could be inferred that the increased humic compounds further bound enzyme sites in a noncompetitive inhibition mechanism. Although these results are very preliminary and further study is underway, it is important to note that complex humic-enzyme interactions can function in fresh waters at natural concentrations of dissolved organic matter.

A number of other studies have suggested a cause-and-effect metabolic link between loadings of recalcitrant dissolved organic matter (DOM) and suppressed microbial metabolism. For instance, the presence of high-molecular-

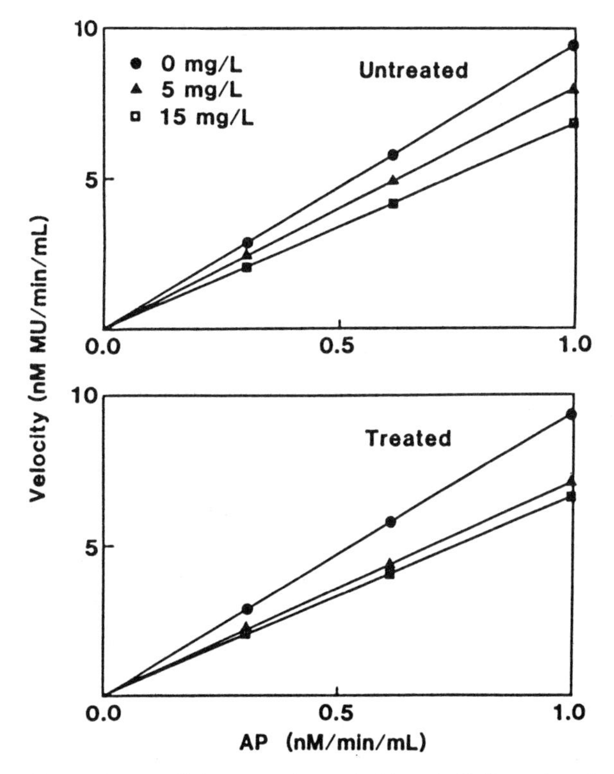

Figure 2.11 Effects of pyrogallol on reaction velocities of alkaline phosphatase of periphytic microflora that had not been exposed (untreated) and had been exposed (treated) to a 20% increase in dissolved organic carbon from organic acids of *Typha latifolia* leachate.

weight (>1,000 daltons) DOM in rivers reduced the respiratory metabolism of epilithic algae and bacteria (Ford and Lock, 1987). It was suggested that high-molecular-weight DOM sorbs by ionic exchange to active metabolic sites and alters enzyme activities.

2.4 Humic Loadings to Fresh Waters

Any evaluation of potential effects of dissolved organic compounds upon organisms and their metabolism must examine both the quantitative loading of these compounds to a water body as well as their qualitative chemical sources. Most lakes and rivers of the world are small and shallow. The importance of organic matter synthesized by emergent (wetland) and floating/submersed (littoral) macrophytes usually assumes total dominance over planktonic-synthesized organic matter. The organic matter of macrophytes is chemically different than that of phytoplankton. As such, higher plant organic matter, in dissolved forms, is not only the energetic foundation of most lake and river ecosystems but also interacts chemically with pelagic organisms in regulatory processes. This chemical difference, and the much larger amounts produced by the macrophytes, is crucial to the thesis of this chapter.

The wetland-littoral component By definition, lakes and rivers are basins that receive water and soluble and particulate materials that are either imported with the water or are synthesized within the waters. Most basins are destined to fill by sedimentation because decomposition and exports are nearly always less than allochthonous and autochthonous importations. A seminal point of this discussion is that in most fresh waters, much of the loading of organic matter is dissolved organic matter and emanates from littoral-wetland regions (Figure 2.12). It is this dissolved organic matter, regardless of its origins, that provides the energy to drive the ecosystem's metabolism.

Area and depth of freshwater bodies In order to determine how extensive the land-water interface zone functions are in freshwaters in general, it is necessary to examine numbers comparatively. Good data do not exist, unfortunately; however, estimates are possible. For streams and rivers, the nature of the drainage basin is extremely important to the loading and functions of these waters. What portion of the "drainage basin" is considered to be in the wetland-littoral category is highly variable with geomorphology and flood-stage

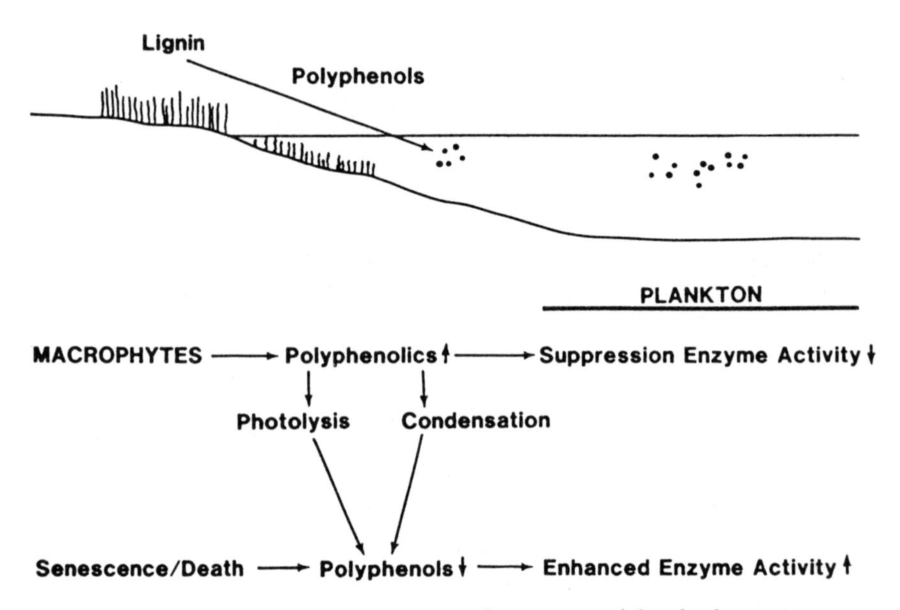

Figure 2.12 Diagram of the common high loading process of dissolved organic matter from the wetland and littoral zones of river and lake ecosystems. The dissolved polyphenolic compounds can suppress enzymatic activities directly. Degradation and condensation with particulate precipitation will cause inactivation of polyphenolic reactivity, thereby enhancing enzyme activities. See text.

conditions. From a functional point of inorganic and organic loadings to the recipient streams, however, the extent of the soil saturation and inundation is also critical. Although quantitative data are meager, it is apparent that riverine loadings of organic matter and nutrient flux regulation are dominated by floodplain macrophyte and attached microbial productivity and metabolism associated with sediments and particulate detritus.

Among reservoirs, the interface zone is less influenced by wetlands and more by dominance of the littoral zone (floating, floating-leaved, and submersed macrophytes, and attendant microflora, both epiphytic and on and within the sediments). The more productive emergent macrophytes of wetlands are often restricted or excluded because artificial variations in water levels are frequently too great for their physiological desiccation/flooding tolerances. Emergent wetlands of reservoirs are therefore often restricted to the inflow regions where water tables at the interface zones exhibit greater stability.

The mechanisms of lake basin formation have been largely geological processes, e.g., glacial mechanisms, that resulted in predominantly shallow lake basins (Hutchinson, 1957). Consequently, most lakes and other standing freshwater bodies are quite shallow, with mean depths much less than 10 m (Figure 2.13). Indeed, in vast northern areas of the Northern Hemisphere, literally millions of lakes of considerable area exist with mean depths of less than three m. Net deposition rates per total volume are nearly always higher in lake basins that are relatively shallow than in deeper lakes. As a consequence, most of these many shallow waters can accommodate high growth of aquatic vegetation over much of their basins. Moreover, the surrounding drainage basins are of

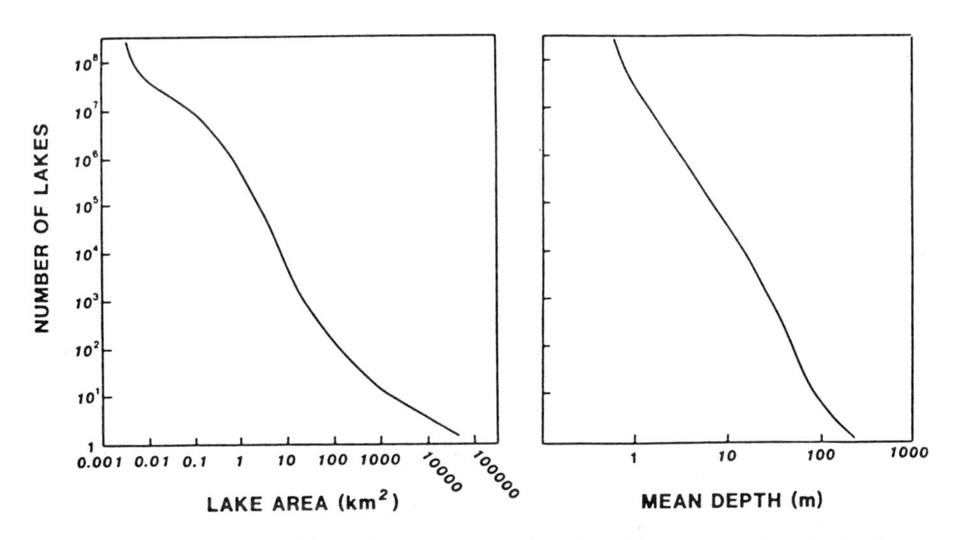

Figure 2.13 Number of lakes of the world as a function of lake area and mean depth (from Wetzel, 1990).

such gentle slopes that emergent wetlands often extend nearly continuously among a large number of lakes.

What then do comparisons of numbers of lakes and other fresh waters show? Examination of the quantity of lakes versus lake area can only be done reasonably on a log-log plot because of the numbers involved. Large lakes are simply very few in numbers (Figure 2.13). Most lakes are less than 10 km^2 in area. Literally millions of lakes and reservoirs are less than 100 ha (<1 km^2) and many more freshwater basins are less than 10 ha. One quickly approaches the question of at what size is a lake or reservoir no longer a lake but rather a shallow, ephemeral depression that holds water for a portion of the year. Figure 2.13 refers to permanently standing waters, and their number is rapidly increasing with the construction of many small reservoirs in agricultural areas.

Examination of the mean depths of lakes and reservoirs also indicates that the deep lakes of the world are very uncommon (Figure 2.13). Most lakes have a mean depth of less than 10 m. As one progresses northward, the mean depth progressively decreases to the point where, in the tundra, mean depths are frequently less than one meter (Hobbie, 1980).

Pelagic/littoral ratios The importance of the examination of numbers of lakes and reservoirs in relation to surface area and mean depth results from the relative percentage of the ecosystem that is available for potential colonization by macrophytes and attendant sessile microflora. As is depicted in Figure 2.14, the ratio of interface zones to pelagic zones, as defined by Hutchinson (1975; see Wetzel, 1983), permits one to estimate the importance of the littoral (L) in relation to the pelagic (P) zone (Figure 2.15). On a global basis, the littoral zone

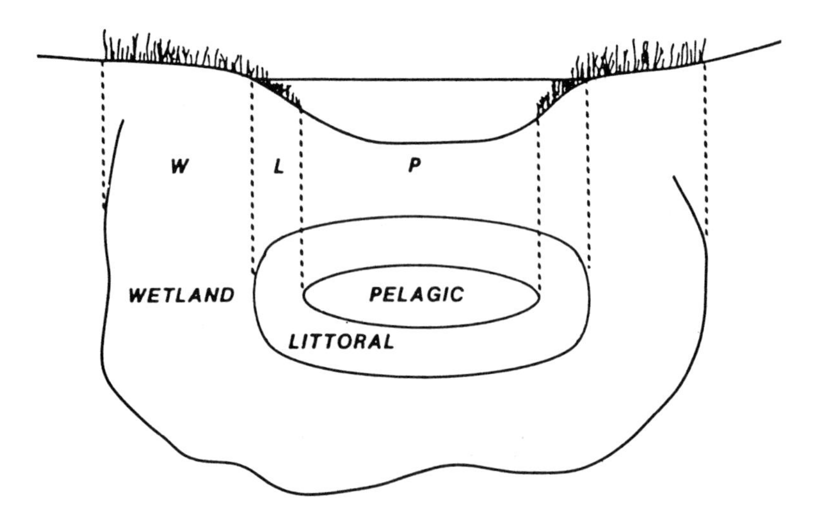

Figure 2.14 Diagram for evaluating the relative areas of wetland, littoral, and pelagic zones of aquatic ecosystems (from Wetzel, 1990).

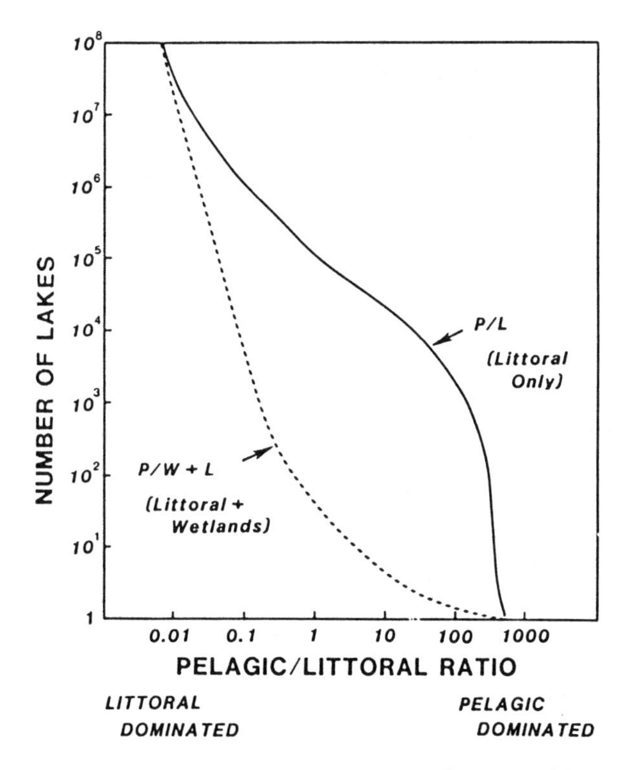

Figure 2.15 Number of lakes of the world as a function of the ratio of the areas of pelagic versus littoral (and wetland) zones (from Wetzel, 1990).

dominates over the pelagic zone (low P/L ratios) for most standing-water ecosystems. Addition of the wetland components to the littoral results in low ratios [P/(W+ L)] for most standing-water ecosystems (Figure 2.15). For flowing waters, floodplains function as the major wetland and littoral component of most river systems.

Organic loading from land-water interface regions The high rate of wetland-littoral productivity and the dominance of anaerobic decomposition of much of its large pool of organic matter results in the formation of large amounts of relatively recalcitrant dissolved organic matter. The transport of organic matter from these interface regions is largely in dissolved form. Particulate organic matter settles out and is not transported extensively. Hydrological flows are generally shallow and surficial, and movements tend to be through surficial detrital zones of intensive microbial metabolism. The simpler, more energy-rich organic substrates are rapidly decomposed, largely by sediment- and particulate-detritus-associated microflora (e.g., Cunningham and Wetzel, 1989). Most of the dissolved organic matter, however, is of plant structural origins and relatively recalcitrant to rapid microbial degradation. These compounds dominate the organic loading and move into the pelagic zones (cf. Wetzel, 1979; 1984).

2.5 Regulation of Enzyme Activity in Aquatic Ecosystems

I suggest that all aquatic ecosystems, and certainly all fresh waters, are functionally a system of stored, immobilized enzymes. This hypothesis is simply that a majority of the enzyme activity is associated with surfaces, the enzymes are largely reversibly immobilized by humic compounds, and that under certain circumstances, biotic access to these enzymes can be gained and their functional capacities restored within the ecosystem.

The premise for these arguments are several. Most enzymes are not dissolved in the water but rather are enzymes associated with living or inert surfaces. Many enzymes, indeed most enzymes, must be complexed or immobilized with clays, humic acid compounds, or carbonates. Mostly hydrophilic, these largely humic-enzyme compounds are quite stable and active, like the synthetic polymer-enzyme analogs that we use analytically to separate and concentrate enzymes. I further suggest that colloidal humus-enzyme compounds protect enzyme moieties and their activity from reactivity with other enzyme processes, and thus protect them from competitive interactions that would collectively lessen their functions in the ecosystem.

I also suggest that one cannot understand enzymatic activities and functions in aquatic ecosystems without examining the microenvironmental conditions associated with humic-enzyme or surface-enzyme complexes. The microenvironmental conditions at the sites of enzyme activities, particularly at living cell surfaces, can be totally different from conditions in bulk water. For example, the pH at the cell surface of an alga or bacterium can vary by two to three units within hours under normal diel cycles. Effective examination and understanding of mechanisms controlling enzyme immobilization and reactivation may require information about these conditions as well as those in the bulk water.

Finally, I would like to suggest that aquatic ecosystems are systems consisting of living cells and immobilized enzymes and function much as does an organ such as a plant leaf or an animal liver. Most regulation of species-species or species-environment interactions takes place at the enzymatic level. The ability of organisms to mobilize complexed enzymes at rates exceeding those of other organisms would instill marked competitive advantages over other organisms. The ecosystem stores the enzymes chemically; the better competitors may be able to reactivate and use them more efficiently than other organisms.

Metabolic couplings and interactions among organisms in different parts of the ecosystem have to be at the chemical level. The ability to complex and store enzyme activities with humic compounds, although energetically expensive and at the risk of loss of some of energetically rich compounds, confers a communicative capacity to the ecosystem. I suggest that it is here that true interactions among ecosystem components occur: enzymatic storage in organic complexes, transport in immobilized forms, and reactivation in displaced sites. Although displacement of enzymes from the organisms that produced them may be construed as their "loss," from the standpoint of the ecosystem, their potential to function is not truly lost.

2.6 Summary Conclusions

Evaluation of the functions and capacities of extracellular enzymatic activities in freshwater ecosystems suggests that altered perspectives are needed. Enzymes are largely associated with surfaces. Methodological limitations restrict the perspicacity of determining whether or not enzymes are truly dissolved ('free'). Extrapolation of known physiological enzymatic responses to apparent responses under differing natural and modified environmental conditions is complicated by complexation reactions.

Determination of the factors controlling differences between potential and realized enzyme activities is a key area of required mechanistic research. True releases of enzymes into the surrounding medium are relatively small; these proteinaceous compounds are readily complexed in both membrane-bound and free states by non-enzymatic chemical reactions or may be lost by biotic decomposition.

The following hypotheses are proposed as important in regulation of biotic metabolism in aquatic ecosystems:

(a) Colloidal organic acid-enzyme complexes protect enzyme moieties from reactivity with other enzymes or chemical processes. These complexes protect the enzymes and allow storage in a suppressed but chemically active state. As a result enzymes can be transported to other parts of the ecosystem. Aquatic ecosystems thus consist of living cells and inactivated or immobilized enzymes, and operate analogously to complex organs of higher plants and animals.

(b) Microenvironmental conditions at sites of enzyme activities, particularly at living cell surfaces, differ greatly from conditions in bulk water.

(c) Biochemical interactions or perhaps better said chemical communication occurs among biotic components of the ecosystem in many ways. Enzymatic storage in organic complexes, transport in inactivated or immobilized forms, and reactivation at displaced sites is suggested as a potentially important means of community regulation, particularly between the littoral and pelagic communities.

Acknowledgments The critical reviews of the manuscript by James Cotner, Jr. and William E. Spencer are gratefully acknowledged. The study was supported in part by subventions of the U.S. Department of Energy (DE-FGO2-87ER60515) and the National Science Foundation.

References

Aaronson, S. 1978. Excretion of organic matter by phytoplankton in vitro. *Limnology and Oceanography* 23: 838.

Aaronson, S. and N.J. Patni. 1976. The role of surface and extracellular phosphatases in the phosphorus requirement of *Ochromonas*. *Limnology and Oceanography* 21: 838–845.

Ammerman, J.W. and F. Azam. 1985. Bacterial 5'-nucleotidase in aquatic ecosystems: A novel mechanism of phosphorus regeneration. *Science* 227: 1338–1340.

Beart, J.E., Lilley, T.H. and E. Haslam. 1985. Plant polyphenols—secondary metabolism and chemical defense: Some observations. *Phytochemistry* 24: 33–38.

Cembella, A.D., Antia, N.J. and P.J. Harrison. 1984. The utilization of inorganic and organic phosphorus compounds as nutrients by eukaryotic microalgae: A multidisciplinary perspective: Part 1. *CRC Critical Reviews in Microbiology* 10: 317–391.

Cembella, A.D., Antia, N.J. and P.J. Harrison. 1985. The utilization of inorganic and organic phosphorus compounds as nutrients by eukaryotic microalgae: A multidisciplinary perspective. Part 2. *CRC Critical Reviews in Microbiology* 11: 13–81.

Chróst, R.J., Münster, U., Rai, H., Albrecht, D., Witzel, P.K. and J. Overbeck. 1989. Photosynthetic production and exoenzymatic degradation of organic matter in the euphotic zone of a eutrophic lake. *Journal of Plankton Research* 11: 223–242.

Chróst, R.J. and J. Overbeck. 1987. Kinetics of alkaline phosphatase activity and phosphorus availability for phytoplankton and bacterioplankton in lake Plußsee (North German eutrophic lake). *Microbial Ecology* 13: 229–248.

Cunningham, H.W. and R.G. Wetzel. 1989. Kinetic analysis of protein degradation by a freshwater wetland sediment community. *Applied and Environmental Microbiology* 55: 1963–1967.

Engasser, J.M. and C. Horvath. 1974a. Inhibition of bound enzymes. I. Antienergistic interaction of chemical and diffusional inhibition. *Biochemistry* 13: 3845–3849.

Engasser, J.M. and C. Horvath. 1974b. Inhibition of bound enzymes. II. Characterization of product inhibition and accumulation. *Biochemistry* 13:3849–3858.

Fogg, G.E. 1977. Excretion of organic matter by phytoplankton. *Limnology and Oceanography* 22: 576–577.

Ford, T.E. and M.A. Lock. 1987. Epilithic metabolism of dissolved organic carbon in boreal forest rivers. *Federation of European Microbiology Societies Microbiology Ecology* 45: 89–97.

Francko, D.A. 1986. Epilimnetic phosphorus cycling: Influence of humic materials and iron on coexisting major mechanisms. *Canadian Journal of Fisheries and Aquatic Sciences* 43: 302–310.

Haslam, E. 1988. Plant polyphenols (syn. vegetable tannins) and chemical defense—a reappraisal. *Journal of Chemical Ecology* 14: 1789–1805.

Haslam, E. and T.H. Lilley. 1988. Natural astringency in foodstuffs—a molecular interpretation. *CRC Critical Reviews in Food Science and Nutrition* 27: 1–40.

Hobbie, J.E. (editor). 1980. *Limnology of Tundra Ponds*. Dowden, Hutchinson and Ross, Inc., Stroudsburg. 514 pp.

Hulett, F.M. 1986. The secreted alkaline phosphatase of *Bacillus licheniformis* MC14: Identification of a possible precursor. pp. 109–127 in Ganesan, A.T.and Hoch, J.A. (editors), *Bacillus Molecular Genetics and Biotechnology Applications*. Academic Press, Inc., New York.

Hutchinson, G.E. 1957. *A Treatise on Limnology. I. Geography, Physics, and Chemistry*. John Wiley and Sons, Inc., New York. 1015 pp.

Hutchinson, G.E. 1975. *A Treatise on Limnology. III. Limnological Botany*. John Wiley and Sons, Inc., New York. 660 pp.

Jones, R.I., K. Salonen, and H. De Haan. 1988. Phosphorus transformations in the epilimnion of humic lakes: Abiotic interactions between dissolved humic materials and phosphate. *Freshwater Biology* 19: 357–369.

Ladd, J.N. 1972. Properties of proteolytic enzymes extracted from soil. *Soil Biology and Biochemistry* 4: 227–237.
Ladd, J.N. 1985. Soil enzymes. pp. 175–221 in Vaughan, D. and Malcolm, R.E. (editors), *Soil Organic Matter and Biological Activity*. M. Nijhoff/W. Junk Publishers, Dordrecht.
Ladd, J.N. and J.H.A. Butler. 1970. The effect of inorganic cations on the inhibition and stimulation of protease activity by soil humic acids. *Soil Biology and Biochemistry* 2: 33–40.
Ladd, J.N. and J.H.A. Butler. 1975. Humus-enzyme systems and synthetic, organic polymer-enzyme analogs. pp. 142–194 in Paul, E.A. and McLaren, A.D. (editors), *Soil Biochemistry*, vol. 4. M. Dekker, Inc., New York.
Mills, C. and J.N. Campbell. 1974. Production and control of extracellular enzymes in *Micrococcus sodonensis*. *Canadian Journal of Microbiology* 20:81.
Münster, U., Einiö, P. and J. Nurminen. 1989. Evaluation of the measurements of extracellular enzyme activities in a polyhumic lake by means of studies with 4-methylumbelliferyl-substrates. *Archiv für Hydrobiologie* 115: 321–337.
Otsuki, A. and R.G. Wetzel. 1972. Coprecipitation of phosphate with carbonates in a marl lake. *Limnology and Oceanography* 17: 763–767.
Otsuki, A. and R.G. Wetzel. 1973. Interaction of yellow organic acids with calcium carbonate in a marl lake. *Limnology and Oceanography* 18: 490–493.
Overbeck, J. and H.D. Babenzien. 1963. Nachweis von frien Phosphatasen, Amylase und Saccharase im Wasser eines Teiches. *Naturwissenschaften* 50: 571–572.
Overbeck, J. and H.D. Babenzien. 1964. Über den Nachweis von freien Enzymen im Gewässer. *Archiv für Hydrobiologie* 60: 107–114.
Perdue, E. M. and E.T. Gjessing. 1990. *Organic Acids in Aquatic Ecosystems*. John Wiley and Sons, Chichester. 345 pp.
Pettersson, K. 1980. Alkaline phosphatase activity and algal surplus phosphorus as phosphorus-deficiency indicators in Lake Erken. *Archiv für Hydrobiologie* 89: 54–87.
Rego, J.V., Billen, G., Fontigny, A. and M. Somville. 1985. Free and attached proteolytic activity in water environments. *Marine Ecology Progress Series* 21: 245–249.
Reichardt, W., Overbeck, J. and L. Steubing. 1967. Free dissolved enzymes in lake waters. *Nature* 216: 1345–1347.
Scheffer, F., Ziechmann, W. and W. Rochus. 1962. Die Wirkung synthetischer Huminsaüren auf Phosphatasen. *Naturwissenschaften* 49: 131–132.
Sharp, J.H. 1977. Excretion of organic matter by marine phytoplankton: Do healthy cells do it? *Limnology and Oceanography* 22: 381–399.
Spencer, C.M., Cai, Y., Martin, R., Gaffney, S.H., Goulding, P.N., Magnolato, D., Lilley, T.H. and E. Haslam. 1988. Polyphenol complexation—some thoughts and observations. *Phytochemistry* 27: 2397–2409.
Stewart, A.J. and R.G. Wetzel. 1981. Dissolved humic materials: Photodegradation, sediment effects, and reactivity with phosphate and calcium carbonate precipitation. *Archiv für Hydrobiologie* 92: 265–286.
Stewart, A.J. and R.G. Wetzel. 1982a. Influence of dissolved humic materials on carbon assimilation and alkaline phosphatase activity in natural algal-bacterial assemblages. *Freshwater Biology* 12: 369–380.
Stewart, A.J. and R.G. Wetzel. 1982b. Phytoplankton contribution to alkaline phosphatase activity. *Archiv für Hydrobiologie* 93: 265–271.
Tipping, E., Backes, C.A. and M.A. Hurley. 1988. The complexation of protons, aluminum and calcium by aquatic humic substances: A model incorporating binding-site heterogeneity and macroionic effects. *Water Research* 22: 597–611.
Walther, K. and L. Fries. 1976. Extracellular alkaline phosphatase in multicellular marine algae and their utilization of glycerophosphate. *Physiologia Plantarum* 36: 118–122.
Wetzel, R.G. 1979. The role of the littoral zone and detritus in lake metabolism. *Archiv für Hydrobiologie Beihefte Ergebnisse Limnologie* 13: 145–161.
Wetzel, R.G. 1981. Longterm dissolved and particulate alkaline phosphatase activity in

a hardwater lake in relation to lake stability and phosphorus enrichments. *Verhandlungen der Internationalen Vereinigung für theoretische und angwandte Limnologie* 21: 337–349.

Wetzel, R.G. 1983. *Limnology*, 2nd edition. Saunders College Publishing, Philadelphia. 860 pp.

Wetzel, R.G. 1984. Detrital dissolved and particulate organic carbon functions in aquatic ecosystems. *Bulletin of Marine Sciences* 35: 503–509.

Wetzel, R.G. 1990. Land-water interfaces: Metabolic and limnological regulators. Edgardo Baldi Memorial Lecture. *Verhandlungen Internationale Vereinigung für theoretische und angwandte Limnologie* 24: 6–24.

Yamane, K. and B. Maruo. 1978. Purification and characterization of extracellular soluble and membrane-bound insoluble alkaline phosphatases possessing phosphodiesterase activities in *Bacillus subtilis*. *Journal of Bacteriology* 134: 100.

3

Environmental Control of the Synthesis and Activity of Aquatic Microbial Ectoenzymes

Ryszard J. Chróst

3.1 Introduction

The majority (>95%) of organic matter in aquatic environments is composed of polymeric, high-molecular-weight compounds (Allen, 1976; Romankevich, 1984; Cole et al., 1984; Thurman. 1985; Münster and Chróst, 1990). Because the passage of organic molecules across the microbial cytoplasmic membrane is an active process requiring specific transport enzymes (permeases), only small (low-molecular-weight) and simple molecules can be directly transferred from the environment into the cell (Rogers, 1961; Payne, 1980a; Geller, 1985). This means that only a small portion of the total dissolved organic matter (DOM) is readily utilizable in natural waters (Münster, 1985; Azam and Cho, 1987; Jørgensen, 1987), and that the majority of DOM cannot be directly transported to microbial cells because of the large size of its molecules.

According to our operational definition, utilizable dissolved organic matter (UDOM) is composed of readily transportable molecules, which may directly cross the cell membrane using microbial active transport systems (Münster and Chróst, 1990). The concentrations of UDOM in aquatic environments, however, are vanishing low (Jørgensen, 1987; Chróst et al., 1989; Münster and Chróst, 1990), often in the pico to nanomolar range, thereby limiting the rate of growth and metabolism of heterotrophic bacteria. Recent studies, however, have indicated that bacteria can contribute significantly to the total microplankton biomass (Chróst, 1988; Chróst et al., 1989), and that the bacterial secondary production is comparable to phytoplankton primary production (Bell and Kuparinen, 1984; Riemann and Søndergaard, 1986).

This apparent "bacterial paradox" (Chróst, 1991) between the low concentration of UDOM and the high biomass and production of aquatic bacteria indicates that heterotrophic bacteria must also be able to efficiently utilize polymeric DOM. In order for polymeric substrates to be available for bacterial metabolism, however, they have to undergo preliminary transformation involving enzymatic step-wise depolymerization and hydrolysis. Heterotrophic bacteria are excellent producers of these hydrolytic enzymes, and their hydrolysis of polymers is an acknowledged rate-limiting step in the microbial utilization of dissolved and particulate organic matter in aquatic environments (Hoppe, 1983; 1986; Haemejko and Chróst, 1986; Chróst et al., 1986; 1989; Hoppe et al., 1988; Chróst, 1989; 1990a; Chróst and Overbeck, 1987; 1990).

Three terms are commonly used for the enzymes involved in the transformation and degradation of polymeric substrates outside the cell membrane. They are: "ectoenzymes" (Chróst, 1990a), "extracellular enzymes" (Priest, 1984), and "exoenzymes" (Hoppe, 1983). In this chapter, the term "ectoenzyme" is used, according to Chróst's definition (Chróst, 1990a), to refer to any enzyme that is secreted and actively crosses the cytoplasmic membrane but remaining associated with its producer. Ectoenzymes are cell-surface-bound or periplasmic enzymes that can react with polymeric substrates outside the cell. An extracellular enzyme occurs in free form dissolved in the water and/or is adsorbed to surfaces others than those of its producer (e.g., detrital particles, clay material). Extracellular enzymes in waters may be actively secreted by intact viable cells, they can be liberated into the environment after cell damage or lysis, and/or they may result from zooplankton "sloppy feeding" on algal cells and from protozoan grazing on bacteria.

Ectoenzymes and extracellular enzymes (in contrast to intracellular enzymes) react outside the cell, and most of them are hydrolases. The ectoenzymes which cleave polymers by splitting the sensitive linkage located in the interior of the substrate molecule and form fragments of intermediate size I propose to call as endoectoenzymes (e.g., amino-endopeptidases act on the centrally located peptide bonds and liberate peptides; Murgier et al., 1976). Those ectoenzymes which attack the substrate by a consecutive splitting of a monomeric product from the end of the molecule are sensu stricto exoenzymes, and I suggest to call them exo-ectoenzymes (e.g., amino-exopeptidases hydrolyze peptide bonds adjacent to terminal α-amino or α-carboxyl groups and liberate free amino acids; Law, 1980).

Microbial cells living in aquatic ecosystems are influenced by a variety of environmental factors which affect the molecular control of their enzyme synthesis. Thus, the signal(s) for appropriate gene expression, and consequent enzyme production, reaches the cell from the surrounding environment. The most important step in the control mechanism of the synthesis of the ectoenzymes in both prokaryotic cells (bacteria) and eukaryotic cells (algae) is exerted primarily at the level of transcription (Priest, 1984). This has obvious advantages to the microorganisms, since it is wasteful to produce transcripts that may not be translated. The regulation of transcription requires modulation of mRNA synthesis upon induction or repression-derepression of an ectoenzyme.

It is obvious, however, that conditions in the aquatic environment are unfavorable for ectoenzymes. First, the substrate concentration is usually very low and highly variable. Many substrates may be insoluble; exist in intimate association with other compounds; and/or be bound to humic substances (Münster and Chróst, 1990), colloidal organic matter, and detritus. Therefore, these conditions are suboptimal for the coupling of an ectoenzyme to its substrate. Next, an ectoenzyme itself may be separated from the maternal cell and bind to suspended particles and humic materials, or it may be exposed to a variety of inhibitors present in the water. Furthermore, an ectoenzyme may also be denaturated by physical and chemical factors in the aquatic environment, or hydrolyzed by proteases. For an ectoenzyme to be of benefit to its producer microorganism, it is obvious that it must avoid destruction long enough to locate its substrate. Moreover, even if an ectoenzyme overcomes these obstacles and binds with its substrate, the physical and chemical conditions of the reaction medium may be unsuitable for catalysis (e.g., suboptimal pH or temperature, presence of inhibitors, absence of activators, suboptimal ionic strength, etc.). Nevertheless, the fact remains that various aquatic microorganisms produce ectoenzymes capable of reacting with many polymeric substrates, and that microbial growth is dependent on the products of ectoenzymatic reactions (Chróst and Overbeck, 1987; Chróst, 1988; 1989; 1990a; Chróst et al., 1989).

The aim of this chapter is to examine the environmental regulation of synthesis and activity of the most active microbial ectoenzymes in aquatic environments. The ectoenzymes that are responsible for the hydrolysis of the major organic constituents (β-linked polysaccharides, proteins, organophosphoric esters) in the DOM pool in lake water, such as β-D-glucosidase (βGlc), leucine-aminopeptidase (Leu-amp), and alkaline phosphatase (APase), are emphasized.

3.2 Experimental Design

Sampling procedure Water samples (4 liters) were taken from the euphotic zone (1 m depth) of the moderately eutrophic lake Plußsee (area 14.3 ha, max. depth 29 m, average depth 9.4 m, naturally eutrophic, Baltic-type lake located in East Holstein, Germany) and from the highly eutrophic Lake Mikołajskie (area 460 ha, max. depth 27 m, average depth 12 m, Baltic-type lake located in the Mazurian Lake District, Poland). Immediately after collection, water samples were prefiltered through 100-μm-pore-size nylon plankton net (to remove large organisms) and transported in plastic (polyethylene) containers to the laboratory within 1 hour. Samples filtered through 0.2-μm-pore-size polycarbonate Nuclepore filters (vacuum filtration <300 Pa) were used for determination of enzyme extracellular activity.

Enzyme assays Alkaline phosphatase (APase) and β-glucosidase (βGlc) activities were measured based on an increase in fluorescence as the nonfluorescent methylumbelliferyl substrates (Table 3.1) were enzymatically hydrolyzed, leading to the production of the highly fluorescent molecule 4-methylumbelliferone (MUF). The amount of substrate cleaved was equivalent to the amount of highly fluorescent MUF anion generated by quenching the reaction with alkaline (pH 10.5) glycine-ammonium hydroxide buffer (0.05 M glycine + 0.2 M NH_4OH; Daniels and Glew, 1984). Quantification was achieved by calibrating the spectrofluorometer (Kontron SFM-25, Switzerland) with a standard solution of MUF (Sigma) prepared in filtered (Whatman GF/F glass fiber filter) lake water buffered with glycine-ammonium hydroxide solution (pH 10.5). MUF fluorescence was measured at 460 nm under 365 nm excitation.

L-Leucine-*p*-nitroanilide (Leu-PNA; Sigma) and L-leucine-4-methyl-coumarinylamide hydrochloride (Leu-MCA; Fluka) were used for spectrophotometric and fluorometric determination of leucine-aminopeptidase (Leu-amp) in Mikołajskie and Plußsee lake water, respectively. *p*-Nitroaniline (final product of the enzymatic hydrolysis of Leu-PNA) was measured in 5-cm cuvettes at 380 nm wavelength. The amount of Leu-PNA hydrolyzed was calculated from the slope of the linear regression of absorbance versus five-fold-concentration increases of *p*-nitroaniline (Sigma) standards dissolved in filtered (GF/F Whatman glass fiber filter) lake water. In fluorometric assay, Leu-amp activity was equal to the amount of highly fluorescent 7-amino-4-methylcoumarin (AMC) produced after Leu-MCA hydrolysis. The spectrofluorometer was calibrated with an AMC standard (Fluka) prepared in filtered (GF/F Whatman)

Table 3.1 Substrates and products used to measure the kinetics of three ectoenzymes

Enzyme	Substrate	Product measured (technique)
Alkaline phosphatase (APase)	4-Methylumbelliferyl phosphate (MUFP)	4-Methylumbelliferone (MUF; fluorometry: excitation 365 nm, emission 460 nm)
β-D-Glucosidase (βGlc)	4-Methylumbelliferyl-β-D-glucopyranoside (MUFGlp)	4-Methylumbelliferone (MUF; fluorometry: excitation 365 nm, emission 460 nm)
Leucine aminopeptidase (Leu-amp)	L-Leucine-4-methyl-7-coumarinylamide hydrochloride (Leu-MCA)	7-Amino-4-methyl-coumarin (AMC; fluorometry: excitation 380 nm, emission 440 nm)
	L-Leucine-*p*-nitroanilide (Leu-PNA)	*p*-Nitroaniline (spectrophotometry: 380 nm)

lake water. AMC fluorescence was determined at 380 nm (excitation) and 440 nm (emission).

Stock solutions of the substrates (Table 3.1) were prepared to a concentration of 5 mM in deionized autoclaved water and stored at a temperature of −25°C. The stock substrate solutions were thawed at room temperature and diluted with deionized water to 0.10, 0.25, 0.50, 1.0, and 2.5 mM immediately before assay. For all enzyme assays, 0.5 ml of substrate solutions were added to 4.5-ml, triplicate water samples, yielding final substrate concentrations in assays of 10, 25, 50, 100, 250, and 500 μM. Autoclaved lake water was used as a blank. Samples were incubated at 20°C for 2–4 h (fluorometric assays) or 12–24 h (spectrophotometric assays) in the dark. MUFGlp and MUFP hydrolysis was terminated by the addition of 0.5 ml glycine-ammonium hydroxide buffer (pH 10.5).

Enzyme kinetic data analysis Varying amounts of substrates were added to the samples to establish enzyme-substrate saturation and to enable the calculation of the kinetic parameters of studied enzymes. The enzymatic reactions followed Michaelis-Menten kinetics, and the plot of the initial velocity of reaction (v) against increased concentrations of substrate ($[S]$) gave a rectangular hyperbola:

$$v = (V_{max} \times [S])/(K_m + [S])$$

The parameters characterizing this equation were calculated from the experimental data. They are: V_{max}, the maximum velocity of enzyme reaction, which is theoretically attained when the enzyme has been saturated by an infinite concentration of substrate ($[S]$); and K_m, the Michaelis constant, which is numerically equal to the concentration of substrate for the half-maximal velocity ($1/2\ V_{max}$).

Since the relationship between the independent variable ($[S]$) and the dependent variable (v) is curvilinear, it is customary to plot the experimental data according to one of three linear transformations (Armstrong, 1983), and V_{max} and K_m are obtained from the slope and intercept. Such graphical methods, however, will theoretically only produce correct values for the kinetic parameters in the absence of analytical error (Dowd and Riggs, 1965). Unfortunately, all the measurements that were obtained were inevitably subject to some degree of imprecision, and therefore use of linearized equations, such as Lineweaver-Burk, Woolf, and Eadie-Hofstee equations, would not in practice give the correct values for the experimental data (Hałemejko and Chróst, 1986; Lundin et al., 1989; Chróst, 1990a).

One solution to this problem was to perform a nonlinear regression analysis on the original experimental data. The kinetic parameters then were calculated from the direct plot of reaction velocity (v) versus substrate ($[S]$) concentration using the IBM PC computer software program "Enzfitter" (Elsevier-Biosoft, U.K.) to determine the best fit of the rectangular hyperbola (Leatherbarrow, 1987).

Permeabilization and lysis of bacterial cells Permeabilized and lysed cell preparations were used for determination of enzyme activity in the study of enzyme location in aquatic bacteria. Bacterial cells, in filtered (1.2-μm-pore-size polycarbonate membrane filters; Nuclepore, U.S.A) lakewater samples, were permeabilized at 20°C for 30 min by their treatment with organic solvents, toluene and chloroform (4%), nonionic detergent Triton X-100 (0.5%), and ionic detergent sodium dodecyl sulfate (SDS; 0.5%). A lysozyme-EDTA procedure was used for rapid, gentle lysis of bacterial cells (Dobrogosz, 1981). Ten-ml lakewater samples (prefiltered through 1.2-μm-pore-size polycarbonate filters) were supplemented with Tris-hydrochloride buffer (final concentration 100 M, pH 8.0), 2.5 mg of EDTA, and 150 μg of lysozyme. Lysis was conducted at room temperature for 15 min.

Other analyses Bacterial biomass was determined by the ATP method, as previously described by Chróst and Overbeck (1987), and converted into the amount of organic carbon (μg org.C l^{-1}) according to Holm-Hansen (1984). Bacterial cell numbers were determined by epifluorescence microscopic method after acridine orange staining. Bacterial cell production and the growth rates were estimated by the [^{3}H-*methyl*]thymidine incorporation method according to Chróst et al. (1988). Chlorophyll$_a$, extracted with 90% methanol, was measured fluorometrically (Wetzel and Likens, 1979) with a Kontron spectrofluorometer SFM-25. Surplus phosphorus (i.e., the amount of phosphorus stored intracellularly in microplankton cells) was determined by the extraction method (Chróst and Overbeck, 1987).

3.3 Ectoenzyme Location in Bacteria

Bacterial community in aquatic environments is composed of Gram-negative and Gram-positive bacteria that have gross morphological differences in the cell wall structure and cell permeability. Gram-negative bacteria usually predominate in aquatic environments (Chróst, 1975) in contrast to Gram-positive bacteria which constitute the majority of microorganisms in soil ecosystems (Burns, 1983). The cell envelopes of most prokaryotic microorganisms (with the exception of *Spiroplasma* sp. and mycoplasmas) are characterized by the presence of two distinct components: an outer cell wall and inner cytoplasmic membrane. The cell wall maintains the shape of the cell and protects the mechanically fragile cytoplasmic membrane from rupture owing to the high osmotic pressure exerted on it by the cell cytoplasm. The cytoplasmic membrane controls the substrate and electron transport processes of the cell and is a site of biosynthesis of extracellular macromolecules (Hancock and Poxton, 1988). The cell wall and the cytoplasmic membrane of the microbial cell serve as physical and functional barriers controlling the transport of solutes to the cytoplasm and the passage of ectoenzymes and metabolites into the surrounding environment. A variety of microbial taxa synthesize and secrete different ectoenzymes, which allow the cells to communicate with their environment. Depending on the cell structure and the permeability of its outer layer, there are a number of possible locations for each ectoenzyme.

Gram-positive bacteria have a relatively simple cell wall structure. The cytoplasmic membrane of Gram-positive bacteria is encased within a thick peptidoglycan/teichoic or teichuronic acid layer, and these components form a mixed matrix (Rogers et al., 1980). In some species, an extracellular, usually polysaccharide, capsule surrounds the cell. Gram-positive cell walls do not normally exhibit any layering or fine structure. The cytoplasmic membrane of Gram-positive bacteria is relatively permeable to higher-molecular-weight compounds. Ectoenzymes may be temporarily restricted by the cell wall, and/or eventually diffuse into the environment. Some ectoenzymes, however, remain attached to the outer surface of the membrane and under certain growth conditions, such enzymes may be naturally released from the cell (Chróst, 1990a).

The envelope of Gram-negative bacteria is more complex (Chróst, 1990a). The cell wall peptidoglycan appears to be less substantial and not as closely associated with the cytoplasmic membrane as the equivalent structure in Gram-positive bacteria. The most notable feature of the Gram-negative envelope is the presence of a second membrane to the exterior of the peptidoglycan. This outer membrane constitutes a barrier making the surface of Gram-negative bacteria less permeable to a wide variety of molecules than that of Gram-positive bacteria. In general, the outer membrane makes the Gram-negative envelope impermeable to hydrophobic compounds and high-molecular-weight hydrophilic molecules (Nikaido and Nakae, 1979). The region of Gram-negative envelope, situated between the cytoplasmic and outer membrane, is known as the periplasmic space. The space constitutes some 20% to 40% of the total cell volume. The restrictive permeability properties of the outer membrane ensure that periplasmic components cannot easily leak into the environment. The periplasm contains a series of nutrient-binding proteins, that are essential components of certain active transport systems. It exhibits a variety of ectoenzymes which are involved in the degradation of metabolizable compounds that are too large or too highly charged to pass through the cytoplasmic membrane. The substrates of the periplasmic ectoenzymes cannot normally be taken up by the cell, but after enzymatic degradation, the reaction products can be translocated to the cytoplasmic interior and metabolized. Therefore, ectoenzymes of Gram-negative bacteria may be located outside the cytoplasmic membrane, either in the periplasm, or attached to the inner or outer surface of the outer membrane. In addition, they are sometimes secreted into the environment. However, the frequency of direct secretion of ectoenzymes into the surrounding environment by the intact living Gram-negative cells is significantly lower in comparison to that of Gram-positive bacteria.

Enzymatic activity in natural waters results both from the cell-surface-bound microbial enzymes (ectoenzymes) and from free, dissolved in the water extracellular enzymes. Extracellular enzymes may originate from the ectoenzyme pool, due to their washout from the cell surface or periplasmic space (see discussion in Chapter 13), and/or from the pool of intracellular enzymes after lysis or cell damage by the grazers. Until now, there is no evidence that extracellular enzymes in aquatic environments are actively secreted by intact living microorganisms, however, the fact remains that their activity is often detected in water and sediment (Maeda and Taga, 1973; Rego et al., 1985; Chróst et al., 1986; Rosso and Azam, 1987; Paul et al., 1987; Chróst, 1989; Mayer, 1989; see also Chapters 15 and 16 in this volume). It seems that one of the responsible mechanisms for the release of extracellular enzymes in aquatic environments are the changes of microbial cell permeability that lead to liberation of the periplasmic and cell surface-bound ectoenzymes.

Permeability of bacterial cells and the consequent release of extracellular enzymes in lakewater samples was strongly affected by the Triton X-100 and EDTA-lysozyme treatment (Figure 3.1). Cell-bound and extracellular enzymes contributed from 85% to 92%, and from 8% to 15% to the total enzyme activity in control samples, respectively. Detergent and lysozyme treatment of the lakewater samples resulted in the significant release of previously cell-bound enzymes to the water, and the enzyme extracellular activity approached 90 to 100% of the total activity. Supplementation of water samples with Triton X-100 and EDTA-lysozyme treatment did not significantly affect activity of alkaline phosphatase, β-glucosidase, and aminopeptidase (Figure 3.2). Toluene and chloroform addition caused slight inhibition (ca. 25%) of enzyme activities. Sodium dodecyl sulfate (SDS), the most commonly used ionic detergent for cell

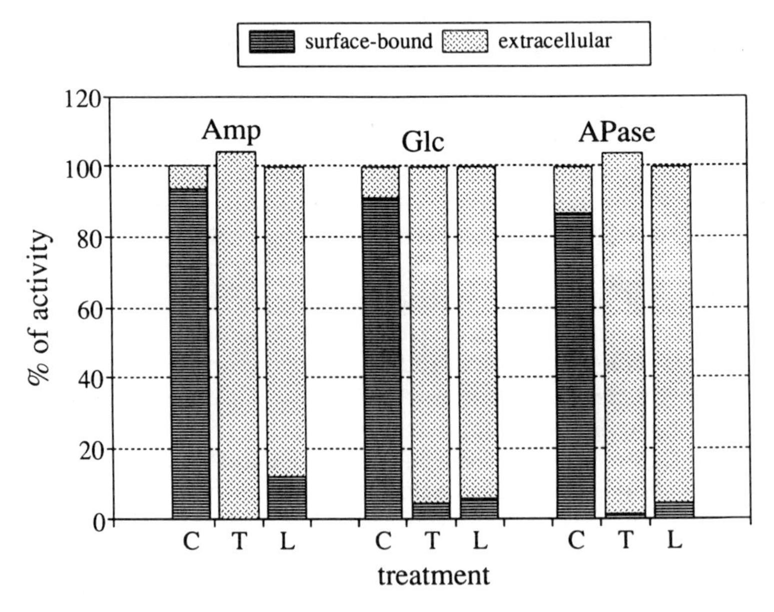

Figure 3.1 Distribution of surface-bound and extracellular activities of bacterial aminopeptidase (Amp), β-glucosidase (Glc), and alkaline phosphatase (APase) in control (C), Triton X-100 (T), and EDTA-lysozyme (L) treated Plußsee water.

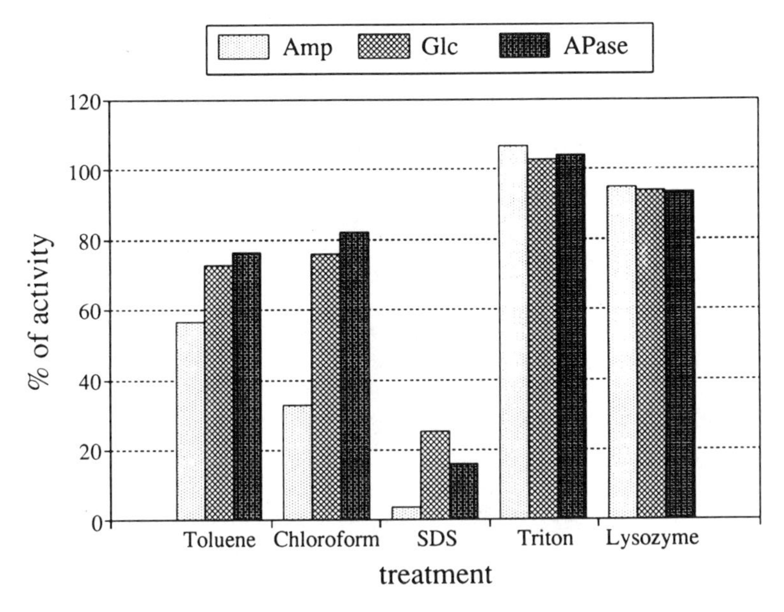

Figure 3.2 Effect of the addition of toluene, chloroform, sodium dodecyl sulfate (SDS), Triton X-100, and lysozyme on the activity of bacterial aminopeptidase (Amp), β-glucosidase (Glc), and alkaline phosphatase (APase) in Plußsee water. Enzyme activity in control samples = 100%.

fractionation (Gerhardt et al., 1981), strongly inhibited (75–95%) activity of the studied enzymes. Because SDS binds strongly to proteins and enzymes resulting in the unfolding and irreversible denaturation of proteins, it can not be applied for cell permeabilization for further enzyme activity determination.

Fractionation of aminopeptidase, β-glucosidase and alkaline phosphatase activities of lakewater showed that from 75% to 88% of the total enzyme activity associated with cell-surface of bacteria (Figure 3.3). Activities of the intracellular and extracellular pool of enzymes were low. Intracellular enzymes contributed from 2% (β-glucosidase) to 8% (aminopeptidase) to the total activity of samples. Activity of the extracellular enzymes, dissolved in the water, constituted from 3% (aminopeptidase) to 18% (alkaline phosphatase) of the total activity.

Close association of ectoenzymes with the cell-surface optimizes modulation of their activity and synthesis and amplifies an efficient utilization of high-molecular-weight-substrates by bacteria. The optimum strategy for aquatic microorganisms is to perform enzymatic hydrolysis of polymers in close proximity to the cell surface where uptake systems are also present. It would be energetically wasteful to maintain hydrolysis in the macroenvironment, as well as, to take up the products of enzyme reaction after they have diffused and become greatly diluted in water. The ectoenzymatic liberation of utilizable substrates in close proximity to the cell creates and sustains high substrate concentrations on the cell surface. Thus, active transport systems of the cell membrane are

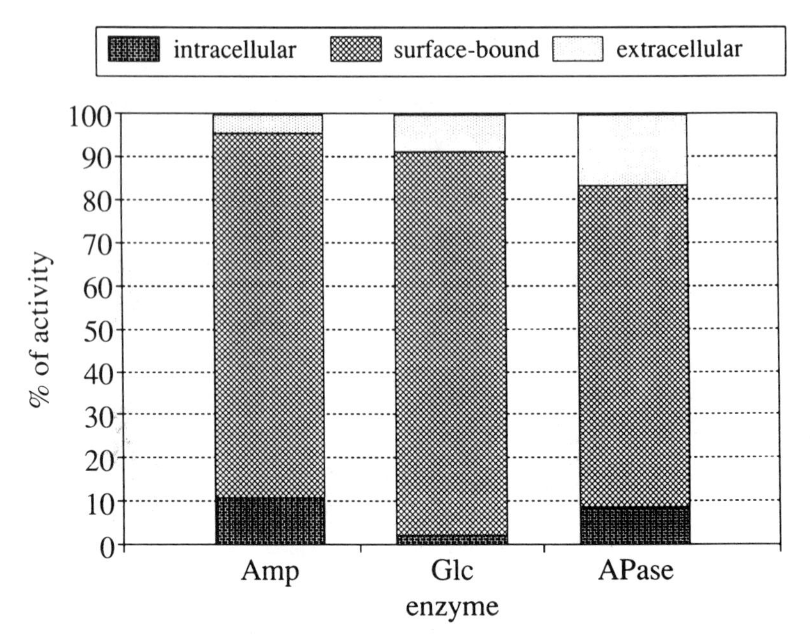

Figure 3.3 Distribution of cell-bound and extracellular activity of bacterial aminopeptidase (Amp), β-glucosidase (Glc), and alkaline phosphatase (APase) in Plußsee water.

exposed to an increased concentration gradient of the substrate, which facilitates its rapid uptake. Recent studies have demonstrated that enzymatic hydrolysis of dissolved organic polymers and the uptake of the low-molecular-weight products of hydrolysis are tightly coupled processes (Hollibaugh and Azam, 1983; Somville and Billen, 1983; Ammerman and Azam, 1985; Hoppe, 1988; Chróst, 1989; Chróst et al., 1989; Tamminen, 1989). The microbial hydrolysis of dissolved polymeric organic compounds with surface-associated ectoenzymes and its coupling with uptake of products is a perfect strategy for the survival of microorganisms in aquatic environments, where readily utilizable substrates are often at low, growth-limiting concentrations. Ectoenzymatic hydrolysis greatly enhances the spectrum of organic and inorganic compounds that can enter microbial metabolism.

3.4 Alkaline Phosphatase

Alkaline phosphatase (orthophosphoric monoester phosphohydrolase; E.C. 3.1.3.1) is one of the most frequently studied and best-known ectoenzymes in aquatic environments. Alkaline phosphatase (APase) is a dimeric molecule of molecular weight about 160,000 daltons and is composed of two subunits. The two subunits interact with negative cooperativity and consequently the enzyme shows "half-of-the-sites" reactivity at physiological pH, but displays Michaelis-

Menten kinetics at alkaline pH (Lazdunski, 1974; McComb et al., 1979). Each subunit contains a tightly bound atom of zinc, which is essential for the structural integrity of the enzyme, and a second, less tightly bound zinc atom, which is involved in the catalytic process (Crofton, 1982). APase activity is stimulated by magnesium, which binds to an effector site on each subunit that is different from the site for zinc (Linden et al., 1977). The magnesium ions exert an allosteric effect on the enzyme to stimulate the dephosphorylation process. Zinc ions can also bind to the magnesium site with greater affinity than the magnesium itself, resulting in loss of activity. Common inhibitors of APase are chelators of divalent ions, such as EDTA (Bretaudiere and Spillman, 1984).

APase catalyzes the hydrolysis of a variety of phosphate esters, including esters of primary and secondary alcohols, sugar alcohols, cyclic alcohols, phenols, and amines, liberating inorganic phosphate (P_i; see Figure 13.1 in Chapter 13). Phosphodiesters are not hydrolyzed. The enzyme also hydrolyses polyphosphates (PP_i). APase is also known to catalyze a variety of transphosphorylation reactions; for instance, the enzyme from *E. coli* catalyzes both the hydrolysis of PP_i and the transfer of PO_4^{3-} group from PP_i (and from a number of nucleoside di- and triphosphates and from mannose-6-phosphate) to glucose, forming glucose-6-phosphate (Barman, 1969).

The presence of APase has been demonstrated in filtrates of algal cultures and in fresh and seawaters, as a dissolved (extracellular) enzyme, and in phytoplankton, bacteria, zooplankton, and protozoans (Aaronson, 1981; Wynne and Gophen, 1981; Stewart and Wetzel, 1982; Hałemejko and Chróst, 1984; Chróst et al., 1984, 1986; Chróst and Overbeck, 1987). Most of the data in the literature indicate that APase is produced in copious amounts when inorganic phosphate becomes limiting in natural waters or in culture medium, suggesting derepression of this enzyme (Siuda, 1984; Chróst and Overbeck, 1987; Siuda and Chróst, 1987). Moreover, measurement of APase activity in algae has been suggested as a procedure for detecting phosphorus deficiency or restricted P_i availability in aquatic ecosystems (Healey and Hendzel, 1980; Pettersson, 1980; Vincent, 1981; Gage and Gorham, 1985).

It is well documented that the synthesis of many ectoenzymes produced by aquatic microorganisms is repressed by the end product that is derived from the substrate and accumulates in the cell or in surrounding environment (Chróst, 1990a). The repression of synthesis of APase by PO_4^{3-} in microalgae and bacteria is probably one of the best known examples (Siuda and Chróst, 1987). In Plußsee, the specific activity of APase significantly decreased when the ambient P_i concentrations were higher than 15 μg $P\text{-}PO_4^{3-}$ l^{-1} (Chróst and Overbeck, 1987). Derepression of APase synthesis in lake microplankton, however, did not occur immediately at the beginning of summer P_i depletion in the photic zone of the lake because of intracellular pools of orthophosphate and polyphosphates (P_{surpl}) stored in the microbial cells. Finally, when all external and internal phosphorus reservoirs were nearly depleted, microplankton began to synthesize APase with high specific activity to hydrolyze phosphate esters as an alternative P source to compensate for the lack of P_i.

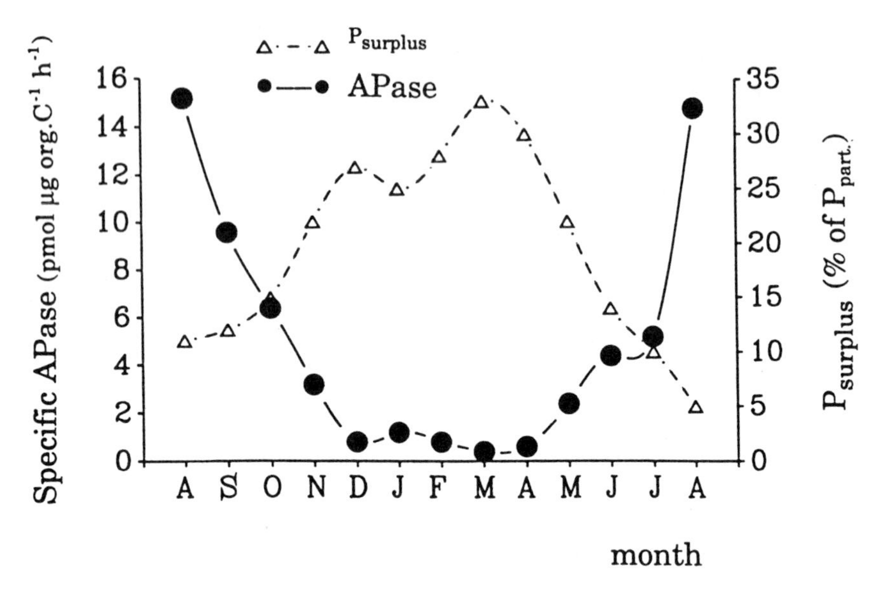

Figure 3.4 Annual course of specific activity of (left) alkaline phosphatase (APase) and (right) surplus phosphate ($P_{surplus}$) stored intracellularly in microplankton in the euphotic zone of Plußsee.

When ambient P_i concentrations in Plußsee lake water were high (after spring and autumn turnover of the lake), microplankton accumulated phosphorus in the intracellular pool, and P_{surpl} constituted from 23–34% (average 30%) of the total particulate phosphorus (P_{part}; Figure 3.4). In parallel with the decrease of P_i concentrations in the water (data not shown), I observed a decrease of P_{surpl} content in microplankton, i.e., part of the phosphorus stored intracellularly was utilized to support the cellular phosphate metabolism of microplankton. Simultaneously to the gradual decrease in P_i and P_{surpl} content, the specific APase activity also increased slowly from ca. 1 to 6 pmol µg org.C^{-1} h^{-1}. However, when P_{surpl} constituted less than 15% of the P_{part} (P_i concentrations were at the same time below 1 µg P-PO_4^{3-} l^{-1}), the microplankton reacted rapidly and synthesized APase with high specific activity (ca. 15 pmol µg org.C l^{-1}; Figure 3.4).

Several authors have assumed that high phosphatase activity can be derepressed or activated by low P_i concentrations (Siuda, 1984). It is not quite correct, however, to conclude that APase synthesis is derepressed directly by low P_i concentrations. The mechanism of APase derepression is regulated by the intracellular phosphate pool in microbial cells, which obviously depends on ambient P_i concentrations. Consequently, the synthesis of APase by microplankton in Plußsee was repressed by high intracellular levels of P_{surpl}. On the other hand, P_i was a strong competitive inhibitor of APase activity in the algal-size fraction of microplankton (Chróst and Overbeck, 1987), but inhibition of

APase activity of bacteria was only slight (Chróst et al., 1986). When PO_4^{3-} was added to the size fractionated water samples, I observed a rapid and marked increase of the apparent Michaelis constant (K_m) value of APase of the algal origin (Table 3.2). The hallmark of competitive inhibition of APase by PO_4^{3-} is that the orthophosphate combines with the enzyme in such a way that it competes with a substrate for binding at the active site, thus decreasing the enzyme affinity for the substrate, and therefore, inhibits the initial velocity of the reaction. Competitive inhibition was reversible and was overcome by increased substrate concentration, therefore the maximum velocity (V_{max}) of the reaction was almost unchanged (the standard deviation for V_{max} in all assays was ±12%).

Bacterial APase, in lake water samples (using a microplankton size fraction of 0.2–1.5 μm) supplemented with different organic solutes naturally present in the water, was induced or repressed, depending on the compound added (Figure 3.5). Supplementation of samples with additional phosphoester substrates (cAMP, ATP, glucose-6-phosphate) induced APase synthesis, and bacteria produced the enzyme with much higher specific activity and affinity for substrate in comparison to the control samples. The highest specific APase activity was observed in samples when glucose-6-phosphate (78 fmol $cell^{-1}$ h^{-1}) and ATP (61 fmol $cell^{-1}$ h^{-1}) were added. cAMP only slightly increased the enzyme specific activity, but markedly decreased the K_m value.

The repression of bacterial APase synthesis was observed, however, when samples were supplemented with easily utilizable, natural substrates for bac-

Table 3.2 Competitive inhibition of alkaline phosphatase activity by PO_4^{3-} added to the crude extract of enzyme of size-fractionated Plußsee microplankton samples[a]

APase source	Initial K_m (μmol l^{-1})	PO_4^{3-} added (μmol l^{-1})	K_m after treatment (μmol l^{-1})	Change of K_m (%)
Summer stratification (August)				
Algae	25 (±0.8)	1	35 (±1.0)	+40
		3	48 (±3.2)	+92
		5	66 (±2.8)	+164
Bacteria	20 (±1.1)	1	19 (±1.6)	−5
		3	22 (±3.0)	+10
		5	27 (±1.5)	+35
Spring homothermy (April)				
Algae	45 (±2.1)	1	58 (±3.1)	+29
		5	85 (±1.7)	+89
		10	125 (±2.3)	+178
Bacteria	26 (±1.5)	1	24 (±2.2)	−8
		5	31 (±0.7)	+19
		10	43 (±1.2)	+65

[a]Homogenates of APase of algal (microplankton size fraction > 3.0 μm) origin and bacterial (microplankton size fraction of 0.2–1.5 μm) origin were extracted with an n-butanol + chloroform mixture, according to Morton (1954).

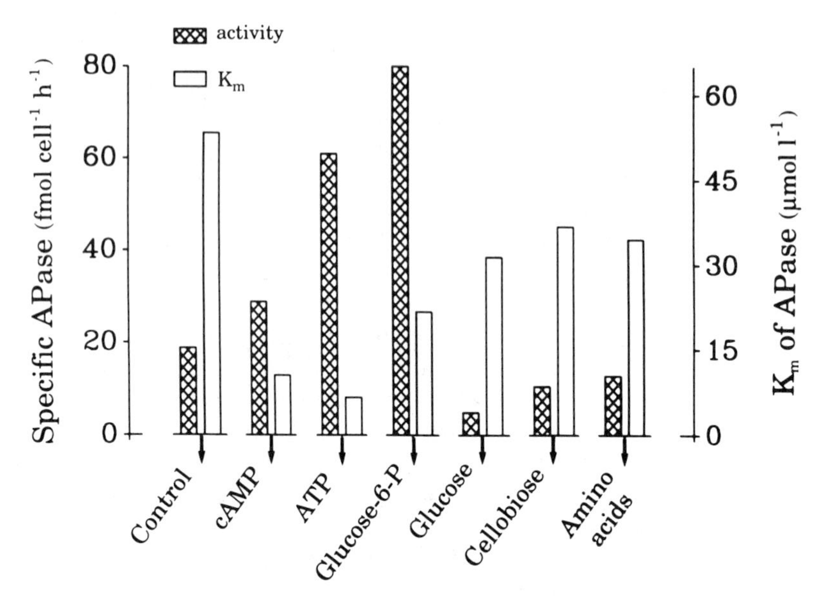

Figure 3.5 Specific activities of bacterial alkaline phosphatase (APase) and the Michaelis constants (K_m) in Plußsee water supplemented with various organic compounds. Water samples (200 ml) were prefiltered through 1.5-µm-pore-size polycarbonate filters (Nuclepore), supplemented with organic substrates (100 µM) and incubated at the in situ temperature (15°C). APase kinetics were measured after 24 h of incubation.

teria in Plußsee, such as glucose, cellobiose, and amino acids (Chróst et al., 1989). APase specific activity decreased significantly, and also the affinity of the enzyme for MUFP decreased in comparison to the samples enriched with phosphoester compounds. The catabolic repression of APase by readily assimilable, low-molecular-weight substrates has not been reported in the literature. Comparison of the results of experimental variants, where glucose-6-phosphate and glucose were added (Figure 3.5), indicated that aquatic bacteria can produced APase to supply their metabolism with organic carbon substrates (glucose, in this example), which (apart from PO_4^{3-} ions) are also final products of phosphoester hydrolysis.

3.5 β-Glucosidase

β-Glucosidase (βGlc; β-D-glucoside glucohydrolase, E.C. 3.2.1.21) was found to be produced by heterotrophic microorganisms (predominately by bacteria and fungi) in waters and sediments of both freshwater and marine environments (Hoppe, 1983; Somville, 1984; King, 1986; Meyer-Reil, 1987; Chróst et al., 1989). The enzyme has high glycone specificity but low aglycone specificity. Aryl glycosides are in general better substrates than alkyl glycosides. βGlc is a broad-specificity enzyme that catalyzes the hydrolysis of β-linked (1→2, 1→3,

1→4, 1→6) disaccharides of glucose, celluhexose, and carboxymethylcellulose (Barman, 1969).

Temperature of incubation had a pronounced effect on the kinetic parameters of βGlc in Plußsee water samples. At the optimal temperature of 28°C, the enzyme had lowest K_m values, i.e., highest affinity for substrate. βGlc also had high substrate affinity at temperatures between 15 and 30°C, however, both low (2–15°C) and high temperatures of incubation significantly decreased the affinity of enzyme for substrate. The V_{max} of βGlc in lake water was strongly dependent on the pH of incubation; activity increased smoothly up to the optimal pH of 8.0 and rapidly decreased at pHs above 8.5. The pH optimum of βGlc tended to reflect the in situ value of the aquatic environment (King, 1986; Chróst, 1989).

During the spring phytoplankton bloom, the specific V_{max} of βGlc showed a distinct temporal pattern of activity. β-Glucosidase activity was low (mean 255±22 pmol μg org.C^{-1} h^{-1}) during the first stage of bloom development when algae grew rapidly (Figure 3.6). The specific activity of the enzyme began to increase gradually during the bloom breakdown and reached the highest values (1,147±132 pmol μg org.C^{-1} h^{-1}) during the late stage of phytoplankton collapse. The increase in β-glucosidase activity was proportional to the increase in biomass and production of bacteria. These results confirmed previous reports that most of the βGlc activity (>95%) in seawater and freshwater samples was associated with the bacterial-size fraction (Hoppe, 1983; Somville, 1984; Chróst and Overbeck, 1990). The above results of field studies suggested that β-glucosidase in lake water was under the control of complex regulatory mechanisms. These regulatory mechanisms may include: (i) induction/derepression

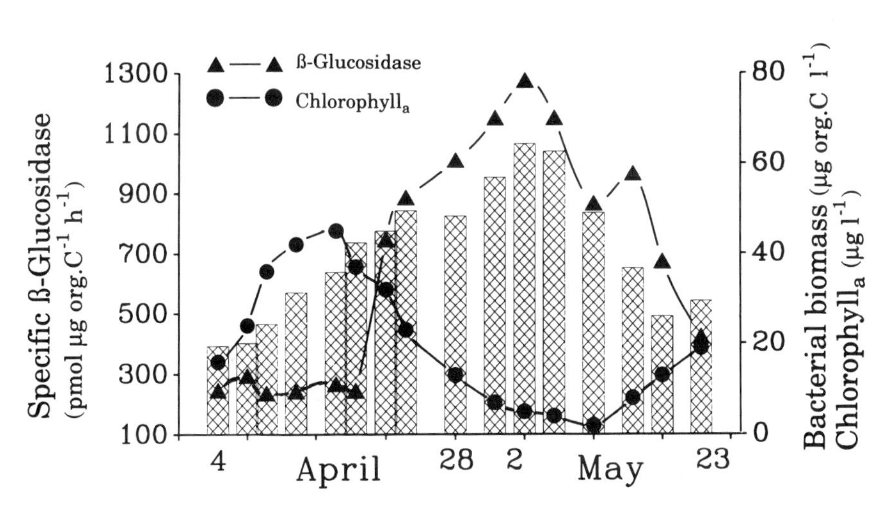

Figure 3.6 Specific activity of β-glucosidase (βGlc calculated per μg org.C of bacterial biomass), the chlorophyll$_a$ content, and the bacterial biomass (bars) during the course of a spring phytoplankton bloom in the euphotic zone of Plußsee.

of synthesis, (ii) repression of synthesis, and/or (iii) inhibition of activity (Chróst, 1989; 1990a).

To verify some of the postulated regulatory mechanisms of βGlc synthesis and activity, I supplemented lake water samples with different organic compounds, naturally present in the water (e.g., potential products of phytoplankton decomposition). Depending on the compound added, results of these experiments revealed the stimulation or inhibition of βGlc synthesis and activity in natural bacterial assemblages in Plußsee water (Figure 3.7). Phosphorylated esters (cAMP, ATP, glucose-6-phosphate) did not affect the specific activity of βGlc. These compounds decreased the affinity of the enzyme for substrate (K_m values increased). The lowest specific activity of βGlc was found in samples supplemented with glucose and amino acids. Addition of cellobiose (a disaccharide composed of two subunits of β-linked D-glucose; and a by-product of cellulose degradation in the waters that is a very suitable substrate for βGlc) strongly induced βGlc synthesis, and bacteria produced the enzyme with high specific activity and affinity.

Readily utilizable bacterial substrates added to water samples presumably acted as repressors, and they caused strong catabolic repression of enzyme synthesis (Chróst and Overbeck, 1990). Moreover, in parallel to the repression of synthesis of "new enzyme protein", D-glucose inhibited the initial velocity of enzyme reaction, because it drastically decreased the affinity of βGlc for substrate (K_m value increased twofold). Competitive inhibition of βGlc activity in

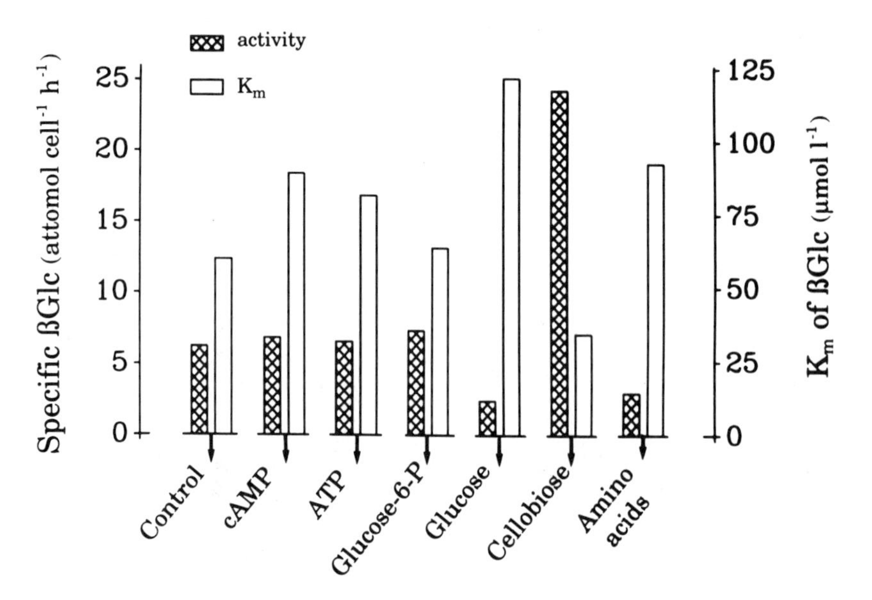

Figure 3.7 Specific activities of bacterial β-glucosidase (βGlc) and the Michaelis constants (K_m) in Plußsee water supplemented with various organic substrates (see Figure 3.5 for assay conditions).

lake water by D-glucose (which is the end product of βGlc catalysis) has been well documented in previous reports (Chróst, 1989; 1990a; Chróst and Overbeck, 1990).

3.6 Leucine Aminopeptidase

Leucine aminopeptidase (L-leucyl-peptide hydrolase, E.C. 3.4.1.1) is widely distributed in aquatic habitats (Somville and Billen, 1983; Halemejko and Chróst, 1986; Chróst et al., 1986; 1989; Hoppe et al., 1988; Mayer, 1989). Most studies of aquatic environments have indicated that ectoenzymatic activity of leucine aminopeptidase (Leu-amp) is always associated with heterotrophic bacteria (Rego et al., 1985; Rosso and Azam, 1987; Jacobsen and Rai, 1988; see also Chapter 7 and 9). Leu-amp is a zinc-containing metalloenzyme (molecular weight ca. 300,000 daltons) requiring Mg^{2+} and Mn^{2+} ions for activation. Ions of heavy metals (Cu^{+}, Cd^{2+}, Hg^{2+}, and Pb^{2+}) are completely inhibitory (Smith and Hill, 1960). The enzyme hydrolyses a large number of peptides and amino acid amides of the L-configuration, however, L-leucyl-peptides and L-leucyl-amides are the best substrates. Leu-amp from lake water samples had a temperature optimum at 20°C. The enzyme displayed a rapid increase of activity between 15 and 20°C and a strong drop of activity at temperatures above 20°C. The activity of Leu-amp increased with an increase of pH from 5 to 7.5 and decreased rapidly at pH 8 to 9 (Halemejko and Chróst, 1986).

Environmental studies of Leu-amp activity in lake waters indicated that the enzyme had the highest V_{max} at the end of spring and summer phytoplankton blooms. Leu-amp was actively produced when the concentration of proteins (products of algal cells degradation) increased markedly in the water (Halemejko and Chróst, 1986; Chróst et al., 1989), indicating the induction of the enzyme synthesis. During the period of active growth of spring algal populations in Plußsee and high photosynthetic activity, the V_{max} of bacterial Leu-amp was low, and the enzyme displayed high K_m values (i.e, low affinity for substrate; Figure 3.8). However, when the phytoplankton bloom began to decline (at the end of first week of April), the V_{max} of Leu-amp rapidly increased in the water. At the same time, the enzyme had high affinity for substrate (K_m values were almost twice smaller than at the maximum of bloom) and a high specific activity (Table 3.3). I hypothesize that synthesis of bacterial Leu-amp was repressed, and/or the enzyme activity was inhibited, because a variety of readily utilizable DOM (UDOM) constituents readily supported bacterial metabolism during the active growth of phytoplankton (Chróst, 1981; 1986; Münster and Chróst, 1990). Bacteria directly utilized low-molecular-weight organic substrates (e.g., amino acids), therefore enzymatic degradation and catabolism of peptides and proteins was unnecessary.

This hypothetical explanation of the mechanism of control of Leu-amp synthesis and activity was supported by data obtained from additional laboratory experiments. The specific activity of Leu-amp drastically decreased (Figure 3.9)

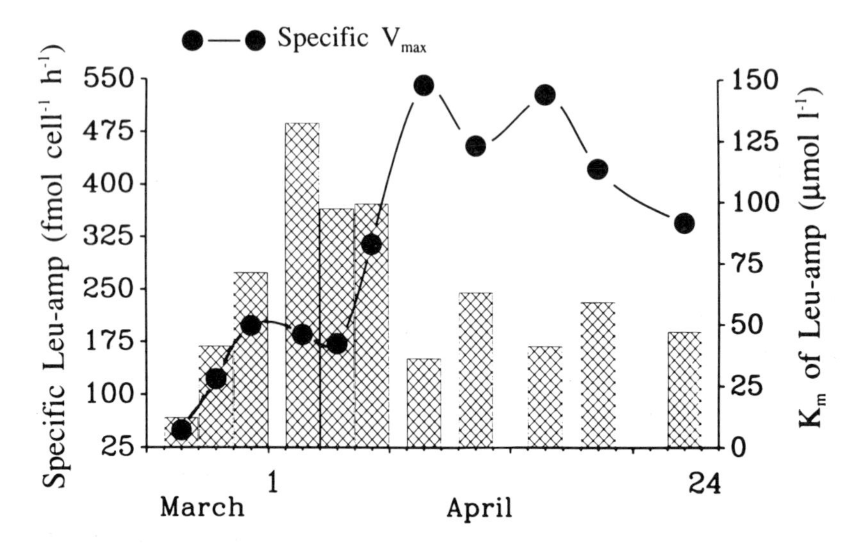

Figure 3.8 Specific activity of bacterial leucine aminopeptidase (Leu-amp) and the Michaelis constants (K_m; bars) in the surface waters of Plußsee during the course of an early spring phytoplankton bloom.

Table 3.3 Specific activity of leucine aminopeptidase produced by bacteria in the euphotic zone of Plußsee during the spring phytoplankton bloom and its breakdown

	Specific activity (fmol cell^{-1} h^{-1})		Number of observations
	Range	Mean	
Bloom	103–255	205 (±22)	6
Bloom breakdown	255–420	375 (±34)	5

when natural bacterial assemblages grew in Plußsee water samples supplemented with readily transportable and utilizable low-molecular-weight substrates (acetate, glucose, amino acids). Fastest bacterial growth rate was found in samples where glucose was added; however, the highest cell production was found after addition of amino acids (Figure 3.10). Leu-amp specific activity of bacteria, grown in samples supplemented with acetate and glucose, decreased markedly only after 20 h and 44 h of incubation. Samples where a mixture of amino acids was added, however, displayed rapid decrease in Leu-amp activity just after 4 h, and further lowering of activity was observed after 20 and 44 h. The affinity of Leu-amp for its substrate in samples supplemented with acetate and glucose was almost unchanged in comparison to the control, as indicated by the K_m values (Figure 3.11). Augmentation of water samples with amino acids, however, not only repressed "new enzyme protein" synthesis (Figure

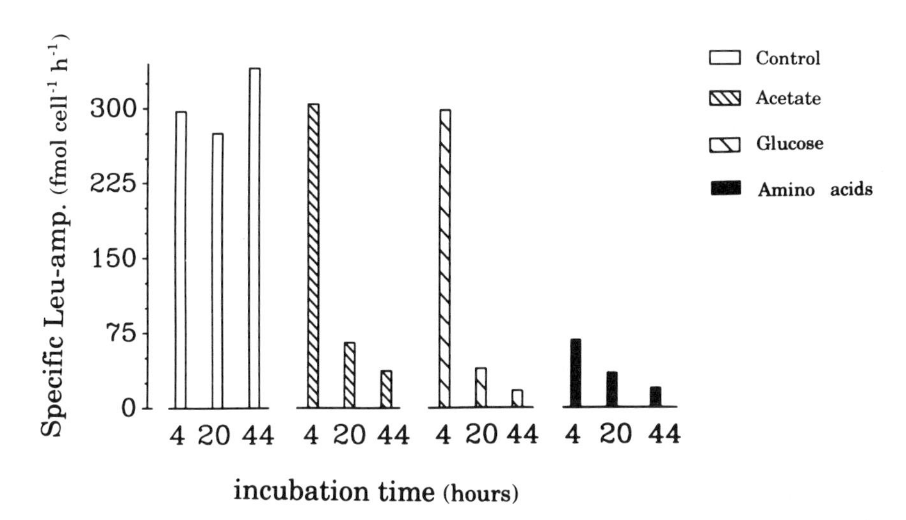

Figure 3.9 Effect of the addition of low-molecular-weight, readily utilizable substrates on the specific activity of bacterial leucine-aminopeptidase (Leu-amp) in Plußsee water. Water samples (200 ml) were: prefiltered through 1.0-μm-pore-size polycarbonate membrane filters (Nuclepore) and supplemented with 100 μM of Na-acetate, D-glucose, or an amino acid mixture (leucine, glycine, serine, alanine; 1 mM solution of each), incubated at the in situ temperature (19°C). Leu-amp kinetics were measured after 4, 20, and 44 h of incubation.

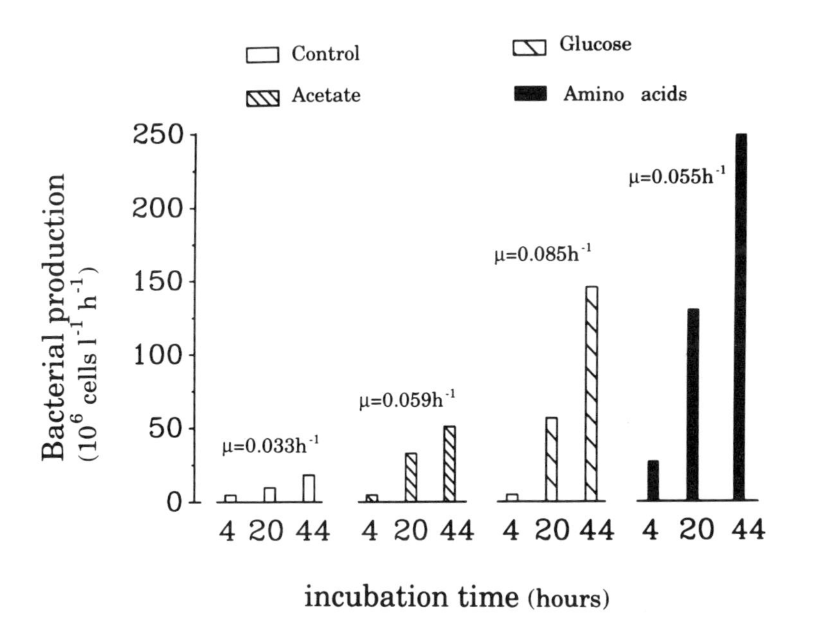

Figure 3.10 Bacterial cell production and growth rates (μ) in Plußsee water samples supplemented with readily utilizable substrates for bacteria (see Figure 3.9 for assay conditions).

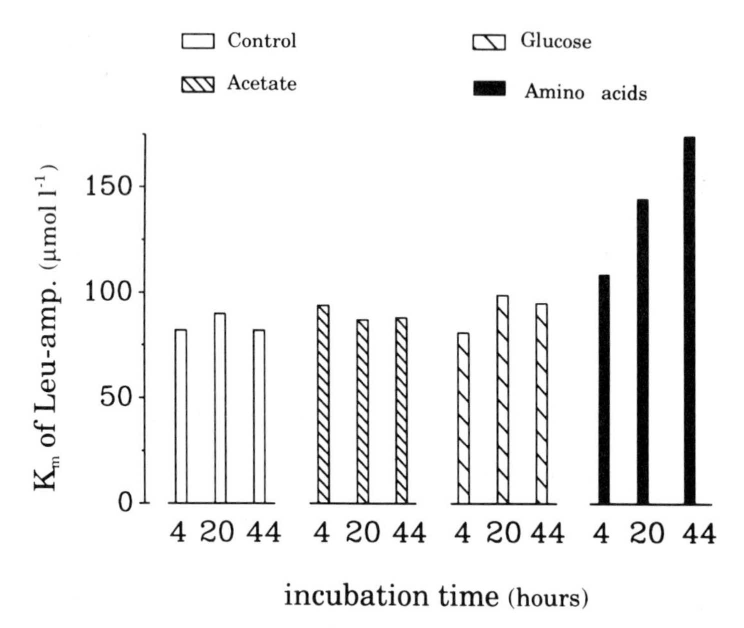

Figure 3.11 Michaelis constants (K_m) of bacterial leucine aminopeptidase (Leu-amp) produced in Plußsee water samples supplemented with readily utilizable substrates for bacteria (see Figure 3.9 for assay conditions).

3.9), but also decreased Leu-amp affinity for substrate (significant increase of K_m values was found) due to the inhibition of the initial velocity of enzyme reaction. The inhibition had a competitive nature.

A complex pattern of bacterial Leu-amp response to dipeptides and protein was observed when Plußsee water samples were supplemented with leucine-leucine, alanine-serine, or albumin (Figure 3.12). The addition of dipeptides, which are directly transported through bacterial cytoplasmic membrane (Payne, 1980b), slightly decreased specific activity of the enzyme synthesized by bacteria in comparison to control samples, where specific (per cell) activity slightly increased after 24 and 48 h of incubation. When bacteria grew in samples enriched with albumin (approx. mol. weight 66,000), which is not readily transportable, the acceleration of growth rate (μ = 0.026 h^{-1}) and rapid induction of Leu-amp synthesis in bacterial cells were observed (Figure 3.12). Newly synthesized Leu-amp also had a higher affinity for substrate, i.e., K_m values decreased after 24 h of incubation (Figure 3.13). At this stage of bacterial growth and Leu-amp synthesis, when water samples were fortified with readily utilizable glucose, the enzyme specific activity gradually decreased after 48, 60, and 72 hours. This catabolic repression of Leu-amp synthesis was coupled to the inhibition of enzyme's initial velocity because of the decline in Leu-amp affinity for substrate, i.e., a strong increase of K_m values after glucose supplementation was found (Figure 3.13).

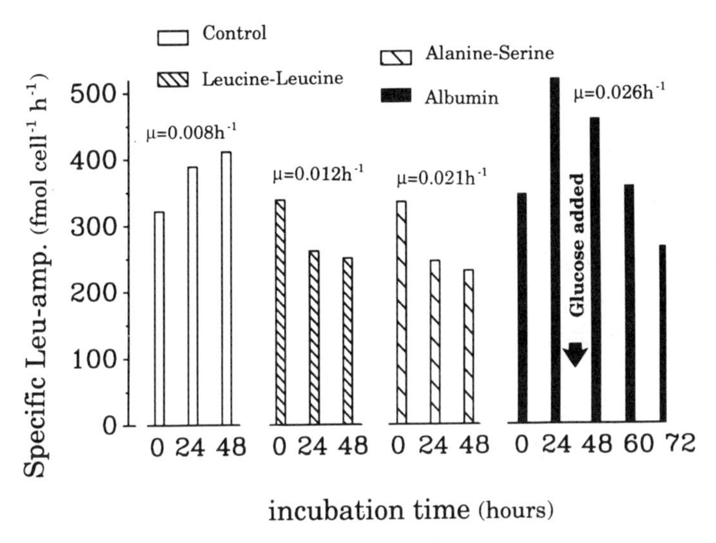

Figure 3.12 Effect on the specific activity of bacterial leucine aminopeptidase (Leu-amp) of Plußsee water supplementation with dipeptides and albumin. Water samples (200 ml) prefiltered through 2.0-μm-pore-size polycarbonate membrane filters (Nuclepore) were supplemented with dipeptides (50 μM of leucine-leucine or alanine-serine) or albumin (5 mg l^{-1}), and incubated at the in situ temperature (20°C). After 24 h, incubation samples previously enriched with albumin were additionally supplemented with glucose (100 μM).

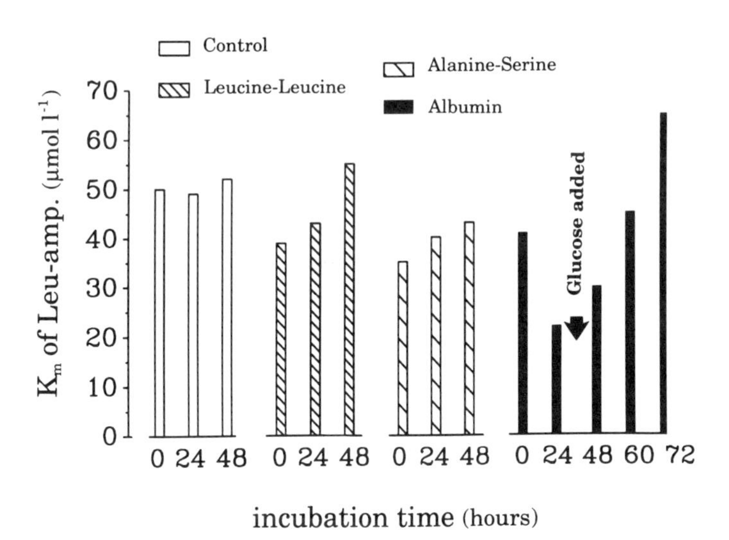

Figure 3.13 Michaelis constants (K_m) of bacterial leucine aminopeptidase (Leu-amp) in Plußsee water samples supplemented with dipeptides or albumin (see Figure 3.12 for assay conditions).

Table 3.4 Effect of nitrate and ammonium ions on leucine aminopeptidase kinetic parameters in lake water

Sample	V_{max} (nmol l^{-1} h^{-1})	K_m (μM)
Lake Plußsee (meso-eutrophic)		
Control	627 (±13)	98 (±7)
+ NO_3^- (50 μM)	366 (±14)	74 (±9)
+ NH_4^+ (50 μM)	325 (±11)	93 (±3)
Lake Mikołajskie (highly eutrophic)		
Control	103 (±8)	100 (±10)
+ NH_4^+ (100 μM)	76 (±5)	118 (±16)

It has been shown that proteinase production by bacteria grown in chemostat is maximal under conditions of nitrogen limitation (Wouters and Bieysman, 1977). My experiments demonstrated that supplementation of lake water samples with inorganic nitrogen inhibited Leu-amp activity produced by natural bacterial assemblages (Table 3.4). Addition of NH_4^+ and NO_3^- ions significantly decreased V_{max} of enzyme reaction; however, the Michaelis constant (K_m) was almost unchanged, indicating a noncompetitive nature of inhibition. Stronger inhibition of Leu-amp V_{max} by inorganic nitrogen was observed in moderately eutrophic Plußsee (stronger N-limitation) than in highly eutrophic Lake Mikołajskie (extremely high level of inorganic nitrogen compounds), which also displayed a low V_{max} of Leu-amp activity in the control samples.

The results of the experiments presented on Leu-amp kinetics in lake water demonstrated that the enzyme synthesis and activity were under different control mechanisms dependent on the physico-chemical conditions of studied habitats. There is ample evidence for general catabolite repression of proteolytic enzymes synthesis by readily utilizable sources of carbon (Boethling, 1975; Litchfield and Prescott, 1976), as well as more specific, end-product repression by amino acids (Glenn, 1976). However, control of the aminopeptidases appears to be distinct and more complex. In some bacteria, amino acids, peptides, and/or proteins seem to induce aminopeptidase synthesis (Daatselaar and Harder, 1974; Litchfield and Prescott, 1976). It is not known specifically how aminopeptidase induction operates, especially since amino acids were reported to act as inducers in some bacteria, rather than acting in their more predictable role as end-product inhibitors.

In natural environments, aminopeptidases coexist with other proteolytic enzymes and create the microbial enzyme systems required for utilization of exogenous proteinaceous materials. The utilization of exogenous peptides and proteins by bacteria requires their degradation to small peptides and amino acids before cellular uptake, and the direct uptake of utilizable proteins and large peptides probably does not occur as such in bacteria (Law, 1980). The role of aminopeptidases in protein and peptide utilization by bacteria is clearly seen in the context of advances in our understanding of peptide and amino acid transport. Most of known aminopeptidases act as exo-ectoenzymes and presumably contribute to the protein-utilizing abilities of bacteria by liberating free amino acids directly from peptides that are too large for direct transport into the organism. The scheme for the use of proteinaceous materials by bacteria producing proteolytic enzymes would be expected to follow a pattern of extracellular protein and polypeptide degradation to amino acids and peptides whose maximum transportable size appears to be limited to about seven amino acid residues. Before further utilization, transported peptides are hydrolyzed to their constituent amino acids by intracellular peptidases of wide specificity (Law, 1980). Amino acids serve not only as components for intracellular protein synthesis but they are also catabolized as carbon and nitrogen sources.

3.7 Control of Ectoenzyme Synthesis and Activity—Conclusions

On the basis of the experiments and in situ studies described earlier, I propose a conceptual model for the interactions of microbial ectoenzymes with simple, readily utilizable dissolved organic matter (UDOM) and with complex, polymeric DOM, and the mechanisms of ectoenzyme synthesis and activity regulation in aquatic environments (Figure 3.14). During the active growth of phytoplankton, algal populations excrete a variety of low-molecular-weight primary metabolites and photosynthetic products into the water, including readily utilizable substrates (Chróst and Faust, 1983; Fogg, 1983), which support bacterial growth and metabolism (Chróst, 1986). These UDOM compounds (e.g., monosaccharides, amino acids, organic acids, simple esters, etc.; Münster and Chróst, 1990) inhibit the activity (end-product inhibition) and repress the synthesis (catabolite repression) of ectoenzymes in heterotrophic bacteria. However, senescent algae release high-molecular-weight secondary metabolites (Fogg, 1966) and/or liberate, through autolysis or due to viral lytic activity on algal cells (Suttle et al., 1990), a high amount of polymeric DOM, such as polysaccharides, proteins, lipids, and nucleic acids (Hellebust, 1974; Karl and Bailiff, 1989; Paul et al., 1990). Low concentrations of UDOM and a high content of polymers (substrates for ectoenzymes) in the water at the same time derepress and induce synthesis of ectoenzymes in aquatic bacteria, respectively.

Most of the ectoenzymes synthesized by aquatic microorganisms are inducible catabolic enzymes (produced by a variety of aquatic microbes), which are actively secreted outside the cytoplasmic membrane and are associated with their producer's cell-surface or located in the periplasmic space of Gram negative bacteria. They are involved in the degradation of nonpenetrating cell membrane substrates and polymers, which are not continuously available in the water. Therefore, the constant synthesis of ectoenzymes in the absence of substrates is unnecessary, because it requires the expenditure of energy that otherwise may be channeled into useful activities. The efficient induction of ectoenzymes is more complicated than that of intracellular enzymes. First, many of the ectoenzyme substrates are polymeric compounds, and they are too large

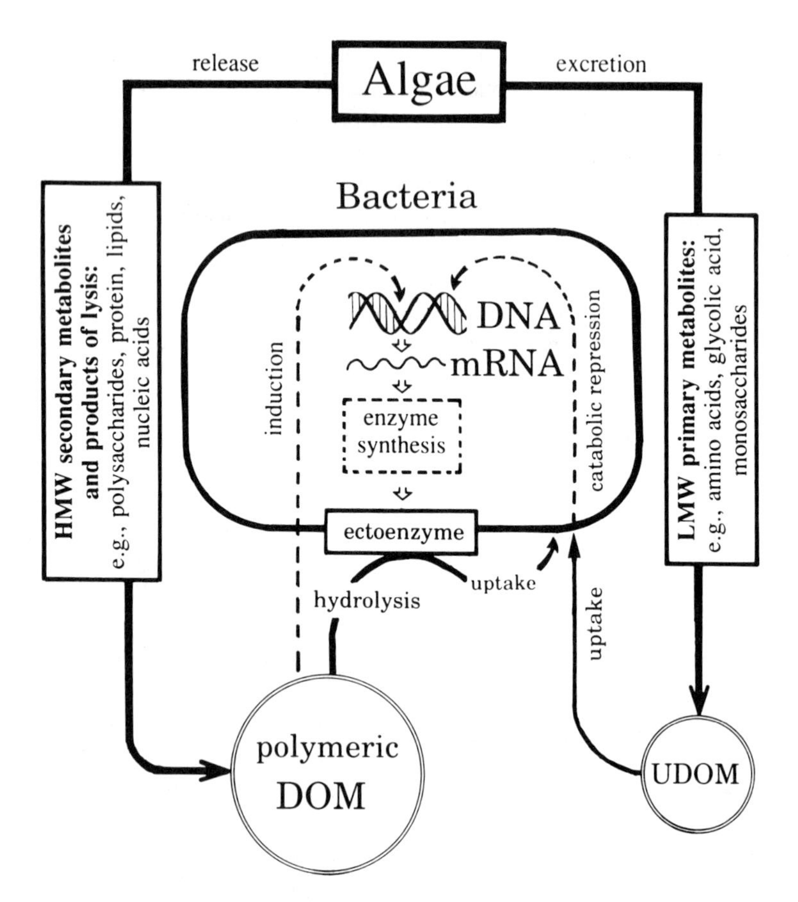

Figure 3.14 A conceptual model for bacterial interactions with readily utilizable substrates (UDOM) and polymeric DOM, and the mechanisms of regulation of ectoenzymes. LMW and HMW are low-molecular-weight and high-molecular-weight algal metabolites, respectively (modified from Chróst, 1989; 1990a).

to enter the cell and serve as an inducer of synthesis. Second, for an ectoenzyme to be secreted at appropriate rates, the microorganism must be able to monitor the activity of the ectoenzyme outside the cell. It has been proposed that these problems are overcome by a low constitutive rate of ectoenzyme secretion. If the substrate is present, then the low-molecular-weight product accumulates to a certain level, enters the cell, and serves as the inducer (Priest, 1984). When environmental conditions inhibit an ectoenzyme activity (e.g., unsuitable pH; for some ectoenzymes, the presence of H_2S; absence of activating cations Mg^{2+}, Zn^{2+}), the induction of its synthesis will not occur because the product of catalysis will not be generated.

The nutritional status of the environment regulates the rates of microbial ectoenzyme synthesis and activity. Ectoenzyme synthesis is under a catabolite repression/derepression control mechanism. The production of virtually all ectoenzymes in most aquatic microorganisms is repressed when grown on sources of readily utilizable dissolved organic matter. This mode of regulation is called catabolite repression. The synthesis of ectoenzymes only become derepressed once the concentration of readily utilizable substrates in the water falls below a critically low level. Using the repression strategy for ectoenzyme synthesis, microorganisms can avoid the wasteful production of inducible enzymes, which are not useful when their growth is not limited by carbon, nitrogen, or phosphorus sources (Bengtsson, 1988; Chróst et al., 1989; Chróst, 1990a). It is also well documented that synthesis of many ectoenzymes in aquatic microorganisms is repressed by the end product that is derived from the substrate and accumulates in the cell or in surrounding environment. Thus, ectoenzyme synthesis is tightly coupled to the availability of UDOM in natural waters. Despite the widespread occurrence of catabolite repression, with the exception of enteric bacteria, the molecular details of repression are poorly understood. Some studies have indicated that cyclic AMP (cAMP), together with its receptor protein, may play a central role in control of catabolite repression (Botsford, 1981; Francko, 1984).

Even if ectoenzyme is already produced, its activity may be rapidly inhibited by a variety of inhibitors present in aquatic environment. Two types of ectoenzyme activity inhibition are known—competitive and noncompetitive. Competitive inhibition occurs when an inhibiting compound is structurally similar to the natural substrate and by mimicry, binds to the enzyme. In doing so, it competes with an enzyme's natural substrate for the active substrate-binding site. The hallmark of competitive inhibition of many ectoenzymes (e.g., alkaline phosphatase, β-glucosidase, aminopeptidase) is that it decreases the affinity of an ectoenzyme (an increase in the value of the Michaelis constant is observed) for the substrate, and therefore, inhibits the initial velocity of the reaction (Hoppe, 1983; Halemejko and Chróst, 1986). It can be also demonstrated that the end product of ectoenzyme action is a competitive inhibitor of its initial activity and the ectoenzyme's affinity for its substrate. Such a situation was found in the study of βGlc and APase in Plußsee lake water (Chróst, 1989; 1990a; Chróst and Overbeck, 1987). Competitive inhibition is reversible and can be overcome by increased substrate concentration, and therefore the maximum velocity (V_{max}) of the reaction is unchanged. Noncompetitive inhibition is generally characterized as an inhibition of enzymatic activity by compounds that bear no structural relationship to the substrate. Therefore, the inhibition cannot be reversed by increasing the concentration of the substrate. It may be reversed only by removal of the inhibitor. Unlike competitive inhibitors, reversible noncompetitive inhibitors cannot interact at the active site, but bind to some other portion of an enzyme-substrate complex (Armstrong, 1983). This type of inhibition encompasses a variety of different inhibitory mechanisms and is therefore not amenable to a simple description. Noncompetitive inhibition of the activity of exoproteases by Cu^{2+} ions (Little et al., 1979), and inhibition of α-glucosidase, β-glucosidase, N-acetyl-glucosaminidase, and alkaline phosphatase by H_2S in natural waters has been described (Hoppe, 1986).

The microorganisms that produce ectoenzymes are presumably superior competitors for organic and inorganic nutrients in aquatic environments. Through the production of ectoenzymes, they are capable of utilizing a variety of polymeric compounds, which are not otherwise utilizable, that predominate in natural waters. Ectoenzymes react on polymeric substrates outside the cell. They catalyze hydrolytic conversion of high-molecular-weight compounds to low-molecular-weight subunits—the final products of ectoenzyme reaction. Thus ectoenzymes transform nonutilizable compounds of the DOM pool to readily transportable and utilizable substrates for microorganisms. Because these very efficient degradative enzyme systems are cell-bound or located in periplasm, they make UDOM available near the cell surface for substrate-transporting systems. This gives an increased chance of survival of microorganisms in aquatic environments when easily available energy and nutrient sources become limiting (Chróst and Overbeck, 1987, 1990; Chróst, 1991) and also enables them to increase their growth and biomass production (Chróst et al., 1989). Obviously the production of ectoenzymes close to the cell, which degrade substrate resistant to attack by other microbial species may allow the producer to dominate an ecosystem through its nutritional advantage. Moreover, microorganisms producing the ectoenzymes form a vital part of aquatic ecosystems in which they not only make carbon, phosphorus, and nitrogen available to themselves by degradation, but also interact with organisms that do not produce ectoenzymes, supplying them with nutrients (Chróst, 1986; 1988; 1990b).

Acknowledgements I thank my student, M.Sc. A. Gajewski, for his excellent assistance with experiments on Leu-amp activity in Lake Mikołajskie. This work was supported by the Max-Planck Gesellschaft zur Förderung der Wissenschaften (F.R.G.) and the research project C.P.B.P. 04.02.3.1.1 (Poland).

References

Aaronson, S. 1981. *Chemical Communication at the Microbial Level*, vol. 1. CRC Press Inc., Boca Raton. 184 pp.

Allen, H.L. 1976. Dissolved organic matter in lakewater: characteristics of molecular weight size fractions and ecological implications. *Oikos* 27: 64–70.

Ammerman, J.W. and Azam, F. 1985. Bacterial 5′-nucleotidase in aquatic ecosystems: A novel mechanism of phosphorus regeneration. *Science* 227: 1338–1340.

Armstrong, F.B. 1983. *Biochemistry*, 2nd edition. Oxford University Press, New York. 653 pp.

Azam, F. and B.C. Cho. 1987. Bacterial utilization of organic matter in the sea. pp. 261–281 in Fletcher, M., Gray, T.R.G. and Jones, J.G. (editors), *Ecology of Microbial Communities*, Cambridge University, Cambridge.

Barman, T.E. 1969. *Enzyme Handbook*, vol. 2. Springer Verlag, Berlin. 928 pp.

Bell, R.T. and J. Kuparinen. 1984. Assessing phytoplankton and bacterioplankton pro-

duction during early spring in lake Erken, Sweden. *Applied and Environmental Microbiology* 48: 1221–1231.
Bengtsson, G. 1988. The impact of dissolved amino acids on protein and cellulose degradation in stream waters. *Hydrobiologia* 164: 97–102.
Boethling, R.S. 1975. Regulation of extracellular protease secretion in *Pseudomonas maltophilia*. *Journal of Bacteriology* 123: 954–961.
Botsford, J.L. 1981. Cyclic nucleotides in prokaryotes. *Microbiological Reviews* 45: 620–645.
Bretaudiere, J.P. and T. Stillman. 1984. Alkaline phosphatases. pp. 75–82 in Bergmeyer, H.U. (editor), *Methods of Enzymatic Analysis*, vol. 4, Verlag Chemie, Weinheim.
Burns, R.G. 1983. Extracellular enzyme-substrate interactions in soil. pp. 249–298 in Slater, J.H., Whittenbury, R., and Wimpenny, J.W.T. (editors), *Microbes in Their Natural Environments*. Cambridge University Press, London.
Chróst, R.J. 1975. Inhibitors produced by algae as an ecological factor affecting bacteria in water ecosystems. I. Dependence between phytoplankton and bacteria development. *Acta Microbiologica Polonica* 7(B): 125–133.
Chróst, R.J. 1981. The composition and bacterial utilization of DOC released by phytoplankton. *Kieler Meeresforschung Sonderheft* 5: 325–332.
Chróst, R.J. 1986. Algal-bacterial metabolic coupling in the carbon and phosphorus cycle in lakes. pp. 360–366 in Meguar, F. and Gantar, M. (editors), *Perspectives in Microbial Ecology*, Slovene Society for Microbiology, Ljubljana.
Chróst, R.J. 1988. Phosphorus and microplankton development in a eutrophic lake. *Acta Microbiologica Polonica* 37: 205–225.
Chróst, R.J. 1989. Characterization and significance of β-glucosidase activity in lake water. *Limnology and Oceanography* 34: 660–672.
Chróst, R.J. 1990a. Microbial ectoenzymes in aquatic environments. pp. 47–78 in Overbeck, J. and Chróst, R.J. (editors), *Aquatic Microbial Ecology: Biochemical and Molecular Approaches*, Springer Verlag, New York. 190 pp.
Chróst, R.J. 1990b. Can bacteria affect the phytoplankton succession in lacustrine environments? pp. 15–20 in Burhardt, L. (editor), *Evolution of Freshwater Lakes*, Adam Mickiewicz University Press, Poznań.
Chróst, R.J. 1991. Ectoenzymes in aquatic environments: microbial strategy for substrate supply. Verhandlungen der Internationalen Vereinigung für Theoretische und Angewandte Limnologie 24: 936–942.
Chróst, R.J. and M.A. Faust. 1983. Organic carbon release by phytoplankton: its composition and utilization by bacterioplankton. *Journal of Plankton Research* 5: 477–493.
Chróst, R.J., Münster, U., Rai, H., Albrecht, D., Witzel, P.K. and J. Overbeck. 1989. Photosynthetic production and exoenzymatic degradation of organic matter in euphotic zone of an eutrophic lake. *Journal of Plankton Research* 11: 223–242.
Chróst, R.J. and J. Overbeck. 1987. Kinetics of alkaline phosphatase activity and phosphorus availability for phytoplankton and bacterioplankton in lake Plußsee (north German eutrophic lake). *Microbial Ecology* 13: 229–248.
Chróst, R.J. and J. Overbeck. 1990. Substrate-ectoenzyme interaction: significance of β-glucosidase activity for glucose metabolism by aquatic bacteria. *Archiv für Hydrobiologie Beihefte Ergebnisse Limnologie* 34: 93–98.
Chróst, R.J., Siuda, W., Albrecht, D. and J. Overbeck. 1986. A method for determining enzymatically hydrolyzable phosphate (EHP) in natural waters. *Limnology and Oceanography* 31: 662–667.
Chróst, R.J., Siuda, W. and G.Z. Hałemejko. 1984. Longterm studies on alkaline phosphatase activity (APA) in a lake with fish-aquaculture in relation to lake eutrophication and phosphorus cycle. *Archiv für Hydrobiologie Supplement* 70: 1–32.
Chróst, R.J., Wcisło, R. and G.Z. Hałemejko. 1986. Enzymatic decomposition of organic matter by bacteria in an eutrophic lake. *Archiv für Hydrobiologie* 107: 145–165.
Chróst, R.J., Wcisło, R. and J. Overbeck. 1988. Evaluation of the [^{3}H]thymidine method

for estimating bacterial growth rates and production in lake water: Re-examination and methodological comments. *Acta Microbiologica Polonica* 37: 95–112.

Cole, J.J., McDowell, W.H. and G.E. Likens. 1984. Sources and molecular weight of dissolved organic carbon in an oligotrophic lake. *Oikos* 42: 1–9.

Crofton, P.M. 1982. Biochemistry of alkaline phosphatase isoenzymes. *CRC Critical Reviews in Clinical and Laboratory Sciences* 16: 161–194.

Daatselaar, M.C.C. and W. Harder. 1974. Some aspects of the regulation of the production of extracellular proteolytic enzymes by a marine bacterium. *Archiv für Hydrobiologie* 101: 21–34.

Daniels, L.B. and R.H. Glew. 1984. β-Glucosidases in tissue. pp. 217–226 in Bergmeyer, H.U. (editor), *Methods of Enzymatic Analysis*, vol. 4, Verlag Chemie, Weinheim.

Dobrogosz, W.J. 1981. Enzymatic activity. pp. 365–392 in Gerhardt, P., Murray, R.G.E., Costilow, R.N., Nester, E.W., Wood, W.A., Krieg, N.R. and Phillips, G.B. (editors), *Manual of Methods for General Bacteriology*. American Society for Microbiology, Washington DC.

Dowd, J.E. and D.S. Riggs. 1965. A comparison of estimates of Michaelis-Menten kinetic constants from various linear transformations. *Journal of Biological Chemistry* 240: 863–869.

Fogg, G.E. 1966. The extracellular products of algae. *Oceanography and Marine Biology Annual Reviews* 4: 195–205.

Fogg, G.E. 1983. The ecological significance of extracellular products of phytoplankton photosynthesis. *Botanica Marina* 26: 3–14.

Francko, D. 1984. Phytoplankton metabolism and cyclic nucleotides. II. Nucleotide-induced perturbations of alkaline phosphatase activity. *Archiv für Hydrobiology* 100: 409–421.

Gage, M.A. and E. Gorham. 1985. Alkaline phosphatase activity and cellular phosphorus as an index of the phosphorus status of phytoplankton in Minnesota lakes. *Freshwater Biology* 15: 227–233.

Geller, A. 1985. Degradation and formation of refractory DOM by bacteria during simultaneous growth on labile substrates and persistent lake water constituents. *Swiss Journal of Hydrology* 47: 27–44.

Glenn, A.R. 1976. Production of extracellular proteins by bacteria. *Annual Reviews in Microbiology* 30: 41–62.

Halemejko, G.Z. and R.J. Chróst. 1984. The role of phosphatases in phosphorus mineralization during decomposition of lake phytoplankton blooms. *Archiv für Hydrobiologie* 101: 489–502.

Halemejko, G.Z. and R.J. Chróst. 1986. Enzymatic hydrolysis of proteinaceous particulate and dissolved material in an eutrophic lake. *Archiv für Hydrobiologie* 107: 1–21.

Hancock, I.C. and I.R. Poxton. 1988. *Bacterial Cell Surface Techniques*. John Wiley and Sons, Chichester, 329 pp.

Healey, F.P. and L.L. Hendzel. 1980. Physiological indicators of nutrient deficiency in lake phytoplankton. *Canadian Journal of Fisheries and Aquatic Sciences* 37: 442–453.

Hellebust, J.A. 1974. Extracellular products. pp. 838–863 in W.D.P. Stewart (editor), *Algal Physiology and Biochemistry*. Blackwell, Oxford.

Hollibaugh, J.T. and Azam, F. 1983. Microbial degradation of dissolved proteins in seawater. *Limnology and Oceanography* 28: 1104–1116.

Holm-Hansen, O. 1984. Composition and nutritional mode of nanoplankton. *Archiv für Hydrobiologie Beihefte Ergebnisse Limnologie* 19: 125–129.

Hoppe, H.G. 1983. Significance of exoenzymatic activities in the ecology of brackish water: measurements by means of methylumbelliferyl substrates. *Marine Ecology Progress Series* 11: 299–308.

Hoppe, H.G. 1986. Degradation in sea water. pp. 453–474 in Rehm, H.J. and Reed, G. (editors), *Biotechnology*, vol. 8, VCH Verlagsgesellschaft, Weinheim.

Hoppe, H.G., Kim, S.J. and K. Gocke. 1988. Microbial decomposition in aquatic environments: combined processes of extracellular enzyme activity and substrate uptake. *Applied and Environmental Microbiology* 54: 784–790.

Jacobsen, T.R. and H. Rai. 1988. Determination of aminopeptidase activity in lakewater by a short term kinetic assay and its application in two lakes of differing eutrophication. *Archiv für Hydrobiologie* 113: 359–370.

Jørgensen, N.O.G. 1987. Free amino acids in lakes: concentrations and assimilation rates in relation to phytoplankton and bacterial production. *Limnology and Oceanography* 32: 97–111.

Karl, D.M. and M.D. Bailiff. 1989. The measurement and distribution of dissolved nucleic acids in aquatic environments. *Limnology and Oceanography* 34: 543–558.

King, G.M. 1986. Characterization of β-glucosidase activity in intertidal marine sediments. *Applied and Environmental Microbiology* 51: 373–380.

Law, B.A. 1980. Transport and utilization of proteins by bacteria. pp. 381–409 in Payne, J.W. (editor), *Microorganisms and Nitrogen Sources*, John Wiley and Sons, New York.

Lazdunski, M. 1974. "Half of the sites" reactivity and the role of subunit interactions in enzyme catalysis. pp. 81–140 in Kaiser, E.T. and Kezdy, F.J. (editors), *Progress in Bioorganic Chemistry*, vol. 3, John Wiley & Sons, New York.

Leatherbarrow, R.J. 1987. *Enzfitter. A Non-linear Regression Data Analysis Program for the IBM PC*. Elsevier-Biosoft, Cambridge. 91 pp.

Linden, G., Chappelet-Tordo, D. and M. Lazdunski. 1977. Milk alkaline phosphatase, stimulation by Mg^{2+} and properties of the Mg^{2+} site. *Biochimica et Biophysica Acta* 483: 100–106.

Litchfield, C.D. and J.M. Prescott. 1976. Regulation of proteolytic enzyme production by *Aeromonas proteolytica*. II. Extracellular aminopeptidase. *Canadian Journal of Microbiology* 16: 23–27.

Little, J.E., Sjogren, R.E. and G.R. Carson. 1979. Measurement of proteolysis in natural waters. *Applied and Environmental Microbiology* 37: 900–908.

Lundin, A., Arner, P. and J. Hellmer. 1989. A new linear plot for standard curves in kinetic substrate assays extended above the Michaelis-Menten constant: application to a luminometric assay of glycerol. *Analytical Biochemistry* 177: 125–131.

Maeda, M. and N. Taga. 1973. Deoxyribonuclease activity in seawater and sediment. *Marine Biology* 20: 58–63.

McComb, R.B., Bowers, G.N., Jr. and S. Posen. 1979. *Alkaline Phosphatase*. Plenum Press, New York. 358 pp.

Mayer, L.M. 1989. Extracellular proteolytic enzyme activity in sediments of an intertidal mudflat. *Limnology and Oceanography* 34: 973–981.

Meyer-Reil, L.A. 1987. Seasonal and spatial distribution of extracellular enzymatic activities and microbial incorporation of dissolved organic substrates in marine sediments. *Applied and Environmental Microbiology* 53: 1748–1755.

Morton, R.K. 1954. The purification of alkaline phosphatases of animal tissues. *Biochemical Journal* 57: 595–603.

Murgier, M., Pelissier, C., Lazdunski, A. and C. Lazdunski. 1976. Existence, location and regulation of the biosynthesis of amino-endopeptidase in Gram-negative bacteria. *European Journal of Biochemistry* 65: 517–520.

Münster, U. 1985. Investigations about structure, distribution and dynamics of different organic substrates in the DOM of lake Plußsee. *Archiv für Hydrobiologie Supplement* 70: 429–480.

Münster, U. and R.J. Chróst. 1990. Origin, composition and microbial utilization of dissolved organic matter. pp. 8–46 in Overbeck, J. and Chróst, R.J. (editors), *Aquatic Microbial Ecology: Biochemical and Molecular Approaches*. Springer Verlag, New York. 190 pp.

Nikaido, H. and Nakae, T. 1979. The outer membrane of Gram-negative bacteria. *Advances of Microbial Physiology* 20: 163–250.

Paul, J.H., Jeffrey, W.H. and M.F. DeFlaun. 1987. Dynamics of extracellular DNA in the marine environment. *Applied and Environmental Microbiology* 53: 170–179.

Paul, J.H., Jeffrey, W.H. and J.P. Cannon. 1990. Production of dissolved DNA, RNA, and protein by microbial populations in a Florida reservoir. *Applied and Environmental Microbiology* 56: 2957–2962.

Payne, J.W. 1980a. *Microorganisms and Nitrogen Sources*. John Wiley and Sons, New York. 764 pp.

Payne, J.W. 1980b. Transport and utilization of peptides by bacteria. pp. 211–256 in Payne, J.W. (editor), *Microorganisms and Nitrogen Sources*. John Wiley and Sons, New York.

Pettersson, K. 1980. Alkaline phosphatase activity and algal surplus phosphorus as phosphorus-deficiency indicators in Lake Erken. *Archiv für Hydrobiologie* 89: 54–87.

Priest, F.G. 1984. *Extracellular Enzymes*. Van Nostrand Reinhold (UK) Co. Ltd., Wokingham. 79 pp.

Rego, V.J., Billen, G., Fontigny, A. and M. Somville. 1985. Free and attached proteolytic activity in water environments. *Marine Ecology Progress Series* 21: 245–249.

Riemann, B. and M. Søndergaard. 1986. *Carbon Dynamics in Eutrophic, Temperate Lakes*. Elsevier, Amsterdam. 345 pp.

Rogers, H.J. 1961. The dissimilation of high molecular weight organic substrates. pp. 261–318 in Gunsalus, I.C. and Stanier, R.Y. (editors), *The Bacteria*, vol. 2, Academic Press, New York.

Rogers, H.J. Perkins, H.R. and Ward, J.B. 1980. *Microbial Cell Wall and Membranes*. Chapman and Hall, London. 258 pp.

Romankevich, E.A. 1984. *Geochemistry of Organic Matter in the Ocean*. Springer Verlag, Tokyo. 478 pp.

Rosso, A.L. and F. Azam. 1987. Proteolytic activity in coastal oceanic waters: depth distribution and relationship to bacterial populations. *Marine Ecology Progress Series* 41: 231–240.

Siuda, W. 1984. Phosphatases and their role in organic phosphorus transformation in natural waters. A review. *Polskie Archiwum Hydrobiologii* 31: 207–233.

Siuda, W. and R.J. Chróst. 1987. The relationship between alkaline phosphatase (APA) activity and phosphate availability for phytoplankton and bacteria in eutrophic lakes. *Acta Microbiologica Polonica* 36: 247–257.

Smith, E.L. and R.L. Hill. 1960. Leucine aminopeptidase. pp. 37–62 in Boyer, P.D., Lardy, H. and Myrbäck, K. (editors), *The Enzymes*, vol. 4, Academic Press, New York.

Somville, M. 1984. Measurement and study of substrate specificity of exoglucosidase activity in eutrophic water. *Applied and Environmental Microbiology* 48: 1181–1185.

Somville, M. and G. Billen. 1983. A method for determining exoproteolytic activity in natural waters. *Limnology and Oceanography* 28: 190–193.

Stewart, A.J. and R.G. Wetzel. 1982. Phytoplankton contribution to alkaline phosphatase activity. *Archiv für Hydrobiologie* 93: 265–271.

Suttle, C.A., Chan, A.M. and M.T. Cottrell. 1990. Infection of phytoplankton by viruses and reduction of primary productivity. *Nature* 347: 467–469.

Tamminen, T. 1989. Dissolved organic phosphorus regeneration by bacterioplankton: 5′nucleotidase activity and subsequent phosphate uptake in a mesocosm enrichment experiment. *Marine Ecology Progress Series* 5(: 89–100.

Thurman, E.M. 1985. *Organic Geochemistry of Natural Waters*. Nijhoff/Junk, Boston. 687 pp.

Vincent, W.V. 1981. Rapid physiological assays for nutrient demand by the plankton. II. Phosphorus. *Journal of Plankton Research* 3: 699–710.

Wetzel, R.G. and G.E. Likens. 1979. *Limnological Analyses*. Saunders, Philadelphia. 395 pp.

Wouters, J.T.M. and P.J. Bieysman. 1977. Production of some exocellular enzymes by *Bacillus licheniformis* 749/C in chemostat cultures. *Federation of European Microbiological Societies Letters* 1: 109–112.
Wynne, D. and M. Gophen. 1981. Phosphatase activity in freshwater zooplankton. *Oikos* 37: 369–376.

4

Microbial Extracellular Enzyme Activity: A New Key Parameter in Aquatic Ecology

Hans-Georg Hoppe

4.1 Introduction

Three general pathways of organic matter degradation exist in natural aquatic environments. These are based on predation, particle feeding, and dissolved organic matter (DOM) uptake. Bacteria are involved in the latter two in that they are able to hydrolyze nonliving particles thereby competing with particle feeders, and take up small organic molecules, which is their exclusive domain. Particle hydrolysis is mediated by extracellular enzymes in the intestines of animals and by the enzymatic activity of attached bacteria. Therefore, successful competition for organic matter among tropic levels is also a question of extracellular enzymatic efficiencies. The decomposition of dissolved organic macromolecules is mediated mainly by the enzymes of free-living bacteria, which subsequently incorporate the small molecules resulting from enzymatic hydrolysis. Therefore, bacterial activity has a strong influence on the concentration and speciation of dissolved organic molecules in the water. A major fraction of the DOM-pool in the water can be expected to consist of dissolved macromolecules since extracellular hydrolysis is a relatively slow process in comparison to the uptake of low-molecular-weight organic matter (LMWOM). The efficiency of animals feeding on particles may vary considerably, depending on various factors. Microbial hydrolysis of particles will depend greatly on the chemical composition and the size of the particles. Thus competition for organic particles between animals and microbes will be determined on the one hand, by the slow but continuous microbial component and, on the other hand, by pulse-feeding activities of animals (Joint and Morris, 1982). In detail, it has been pointed out that efficiencies of microbial decomposition of fast-sinking large particles and nonsinking small particles may be different (Cho and Azam, 1988; Karl et al., 1988).

The input of substances into the pool of DOM is, by necessity, mediated by the hydrolytic activity of bacteria attached to particles and, to a certain degree, by autolysis of cells, sloppy feeding, and excretion of plants and animals. The attached bacteria, however, contribute to the DOM pool and benefit the free-living bacteria only when they do not entirely use up the products of particle hydrolysis themselves. Overproduction of hydrolysis products, as demonstrated in this report, causes a considerable shift in organic materials from the particulate to the dissolved state. This process is mediated by the different enzymatic strategies of attached and free-living microorganisms and provides the basis for their nutritional demands. Therefore, it is suggested that the microbial loop model (Azam et al., 1983) be amended to include the initial and obligatory hydrolytic steps of microbial nutrient generation, which would serve to set the range of model dimensions (Chróst, 1990). The microbial loop model (Figure 4.1) describes the role of bacteria and bacteriovorous organisms in organic matter transfer and mineralization, these processes, however, strongly depend on the enzymatic capacities of the microbial community.

Microbial attachment is a broad field of research, many aspects of which have already been investigated, ranging from adhesion mechanisms to metabolic properties of attached bacteria (Hoppe, 1984) and their chemotactic response (Jackson, 1989). However, information on enzymatic activity of bacteria on surface materials is scarce, despite the fact that it is probably one of the most important functions of surfaces in the aquatic environment. Therefore it

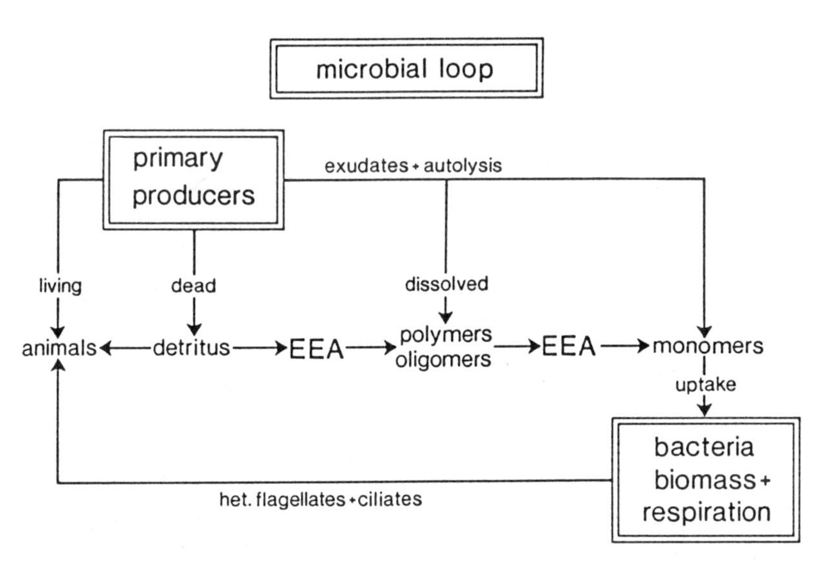

Figure 4.1 Bacterial loop model supplemented with pathways of organic matter transformation from primary producers to bacteria via extracellular enzymatic activity (EEA).

is the aim of this study to provide some preliminary data about bacterial enzymatic activity on organic/inorganic surfaces and to support the theory that free bacteria may benefit from the hydrolysis activity of the attached bacteria. Studies concerning the relationship between extracellular enzyme activities of attached and free-living bacteria were conducted using natural detritus materials and submerged glass slides from a station near the coast of the Kiel Fjord. An example of a depth profile from an offshore North Atlantic station is included, which demonstrates a shift of the balance between coupled hydrolysis and substrate uptake activity along the depth profile.

Since microbial extracellular enzyme activity is integrated into the cycling processes of organic and inorganic matter, it is difficult to directly recognize and measure its dimensions (Chróst et al., 1989). Therefore, it was necessary to label the high-molecular-weight organic matter (HMWOM) pool in question with a tracer. The tracer was chosen to have some characteristics in common with its naturally occurring analogues. For this purpose, fluorogenic combined molecules have come into use in recent years. Table 4.1 lists some of the systems and tracers that have been used. These fluorogenic combined molecules constitute an organic molecule, a specific binding, and a fluorochrome, which is activated after cleavage of the binding. Such molecules contain peptide, glucoside or ester bonds, and their fluorochrome is frequently a methylumbelliferone (MUF), methylcoumarinylamide (MCA), β-naphthylamide, or methylfluorescein. The detection limit of these substances is in the nanomolar range, and therefore the fluorogenic model substrates are well adapted for short-term measurements of extracellular enzyme activities (EEA) under natural conditions, even down to depths of 1,000 m in the sea. The use of these substances is not yet fully researched and has several drawbacks as well as advantages, which have been addressed previously by several authors (Hoppe, 1983; Somville, 1984; Chróst and Krambeck, 1986; Snyder et al., 1986). Some uncertainties arise from the fact that, experimentally, no clear distinction can be made between free enzymes, enzymes associated with cells, and those adsorbed to particles or resulting from autolysis or sloppy feeding. In recent years, the importance of extracellular enzymatic processes in aquatic habitats has received increasing recognition and the application of fluorogenic model substrates has become a powerful tool for the investigation of these processes (Table 4.1).

4.2 Methods and Experimental Design

Standard methods Total bacterial numbers were determined by acridine orange staining of the bacteria on black 0.2-μm-pore-size Nuclepore filters and epifluorescence microscopy (Zimmermann, 1977; Hobbie et al., 1977). Bacteria attached to submerged glass slides were directly stained on the slides and then counted after removal of excess staining solution. Bacterial sizes were estimated using Patterson grids.

[^{3}H]leucine uptake of natural bacterial populations was measured using a

Table 4.1 Microbial extracellular enzyme activities (EEA) studies using fluorogenic substrates.

Biotope, organism, source	Purpose	Substrate	Reference
—	Sulfatase and lipase activity, biochemical test	MUF-sulfate, β-naphthyl-sulfate, MUF-acyl esters, methyl-indoxyl-acyl esters	Guilbault and Hieserman (1969)
Soil	Lipase activity	MUF-acyl esters	Pancholy and Lynd (1972)
Central North Pacific water	Alkaline phosphatase activity in seawater	Methylfluorescein phosphate	Perry (1972)
Lake water	Alkaline and acid phosphatase activity in lakes, methodology	Various fluorogenic substrates for phosphatases	Pettersson and Jansson (1978)
Pure cultures	Viability test of fungal hyphae	Fluorescein diacetate	Ingham and Klein (1982)
Baltic Sea	EEA of brackish water bacteria, methodology	various MUF- and MCA-substrates	Hoppe (1983)
Natural waters	Exoproteolytic activity, methodology	Leucyl-β-naphthylamide	Somville and Billen (1983)
Seafood	Enumeration of *E. coli*	MUF-β-glucuronide	Alvarez (1984)
Eutrophic water	Substrate specificity of exoglucosidase	MUF-α-glucoside MUF-β-glucoside	Somville (1984)
Drinking water, natural waters	EEA in water treatment plants, methodology	MUF-β-glucoside, MUF-phosphate, alanine-β-naphthylamide, fluorescein diacetate	Holzapfel et al. (1985)
Baltic Sea, benthic vegetation	EEA of natural populations of different sites	MUF-β-glucoside MUF-phosphate, Leu-MCA	Mow-Robinson and Rheinheimer (1985)
North Sea, Scheldt estuary	Free and attached proteolytic activity	Leucyl-β-naphthylamide	Vives-Rego et al. (1985)

(Continued next page)

Table 4.1 Continued

Biotope, organism, source	Purpose	Substrate	Reference
Natural waters	Fluorescence correction, methodology	MUF-substrates	Chróst and Krambeck (1986)
Lakes, sea water	Phosphatase activity, methodology	MUF-phosphate	Chróst et al., (1986)
Saprophytes	EEA of saprophytes	Various MUF- and MCA-substrates	Kim and Hoppe (1986)
Marine sediments	β-glucosidase activity in sediments	MUF-β-glucoside	King (1986)
Marine sediments	Relation of EEA and substrate incorporation	MUF-β-glucoside, Leu-MCA	Meyer-Reil (1986)
Pure cultures	In vivo enzyme—substrate fluorescence velocities, review of substrates	Various fluorogenic substrates	Snyder et al. (1986)
Lakewater bacteria and phytoplankton	Alkaline phosphatase activity, kinetics	MUF-phosphate	Chróst and Overbeck (1987)
Coastal seawater	Effect of proteolysis inhibitors on exoproteases	Leucyl-β-naphthylamide	Fontigny et al. (1987)
Marine sediments	Seasonal aspects of EEA	Leu-MCA, MUF-β-glucoside	Meyer-Reil (1987)
Pure cultures	Test for chitinase activity of bacteria	MUF-N-acetyl-glucosaminide	O'Brien and Colwell (1987)
Coastal oceanic water	Proteolytic activity distribution, relation to bacteria, methodology	Leucyl-β-naphthylamide	Rosso and Azam (1987)
Drinking water	Test for fecal coliforms	MUF-galactoside, MUF-glucuronide	Berg and Fiksdal (1988)
Baltic Sea	Relation of EEA and substrate incorporation	Leu-MCA	Hoppe et al. (1988a)
Coral Reef	EEA of pelagic reef-bacteria	Leu-MCA	Hoppe et al. (1988b)

Lakes	Determination of aminopeptidase activity, methodology, field results	Leucyl-β-naphthylamide	Jacobsen and Rai (1988)
Lake water	β-Glucosidase activity, synthesis and repression	MUF-β-D-glucoside	Chróst (1989)
Lake water	Phosphatase, aminopeptidase, β-glucosidase and β-galactosidase activity, degradation of organic matter	Various MUF-substrates	Chróst et al. (1989)
Polyhumic lake	EEA of natural microbial populations, methodology	Various MUF-substrates and Leu-MCA	Münster et al. (1989)
Baltic Sea	Depth profiles of peptidase activity	Leu-MCA	Rheinheimer et al. (1989)
Phragmites australis stems	EEA of epiphytic microbiota	Various substrates	Goulder (1990)
Brackish water	Effect of H_2S on EEA, substrate uptake, and bacterial production	Leu-MCA	Hoppe et al. (1990)
Stream bed sediments	EEA of different type of sediments	MUF-β-glucoside, MUF-phosphate	Marxsen and Witzel (1990)

MUF- = methylumbelliferyl substrates, MCA- = 4-methyl-7-coumarinylamide substrates.

conventional assay for measuring uptake kinetics (Gocke, 1977). [^{3}H-*methyl*]thymidine incorporation as an estimate for bacteria production in the depth profile was measured according to the basic approach of Fuhrman and Azam (1982) using Nuclepore filters for particle collection. Time series incubations were done in PE-vials at 5 nM [^{3}H-*methyl*]thymidine (saturation concentration).

Aminopeptidase activities of bacterial populations were measured by the hydrolysis of leucine methylcoumarinylamide (Leu-MCA) at different concentrations using a Kontron SM 25 fluorometer. A detailed description of the method is given by Hoppe (1983) and Hoppe et al. (1988a). Special procedures developed to measure aminopeptidase activity of bacteria attached to glass slides are described below.

Aminopeptidase activity of bacteria attached to detritus This experiment has already been reported in part by Kim (1985) and Hoppe et al. (1988a). It was designed to simultaneously measure the uptake and the hydrolysis dynamics of free bacteria and bacteria attached to particles (mainly organic detritus). The results are presented here in a form which allows for the comparison with findings from experiments conducted on bacteria attached to inorganic surfaces (glass slides).

The three components of this substrate dilution experiment were: (1) measurement of [^{3}H]leucine uptake kinetics of a natural water sample from the Kiel Fjord (particle bound plus free bacteria) and of its 3-μm filtrate (mainly free bacteria). In both cases, disintegrations per minute (dpm) and leucine uptake were measured at 5.1 ng C [^{3}H]leucine l^{-1} and higher leucine concentrations, which were established by the addition of increasing amounts of unlabelled leucine. The dpms were plotted against the concentration of leucine, and velocities of uptake were calculated; (2) measurement of aminopeptidase kinetics in terms of Leu-MCA hydrolysis at increasing Leu-MCA concentrations; and (3) determination of [^{3}H]leucine-uptake kinetics in the presence of simultaneous Leu-MCA hydrolysis. Dpms of [^{3}H]leucine uptake (5.1 ngC [^{3}H]leucine l^{-1}) were measured in batches with increasing amounts of Leu-MCA. In this case, cold leucine additions were not defined as in the first component, but rather resulted from Leu-MCA hydrolysis. Velocities of leucine uptake under these conditions were calculated by using the dpm/concentration plot of the first component. Velocities of leucine uptake were then compared to velocities of MCA-hydrolysis for the unfiltered water sample and the 3-μm filtrate.

Aminopeptidase activity of bacteria attached to inorganic surfaces Microbial colonization and aminopeptidase activity of inorganic surfaces were investigated on glass slides submerged for seven days in the brackish water of the Kiel Fjord. Some slides were removed from the stock of submerged slides at daily intervals and rinsed with sterile seawater to remove nonattached bacteria. For total bacteria and bacteria biomass investigations, the slides were covered with acridine orange solution for 5 min, resulting in brilliant bacterial staining after the removal of excess dye. Patchiness of bacterial colonization during the first days caused lower statistical confidence levels for bacteria counts per cm^2 than were obtained in the later phases of colonization. In order to estimate the aminopeptidase activity of the attached bacteria, slides were incubated separately in 15 ml of sterile (filtered 0.2-μm) seawater taken from the site of exposure and supplemented with 150 μl of the aminopeptidase substrate Leu-MCA. Subsamples for time series measurement of peptidase activity were withdrawn from these batches at intervals of 30 min. Readings of fluorescence were performed according to Hoppe (1983). Some slides were also incubated in unprocessed water from the site of exposure, further treatments being exactly the same as those described before. At the same time that slides were drawn from the site of exposure, water samples were also taken from the surrounding water and investigated for bacteria and peptidase activity. For the latter, 50-ml samples were supplemented with 500 μl of Leu-MCA, resulting in the same final substrate concentrations as in the slide-incubations.

Depth profile of aminopeptidase activity This experiment was conducted on the JGOFS-Meteor 10 cruise from the Azores to Iceland (57°40.10′ N, 23°35.10′ W), in May 1989. The parameters measured were total number of bacteria, V_{max} and hydrolysis rate (HR, % h^{-1}) of aminopeptidases, and bacterial production by [^{3}H-*methyl*]thymidine incorporation. The application of these methods followed descriptions and references given in this chapter. The depth profile ranged from 5 to 500 m. Changes in the values of the different parameters with regard to depth are compared to each other and their relationships to environmental factors are discussed.

Calculations Total bacterial numbers from slides and water are expressed in terms of bacteria per cm^2 and total bacteria per ml, respectively. Bacterial biomass is expressed as μg bacterial C per cm^2 and μg bacterial C per ml, respectively.

Leu-MCA hydrolysis is calculated as follows: From time series incubation, fluorescence increase per hour and volume were computed by regression analysis. These values are converted to μg C leucine l^{-1} h^{-1} (velocity of Leu-MCA hydrolysis) using a calibration factor (slope of the regression derived from fluorescence readings at increasing concentrations of MCA solution). The values obtained can be attributed exclusively to the peptidase activity of the bacteria attached to the glass slides (32.8 cm^2), because the water used for incubation was originally sterile. Thus the extracellular enzyme activity (EEA) of the attached bacteria per cm^2 could be calculated and compared to the EEAs of the surrounding water.

4.3 Activity of Bacteria Attached to Detritus

The results of the combined radiotracer/fluorescence-tracer experiment described above are presented in a comprehensive form in Table 4.2. The three

Table 4.2 Relationship between enzymatic hydrolysis of Leu-MCA (peptide model substrate) and bacterial leucine incorporation (^{3}H-Leu) with increasing concentrations of peptide model substrate. Values from a particle size fractionation experiment with brackish water.

Leu-MCA concentration (μmol l^{-1})	A: Leu-MCA hydrolysis (μg C l^{-1} h^{-1})	B: Leucine incorporation (μg C l^{-1} h^{-1})	A ÷ B
Total bacterial population			
7.2	0.12	0.04	3
72	0.82	0.09	9
360	2.41	0.11	21
720	3.89	0.12	32
Fraction of <3 μm (mainly free-living bacteria)			
7.2	0.05	0.03	2
72	0.29	0.05	5
360	0.91	0.06	15
720	1.55	0.09	17
Fraction of >3 μm (mainly attached bacteria)			
7.2	0.07	0.01	7
72	0.53	0.04	13
360	1.50	0.05	30
720	2.33	0.03	77

parts of the table show the velocities of bacteria peptide hydrolysis and leucine incorporation at the indicated Leu-MCA concentrations for the unfiltered water sample and for the <3-μm- and >3 μm-size fractions. Bacterial counts revealed that 82% of the total bacteria passed through the 3-μm filter. The remaining part (18%) which was harvested on a 3-μm filter, consisted mainly of bacteria attached to detritus. Of course, not only attached bacteria, but also most of the planktonic algae, small animals, and particles were included in this fraction. Except for the bacteria, these living and nonliving components have only small capacities for the uptake of dissolved organic matter (such as leucine), and their adsorption may also be low in comparison to the active substrate transport mechanisms of the attached bacteria. Aminopeptidase activities in this fraction are also predominantly attributed to the attached bacteria, because extracellular enzyme activities of selected algal species are low (Hoppe, 1986; Chróst et al., 1989). On the other hand, the effect of adsorption of free enzymes (e.g., from autolysis) to particle surfaces has not yet been studied in detail and, therefore, attributing aminopeptidase activity in the >3-μm fraction exclusively to the attached bacteria may overestimate their properties.

Nevertheless, the final results of this experiment (relationships between rates of Leu-MCA hydrolysis and those of leucine uptake) show a very strong tendency towards a stronger hydrolysis in the >3-μm fraction than in the free bacteria fraction. About 60% of the total hydrolytic activity is represented by the fraction of attached bacteria, which is, as stated above, only about 18% of the total bacteria number. This fraction is responsible for only 25–45% of the total leucine incorporation. Thus it can be concluded that the population of bacteria attached to organic particles plays a dominant role in LMWOM generation. However, with respect to substrate uptake and the heterotrophic productivity, it is inferior to the free-living bacteria community, an assertion which has also been made by Azam and Ammerman (1984). According to observations by Fukami et al. (1983), the role of attached bacteria in particle decomposition will be especially important during phases of phytoplankton bloom decline, when attached bacteria may amount to 40% of the total bacteria.

4.4 Activity of Bacteria Attached to Inorganic Surfaces

Microbial colonization and aminopeptidase activity of inorganic surfaces were studied on glass slides submerged for seven days in the brackish water of the Kiel Fjord. The number of bacteria on the glass slides increased from zero to 1.24 million per cm^2 after only one day of exposure. The maximal value of 1.91 million was obtained after 6 days of exposure. This increase was not steady, but rather interrupted by minor decreases, most likely due to grazing or detachment. A reduction of biofilm development by protozoa grazing on bacteria has been documented by Pedersen (1982). The size spectrum of attached cells on the slides shifted significantly during the incubation time from predominantly smaller cells to a greater number of larger ones (Table 4.3). Compared with the bacteria size spectrum of the water surrounding the submersion site, it appeared that, initially, the smaller size classes tended to attach to the glass surface. A similar observation was also made with other surface materials by

Table 4.3 Size class distribution of bacteria on submerged glass slides and in the surrounding water.

Size class[1]	Glass slides: total days of exposure[2]						Water: sampling day[2]	
	1	3	4	5	6	7	1	7
A	0	0	0	0	0	0	0	0
B	25	20	15	11	10	9	0	0
C	41	20	15	21	16	9	0	8
D	12	14	8	18	11	9	26	28
E	6	19	37	36	34	25	23	22
F	2	14	14	69	24	37	6	16
G	0	4	11	65	3	11	4	0
H	0	1	0	2	2	0	0	0
I	0	3	0	0	0	0	0	0

[1]Increasing sizes from A to I are based on the diameter (in μm) of the circles on the Patterson counting grid: A = 0.24, B = 0.34, C = 0.48, D = 0.68, E = 0.96, F = 1.36, G = 1.93, H = 2.73, I = 3.86.
[2]The numbers are counts of bacteria per size class in a defined number of counting fields.

Marshall et al. (1971). Marshall (1985) claimed that initially starved cells attach to surfaces which enlarge their cell volumes in the latter phase of attachment. The counts of bacteria in the surrounding water were approximately 3.9 million per ml, which was typical for the late summer situation in the moderately polluted inner Kiel Fjord.

Aminopeptidase activities of attached bacteria rose continuously over time, more in the experiments in which exposed glass slides were incubated in natural water from the site of exposure than in experiments in which they were incubated in sterile filtered water. Aminopeptidase activity of the natural water from the sampling site alone was quite constant over time, except for an increase on the final day of the experiment (Figure 4.2). The amount of peptidase activities from the surrounding water plus from the slides incubated in sterile water was almost always smaller than that of slides incubated in natural water; this suggests a stimulation of peptide decomposition in the complete system. The values of extracellular peptidase activity increased over the experimental period from 7.4 to 47.5 ngC cm^{-2} h^{-1} for the exposed slides, corresponding activities of the surrounding waters being 27.5 and 35.3 ng C ml^{-1} h^{-1}. Specific peptidase activities per cell were calculated from the counts of bacteria attached

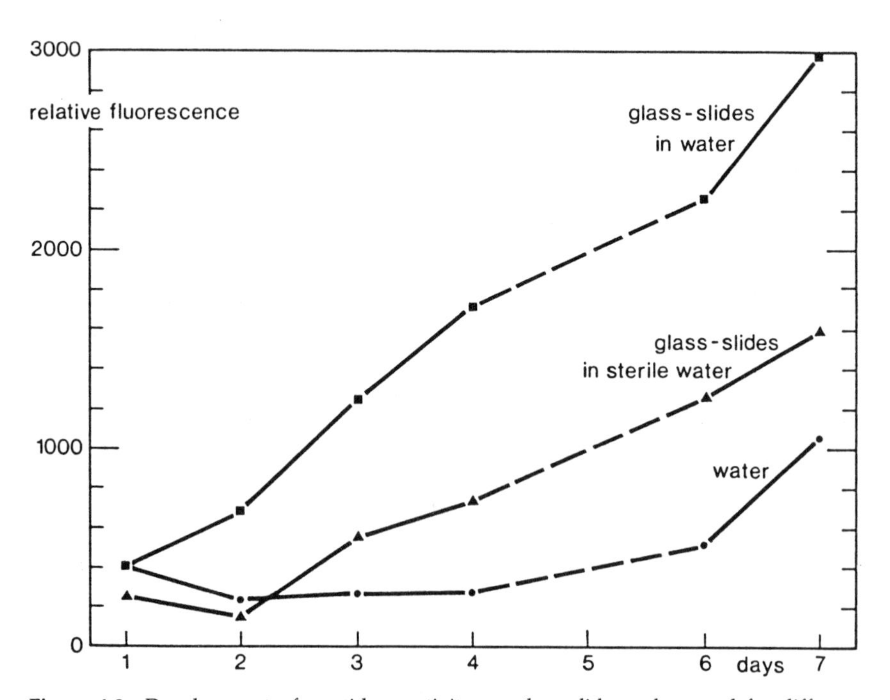

Figure 4.2 Development of peptidase activity on glass slides submerged for different periods of time in brackish water and in the surrounding water. For peptidase activity measurement, the glass slides were incubated together with the model substrate Leu-MCA in 0.2.μm-filter sterilized water from the site of exposure or in natural water.

to the glass slides and those of bacteria in the surrounding water and their corresponding activities. These few data are, of course, very preliminary and include some uncertainties due to the patchy distribution of bacteria on the slides. The specific activities per cell (Table 4.4) vary considerably. At the end of the experiment after seven days the specific activity of the cells attached to the glass slides is about double as high as that of the bacteria in the surrounding water. This may be attributed to the previously mentioned increase in size of the attached bacteria. However, it has to be considered, that the bacteria population in the surrounding water was composed of free bacteria and those attached to organic detritus. This is taken into account in the lower part of Table 4.4 where overall specific activities of bacteria in the surrounding water were diversified for specific activities of free bacteria and bacteria attached to detritus using factors given in subchapter 4.3. Comparison of specific activities of bacteria attached to glass surfaces and those attached to organic detritus exhibit considerable higher values for the latter.

In order to gain a better understanding of the relative changes of peptidase activities of attached bacteria and bacteria in the water during the experiment, direct comparison was attempted (Figure 4.3). It became obvious that the relationship between the activity originating from the bacteria attached to 1 cm^2 of exposed slide and the activity originating from the bacteria in 1 ml of surrounding water was steadily increasing by a factor of up to 1.35. To a minor extent, this increase is due to the specific activity of the attached bacteria, but it can largely be attributed to progressive bacterial colonization of the inorganic surface combined with an increase in cell size.

Table 4.4 Specific peptidase activities per cell (pg C h^{-1}) of bacteria attached to submerged glass slides and of bacteria in the surrounding water

Glass slides: total days of exposure[1]					Water: sampling day	
1	3	4	6	7	1	7
					Total bacteria[2]	
0.006	0.015	0.018	0.020	0.027	0.007	0.013
Free-living bacteria[3]						
—	—	—	—	—	0.004	0.007
Attached bacteria[4]						
—	—	—	—	—	0.021	0.039

[1]Peptidase activity on glass slides is exclusively attributed to attached bacteria.
[2]Peptidase activity of total bacteria in the water is an average attributed to free bacteria as well as bacteria attached to detrital particles.
[3]Free-living bacteria (fraction of <3 μm) constituted 80% of the total number of bacteria and displayed 40% of the total peptidase activity in the water.
[4]Bacteria attached (fraction of >3 μm) to detritus constituted 20% of the total number of bacteria and accounted for 60% of the total peptidase activity in the water.

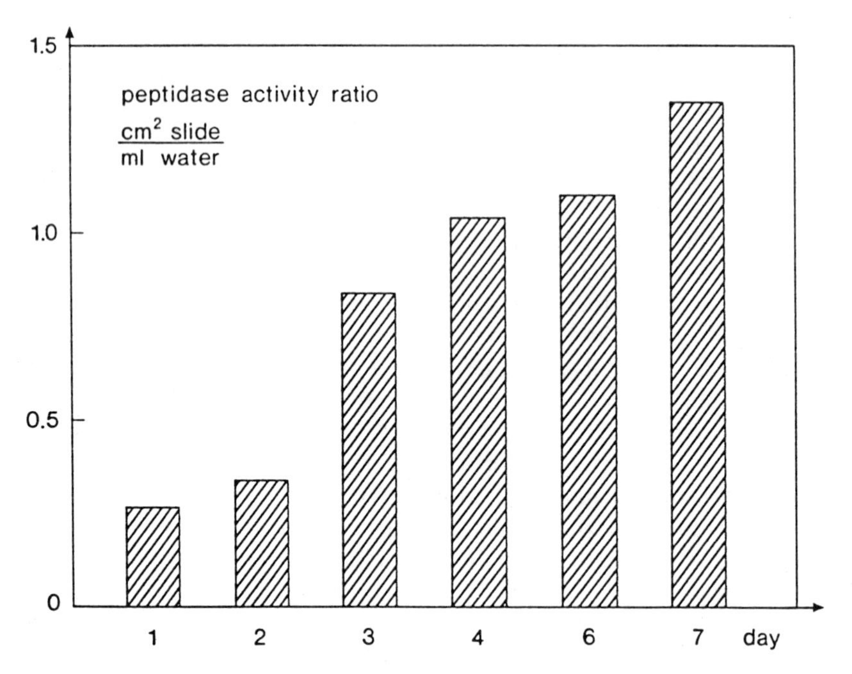

Figure 4.3 The ratio of the peptidase activity originating from the epiflora of 1 cm^2 of submerged slide surface and from 1 ml of the surrounding water over a seven-day exposure period.

4.5 Depth Profile of Aminopeptidase Activity

The depth profile of aminopeptidase activity and other bacterial activities was obtained from a station (57°40.1′ N, 23°35.1′ W) in the North Atlantic. This investigation is described here in order to give an example of changes in hydrolysis and uptake rates in relation to changes in environmental conditions. Some of the hydrographic and biological data from this station may support this interpretation of the microbiological results: water temperature 9.8°C at the surface, 8.8°C below 30 m; boundary of the mixed layer at 20-m depth; highest values for O_2, NO_3^-, SiO_2, PO_4^{3-} are in the mixed layer; stabilization of values for these parameters below 100 m. Maxima of chlorophyll$_a$ and primary production were upwards of 10 m. Chlorophyll$_a$ in the upper 100 m totalled approximately 100 mg m^{-2}, and primary production was approximately 530 mg C m^{-2} d^{-1}. In a drifting sediment trap located at a depth of 100 m, chlorophyll$_a$ accumulation was only approximately 300 μg m^{-2} d^{-1}. The difference between the amount of chlorophyll$_a$ in the living phytoplankton in the water column and the amount of chlorophyll$_a$ in the sediment trap suggests a strong decomposition of materials by way of terminal oxidation or recycling in the upper 100 meters.

Bacterial measurements taken from the depth profile are presented in Table

4.5. Taking into account the fact that the station was 900 miles from the coast, bacterial numbers and production were relatively high in the mixed layer. This also holds true for the peptidase activity, which was almost as high as the annual average peptidase activity found in the moderately polluted Kiel Bay (Baltic Sea). The strong rate of hydrolysis of organic materials (HR about 15% d^{-1}) suggests that materials in the mixed layer were quite prone to bacterial decomposition. In general, the bacterial measurements reflect the favorable nutritional conditions which were created by the spring bloom of phytoplankton in this region.

Also listed in Table 4.5 are the relative changes in bacteriological measurements with depth. Since surface waters are the most productive in terms of primary production and heterotrophic potential, the values of bacterial parameters at any depth were standardized to surface values. Relationships among values from different depths also are obvious using this mode of calculation. The hydrolysis rate (HR, % h^{-1}) of proteinaceous materials decreases the most as a function of depth. This implies that the materials which undergo enzymatic hydrolysis become increasingly refractile, especially below a depth of 80 m, where no photosynthesis occurs and easily degradable substrates are therefore not produced.

The decrease in the rate of peptide hydrolysis (V_{max}) with depth was the

Table 4.5 Surface/depth relationships of bacteria stock and activity parameters at the sampling station in the North Atlantic (part A) and the absolute values of these parameters from the same station (part B)

		Peptidase activity		
	Bacteria (10^6 cells ml^{-1})	V_{max}[1] (μg C l^{-1} h^{-1})	Rate of hydrolysis[2] (% d^{-1})	Thymidine incorporation (pmol l^{-1} h^{-1})
A: Surface/depth relation				
5 m/10 m	1.18	1.04	0.90	1.77
5 m/20 m	1.39	1.33	1.25	1.91
5 m/40 m	1.78	1.92	1.89	7.33
5 m/80 m	2.33	2.65	2.37	8.63
5 m/300 m	5.59	4.04	13.92	9.78
5 m/500 m	16.27	4.37	23.94	14.67
B: Depth				
5 m	1.79	4.24	14.10	4.40
10 m	1.52	4.08	15.70	2.48
20 m	1.29	3.19	11.30	2.30
40 m	1.00	2.21	7.50	0.60
80 m	0.77	1.52	5.90	0.51
300 m	0.32	1.05	1.00	0.45
500 m	0.11	0.97	0.60	0.30

[1]Maximal velocity of peptide hydrolysis.
[2]Hydrolysis rate of peptides in the water.

lowest compared with the distribution of the other parameters. Total numbers of bacteria decreased more rapidly than the hydrolytic potentials which can be attributed to them. This implies that the hydrolytic potential of individual cells increased considerably with depth, most likely due to the lack of easily degradable dissolved substrates. It can even be assumed that, due to changes in environmental conditions affecting sinking bacteria which originate from the surface layer, the decrease in active cells, which are included in the total number estimates, was stronger than evident from the total number relationships.

According to the measurements of chlorophyll$_a$ and primary production, thymidine uptake into nonacid-hydrolyzable cell constituents, a relative measurement of bacteria production, shows a strong decrease starting below the mixed layer at 20 m. This indicates a tight coupling of photoautotrophic and bacterial heterotrophic processes in the mixed surface layer, where easily degradable substances are the basis for bacterial nutrition. This coupling is not necessarily reflected by numbers of bacteria, because grazing activities are also high in this layer. Below the mixed layer, which was distinguished from the underlying water by a low temperature gradient, bacterial production decreased progressively. This pattern of depth distribution shows that rapid bacterial growth is dependent upon easily degradable substrate supply. Increasing recalcitrance of organic materials with depth does not allow for high bacterial productivity, although bacterial numbers may still be relatively high due to the input from sinking particles and less efficient grazing.

4.6 Importance of Extracellular Enzyme Activity in Microbial Activity in Marine Ecosystems

Particle decomposition is considered an important process of dissolved nutrient generation in aquatic environments. This is especially true for the aphotic water column and the sediment, regions in which easily degradable dissolved compounds resulting from phytoplankton exudation are rare. Autolysis of sinking cells and excretion by animals will probably only constitute a minor contribution to the dissolved organic carbon (DOC) pool in the depth. It has been found that dead phytoplankton loses up to 52% of nitrogen content as dissolved organic nitrogen (DON) within 7 h (Garber, 1984). However, these substances will contribute to the pool of easily degradable substrates only in a small zone under the euphotic zone.

Bacteria cannot consume whole particles, but will settle on their surfaces. Therefore, bacteria behavior on particles (organic or inorganic) is closely linked to surface properties. Decomposition of surface molecules (whether they be integrated into the surface or adsorbed to it) is, in most cases, brought about by extracellular enzymes of attached bacteria and, to an unknown extent, also of enzymes adsorbed from the surrounding water. The location of extracellular enzymes (described by Priest, 1984, and Chróst, 1990; see also Chapter 3) can strongly influence the mode and efficiency of surface erosion. Should the enzymes be bound to membranes, or at least remain in the vicinity of their producers, their impact is limited to the microbes dimension. This holds true for attached bacteria, but not so much for free bacteria, where substrates are available around the cells, and clustering of cells around microenvironmental nutrient gradients may be a preferred strategy (Azam and Ammerman, 1984; Mitchell et al., 1985). This hypothesis, i.e., enhancement of bacterial growth in microenvironments, has been challenged by Jackson (1989). Location of enzymes in the cell envelope could also allow for a tight coupling of hydrolysis and the uptake of products of hydrolysis.

If enzymes are secreted in the vicinity of the cells or escape from them by way of diffusion, they may penetrate particle surfaces (if their structures allow) and break down the particle from within. In this case, enzymes would be more prone to degradation and a loose coupling of hydrolysis and uptake could be anticipated. Products of hydrolysis not used up immediately by the attached cells enter the DOC pool of the surrounding water to the benefit of the free bacteria there. It has been shown that products of extracellular hydrolysis of dissolved proteins did not accumulate in the medium (Hollibaugh and Azam, 1983). End products of hydrolysis are expected to be removed immediately by the present microflora (Güde, 1978).

The three examples presented in this chapter are based on such considerations. The first experiment was on enzymatic decomposition of detritus particles (>3-μm). In different size fractionation experiments (Kim, 1985), the highest rates of aminopeptidase activities were occasionally found in the <3-μm fraction. However, activities in the >3-μm fraction also always contributed considerably to the total aminopeptidase activity. In any case, aminopeptidase activity in this fraction was much higher than was to be expected in view of its low contribution to the total number of bacteria.

Activity of dissolved aminopeptidases is low in 0.2-μm filtrates (Hoppe, 1983; Vives-Rego et al., 1985; Rosso and Azam, 1987; Chróst et al., 1989, Boon, 1989), therefore, measured activities in the <3-μm fraction can predominantly be attributed to free bacteria and probably also to small flagellates. In the >3-μm fraction, activities resulting from enzymes associated with bacteria, adsorbed enzymes that originally were free, and enzymes excreted by the attached bacteria cannot be distinguished from each other. Since the activity of free dissolved enzymes is low, their contribution to the total activity on particles may also be low after adsorption. Because excreted enzymes originate mainly from bacteria, it can be hypothesized from the results of this experiment that the combined activities of excreted and cell- associated extracellular aminopeptidase activity of individual bacteria attached to detritus is much higher than the corresponding activity of free bacteria. Also, from the qualitative aspect, attached bacteria seem to have a broader spectrum of substrates which they can hydrolyze than free-living bacteria do (Fukami et al., 1981). As long as no methods are available to detect extracellular enzymatic properties of individual cells, measured activities can only be related to total numbers of bacteria.

When discussing the quantitative aspects of particle hydrolysis, it should be considered that only some fractions of primary production reach the aphotic depth as detritus. According to Joint and Morris (1982), the portion of primary production which settles into the deeper water is probably less than 10%. The time scale for decomposition is highly variable depending on the water depth and sinking speed. Increasing C:N ratios of sedimenting detritus in the deep (Karl et al., 1988) suggest that, after preferential use of proteinaceous constituents, the hydrolysis of the remaining materials must be a rather slow process. Hence, considering particle decomposition to be a key factor in the generation of dissolved nutrients in the deep, their concentrations in deep waters must be low.

The second experiment dealt with bacterial growth and aminopeptidase activity on inorganic solid surfaces. The results of this experiment cannot be generalized because the type of material exposed to the environment and its chemical and physical properties have an influence on surface conditioning and bacterial attachment. Thus, the results are only valid for smooth and plain glass surfaces. It could be shown that aminopeptidase activities of bacteria attached to 1 cm^2 after the first day of exposure were equal to those of 0.27 ml of the surrounding water. This activity increased steadily to equal that of 1.35 ml after 7 days of exposure (Figure 4.3). The increase of this ratio is certainly due to increasing numbers and cell sizes of bacteria and increasing aminopeptidase activity per cell during the period of exposure of the glass slides, while, in the surrounding water, these values remained fairly constant.

When examining individual aminopeptidase activities, it can be stated that per-cell activities observed on the glass slides may exclusively be attributed to attached bacteria. On the other hand, overall activities measured in the surrounding water resulted from the combined activities of free bacteria and bacteria attached to particles (detritus). Because no size fractionation was made in this experiment, results of the first experiment reported in this paper were used to differentiate between the enzymatic activities of the free bacteria and of the attached bacteria in the surrounding water. In this calculation it is assumed that maximally 20% of the bacteria is attached (predominantly to detritus) and represents maximally 60% of the total measured aminopeptidase activity; the remaining 40% of the total activity would be represented by the 80% of free-living bacteria (Table 4.4). It should be noted here that this procedure is, of course, not absolutely correct and that the desire was to generate preliminary ideas about enzyme activities of single bacteria attached to inorganic surfaces, in comparison to free-living bacteria and bacteria attached to organic detritus.

Calculated per cell aminopeptidase activities were 0.004 to 0.007 pg C h^{-1} for free bacteria and 0.021 to 0.039 pg C h^{-1} for attached bacteria in the water around the glass slides. Corresponding activities for the bacteria attached to the glass slides were between 0.006 and 0.027 pg C h^{-1} (Table 4.4). This suggests that, shortly after the beginning of the experiment, bacteria attached to the glass slides were slightly more active than free bacteria from the surrounding water but much less active than bacteria attached to detritus suspended in the surrounding water. After seven days of exposure bacteria attached to the glass slides were much more active than the free-living bacteria but still not as active as the bacteria attached to detritus. Nevertheless, the progressive increase in activity of the bacteria on the glass slides reflects the process of biofilm formation on the slides, which provides a better nutritional supply to the microbial community.

Bacterial extracellular decomposition of macromolecular dissolved or particulate organic matter is of special importance in the water column below the euphotic zone, where easily degradable substrates (exudates) are rare. This situation was investigated in the third experiment, at various depths in the North Atlantic. Just below the euphotic zone, autolyzing cells may still be a source contributing to the pool of dissolved HMWOM, however, at greater depths it can be expected that suspended or sinking organic particles and zooplankton fecal pellets are the main objects of bacterial hydrolytic activity. Relatively high potentials for extracellular enzymatic hydrolysis in the deep may therefore be understood as the microbial population's response to the prevailing nutrient regime, which enables them to make use of the more refractory residues of primary production in the deep waters (Table 4.5). Nevertheless, bacterial numbers and production rates below the euphotic zone are much lower than would be expected considering the potential for hydrolysis. This suggests that potentials for hydrolysis are not fully made use of in the environment. However, they become obvious when larger amounts of degradable organic matter are supplied, as established by the Leu-MCA addition in the experiments. Thus it can be ascertained that hydrolytic capacities are increasingly necessary in the deeper zones, more for survival under conditions of low nutrient availability than for allowing rapid bacterial growth. Very low rates of proteinaceous matter hydrolysis (turnover rate) below 80 m indicate that, despite relatively high potentials for hydrolysis, the production of LMWOM must be low in these waters due to the nature of sedimenting materials. Thus, bacterial numbers and productivities decrease drastically in the deeper aphotic zone. The depth profile presented here is, in principle, very similar to those obtained from the Santa Monica Basin (Rosso and Azam, 1987). However, absolute values of thymidine incorporation and bacteria numbers are considerably higher in the Santa Monica Basin than in the North Atlantic; in contrast, those for peptidase activity in the mixed layer are higher in the North Atlantic than in the Pacific.

Observations of extracellular enzyme activity (EEA) and ideas about the impact of EEA on organic matter conversion have been used to construct a semiquantitative model (Figure 4.4). This model is based on the three most important pools of organic matter in aquatic environments: living and dead particulate organic matter (together called POM) and dissolved organic matter (DOM). The sum of the different fractions of the DOM pool (polymers, oligomers, monomers) is, when integrated over the entire water column, considerably greater than the sum of living and dead POM standing stocks. The fraction of dead organic matter from POM will be quite variable with regard to depth; in the DOM pool it is expected that monomers always constitute a small

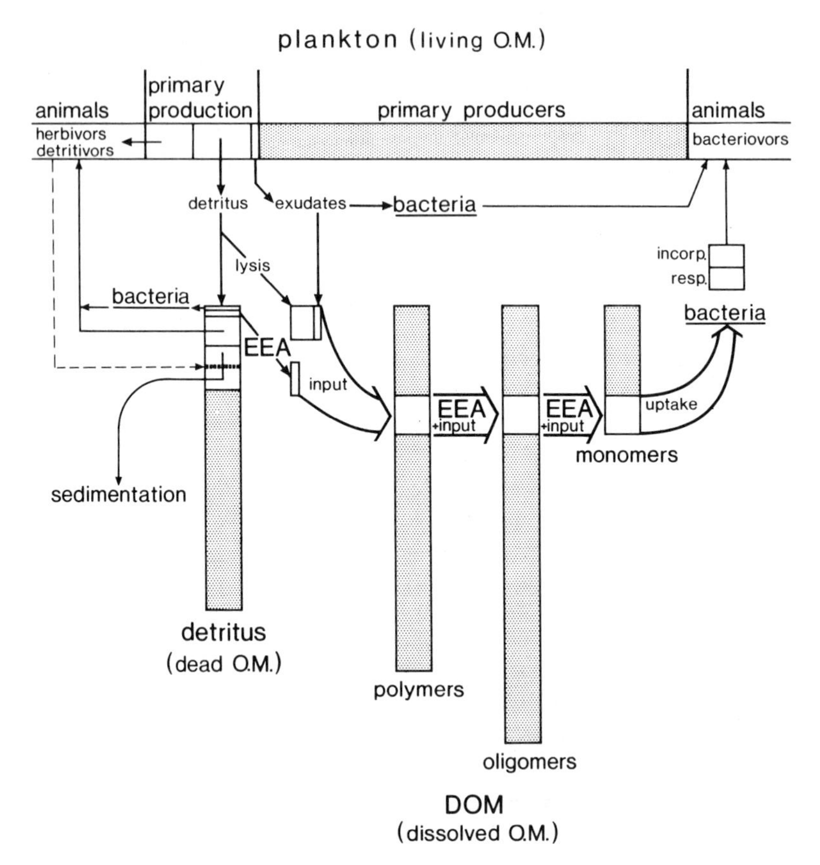

Figure 4.4 Semiquantitative model of organic matter transformation in aquatic ecosystems in a steady state. Blank areas represent production and losses or inputs into the respective stock pools. Shaded areas represent stock pools.

fraction of the total (Coffin, 1989). Standing stock pool sizes of organic matter must be the result of dynamic processes, such as primary production and losses (detritus formation, exudation, lysis, grazing), which serve simultaneously as an input for the resulting pools. In a steady state model, pool sizes are assumed to remain constant with production being balanced by losses to secondary producers and sediment deposits.

According to the model (Figure 4.4), a part of primary production (about 10%) is lost via exudation. The low-molecular-weight substances exudated are almost completely and immediately taken up by bacteria in the euphotic zone. The rest contributes to the DOM pool, which is subject to microbial hydrolysis. A considerable but variable fraction of primary production will be transformed to detritus. Of course, in reality, detritus recruits from the actual standing stock of cells. In the mass balance, however, it is subtracted from gross primary production. Detritus formation is accompanied by the loss of semipermeability of cell membranes and cell lysis. Depending on the total detritus formation, this may cause a considerable input of macromolecular organic matter into the DOC pool in the euphotic zone and in the upper part of the aphotic zone, where fresh detritus starts on its path of sedimentation. Particle fragmentation and solubilization have been said to be an important prerequisite for microbial usage of particulate materials (Karl et al., 1982). The remaining particulate parts of detritus are subject to bacterial colonization, bacterial particle hydrolysis, grazing of animals, and sedimentation. The many different factors involved in particle utilization demonstrate the difficulties which are inherent in the study of particle decomposition.

Nearly all detritus particles, fresh ones as well as those from the detritus standing stock, are potential objects of microbial colonization. Bacterial growth on detritus particles implies the usage of particulate material by way of extracellular hydrolysis and, to a minor extent, also of substrates from the surrounding water. As shown through the example presented in this report, products of extracellular hydrolysis are only partly incorporated by the attached bacteria. A considerable amount of these products, especially macromolecules, escape into the DOC pool, thus becoming exposed to free-living bacterial enzymatic activities. Excretory products and remainders of dead animals will also add to the detritus pool and, in turn, detritus particles with their attached bacteria will be ingested by animals. In a steady state model, losses from the detritus pool (sedimentation, animal grazing and respiration of bacteria by way of EEA) will be compensated for by the input derived from phytoplankton and animals. The input of materials into the DOC pool consists mainly of phytoplankton exudates, products of cell lysis and products of enzymatic hydrolysis from organic particles. While exudates which recycle quickly dominate in the euphotic zone, the latter two factors dominate in the upper and lower aphotic zone, respectively.

Macromolecular materials introduced into the DOC pool are subject to extracellular enzymatic hydrolysis, which is mainly established by free bacteria. Their progressive enzymatic decomposition is responsible for the dynamics of the different molecular size class pools within the DOC pool. The final step of monomer (dimer, trimer) formation is closely coupled with bacterial heterotrophic substrate uptake (Hoppe et al., 1988a; Chróst, 1989). Enzymatic degradation of particles is a relatively fast process based on high concentrations of organic substances in restricted volumes. Enzymatic degradation of dissolved macromolecular organic substances is a relatively slow process because the bulk of these substances is more or less evenly distributed in the water. This phenomenon leads to an accumulation of macromolecular organic substances dissolved in the water. Assuming steady state conditions, the output of monomers from this pool is balanced by bacterial uptake. This implies that real rates for hydrolysis and uptake (not necessarily V_{max}) balance each other, although turnover times are quite different, due to the different pool sizes of polymers, oligomers and monomers.

Acknowledgments This study was supported by DFG grants Ho 715-3/2 and Ho 715-5/1. I thank Mrs. M. Mehrens for excellent technical assistance, and the students of biological oceanography (H. Obstfeld, B. Schirm, G. Schönk) for their enthusiasm during the performance of the experiments. Thanks are also given to Mrs. A. Baum for linguistic correction and typing of the manuscript, and Dipl. Biol. B. Karrasch for the contribution of bacterial production data.

References

Alvarez, R.J. 1984. Use of fluorogenic assays for the enumeration of *Escherichia coli* from selected seafoods. *Journal of Food Science* 49: 1186–1187.

Azam, F., Fenchel, T., Field, J.G., Gray, J.S., Meyer-Reil, L.A. and T.F. Thingstad. 1983. The ecological role of water-column microbes in the sea. *Marine Ecology Progress Series* 10: 257–263.

Azam, F. and J.W. Ammerman. 1984. Cycling of organic matter by bacterioplankton in pelagic marine ecosystem: microenvironmental considerations. pp. 345–360 in Facham, M.J.R. (editor), *Flows of Energy and Materials in Marine Ecosystems*. Plenum Publishing Corp., New York.

Berg, J.D. and L. Fiksdal. 1988. Rapid detection of total and fecal coliforms in water by enzymatic hydrolysis of 4-methylumbelliferyl-α-D-galactoside. *Applied and Environmental Microbiology* 54: 2118–2122.

Boon, P.I. 1989. Organic matter degradation and nutrient regeneration in Australian freshwater: I. Methods for enzyme assays in turbid aquatic environments. *Archiv für Hydrobiologie* 115: 339 359.

Cho, B.C. and F. Azam. 1988. Major role of bacteria in biochemical fluxes in the ocean's interior. *Nature* 332: 441–443.

Chróst, R.J. 1989. Characterization and significance of β-glucosidase activity in lake water. *Limnology and Oceanography* 34: 660–672.

Chróst, R.J. 1990. Microbial ectoenzymes in aquatic environments. pp. 47–78 in Overbeck, J. and Chróst, R.J. (editors), *Aquatic Microbial Ecology: Biochemical and Molecular Approaches*. Springer Verlag, New York. 190 pp.

Chróst, R.J. and H.J. Krambeck. 1986. Fluorescence correction for measurements of enzyme activity in natural waters using methylumbelliferyl-substrates. *Archiv für Hydrobiologie* 106: 79–90.

Chróst, R.J., Münster, U., Rai, H., Albrecht, D., Witzel, K.P. and J. Overbeck. 1989. Photosynthetic production and exoenzymatic degradation of organic matter in the euphotic zone of an eutrophic lake. *Journal of Plankton Research* 11: 223–242.

Chróst, R.J. and J. Overbeck. 1987. Kinetics of alkaline phosphatase activity and phosphorus availability for phytoplankton and bacterioplankton in Lake Plußsee (North German eutrophic lake). *Microbial Ecology* 13: 229–248.

Chróst, R.J., Siuda, W., Albrecht, D. and J. Overbeck. 1986. A method for determining enzymatically hydrolyzable phosphate (EHP) in natural waters. *Limnology and Oceanography* 31: 662–667.

Coffin, R.B. 1989. Bacterial uptake of dissolved free and combined amino acids in estuarine waters. *Limnology and Oceanography* 34: 531–542.

Fontigny, A., Billen, G. and J. Vives-Rego. 1987. Kinetic characteristics and regulation of exoproteolytic enzymes of marine bacterioplankton. *Marine Ecology Progress Series* 25: 127–133.

Fuhrman, J.A. and F. Azam. 1982. Thymidine incorporation as a measure of heterotrophic bacterioplankton production in marine surface waters: evaluation and field results. *Marine Biology* 66: 109–120.

Fukami, K., Simidu, U. and N. Taga. 1981. Fluctuation of the communities of heterotrophic bacteria during the decomposition process of phytoplankton. *Journal of exploration in Biology and Ecology* 55: 171–184.

Fukami, K., Simidu, U. and N. Taga. 1983. Change in a bacterial population during the process of degradation of a phytoplankton bloom in a brackish lake. *Marine Biology* 76: 253–255.

Garber, J.H. 1984. Laboratory study of nitrogen and phosphorus remineralization during the decomposition of coastal plankton and seston. *Estuarine and Coastal Shelf Sciences* 18: 685–702.

Gocke, K. 1977. Comparison of methods for determining the turnover times of dissolved organic compounds. *Marine Biology* 42: 131–141.

Goulder, R. 1990. Extracellular enzyme activities associated with epiphytic microbiota on submerged stems of the red *Phragmites australis. Federation of European Microbiology Societies Microbiology Ecology* 73: 323–330.

Güde, H. 1978. Model experiments on regulation of bacterial polysaccharide degradation in lakes. *Archiv für Hydrobiologie,* (supplement) 55: 157–185.

Guilbault, G.G. and J. Hieserman. 1969. Fluorometric substrate for sulfatase and lipase. *Analytical Chemistry* 41: 2006–2009.

Hobbie, J.E., Daley, R.J. and S. Jasper. 1977. Use of Nuclepore filters for counting bacteria by fluorescence microscopy. *Applied and Environmental Microbiology* 33: 1225–1228.

Hollibaugh, J.T. and F. Azam. 1983. Microbial degradation of dissolved protein in seawater. *Limnology and Oceanography* 28: 1104–1116.

Holzapfel-Pschorn, A., Obst, U. and K. Haberer. 1985. Fluoreszenzspektroskopie zum empfindlichen Nachweis von Enzymaktivitäten in der Trinkwasseraufbereitung. *Gewässerschutz Wasser Abwasser* 79: 352–365.

Hoppe, H.G. 1983. Significance of exoenzymatic activities in the ecology of brackish water: measurements by means of methylumbelliferyl-substrates. *Marine Ecology Progress Series* 11: 299–308.

Hoppe, H.G. 1984. Attachment of bacteria: advantage or disadvantage for survival in the aquatic environment. pp. 283–302 in Marshall, K.C. (editor), *Microbial Adhesion and Aggregation.* Dahlem Konferenzen, Life Sciences Research Report 31, Springer-Verlag, Berlin.

Hoppe, H.G. 1986. Relations between bacterial extracellular enzymatic activities and heterotrophic substrate uptake in a brackish water environment. *Proceedings of 2nd International Colloquium on Marine Bacteriology, Actes de Colloque 3, Brest*: 119–128.

Hoppe, H.G., Kim, S.J. and K. Gocke. 1988a. Microbial decomposition in aquatic environments: combined process of extracellular enzyme activity and substrate uptake. *Applied and Environmental Microbiology* 54: 784–790.

Hoppe, H.G., Schramm, W. and P. Bacolod. 1988b. Spatial and temporal distribution of pelagic microorganisms and their proteolytic activity over a partly destroyed coral reef. *Marine Ecology Progress Series* 44: 95–102.

Hoppe, H.G., Gocke, K. and J. Kuparinen. 1990. Studies on the effect of H_2S on heterotrophic substrate uptake, extracellular enzyme activity and growth of brackish water bacteria. *Marine Ecology Progress Series* 64: 157–167.

Ingham, E.R. and D.A. Klein. 1982. Relationship between fluorescein diacetate-stained hyphae and oxygen utilization, glucose utilization and biomass of submerged fungal batch cultures. *Applied and Environmental Microbiology* 44: 363–370.

Jackson, G.A. 1989. Simulation of bacterial attraction and adhesion to falling particles in an aquatic environment. *Limnology and Oceanography* 34: 514–530.

Jacobsen, T.R. and H. Rai. 1988. Determination of aminopeptidase activity in lakewater by a short term kinetic assay and its application in two lakes of differing eutrophication. *Archiv für Hydrobiologie* 113: 359–370.

Joint, I.R. and R.J. Morris. 1982. The role of bacteria in the turnover of organic matter in the sea. *Oceanography and Marine Biology Annual Reviews* 20: 65–118.

Karl, D.M., Knauer, G.A. and J.H. Martin. 1988. Downward flux of particulate organic matter in the ocean: a particle decomposition paradox. *Nature* 332: 438–441.

Kim, S.J. 1985. Untersuchungen zur heterotrophen Stoffaufnahme und extrazellulären Enzymaktivität von freilebenden und angehefteten Bakterien in verschiedenen Gewässerbiotopen. *Ph.D. Dissertation, University Kiel.* 203 pp.

Kim, S.J. and H.G. Hoppe. 1986. Microbial extracellular enzyme detection on agar plates by means of fluorogenic methylumbelliferyl-substrates. *Proceedings of 2nd International Colloquium on Marine Bacteriology, Actes de Colloque 3, Brest,* 175–183.

King, G.M. 1986. Characterization of a-glucosidase activity in intertidal marine sediments. *Applied and Environmental Microbiology* 51: 373–380.

Marshall, K.C. 1985. Bacterial adhesion in oligotrophic habitats. *Microbiology Sciences* 2: 321–325.

Marshall, K.C., Stout, R. and R. Mitchell. 1971. Selective sorption of bacteria from seawater. *Canadian Journal of Microbiology* 17: 1413–1416.

Marxsen, J. and K.-P. Witzel. 1990. Measurement of exoenzymatic activity in stream-bed sediments using methylumbelliferyl substrates. *Archiv für Hydrobiologie Beihefte Ergebnisse Limnologie* 34: 21–28.

Meyer-Reil, L.A. 1986. Measurements of hydrolytic activity and incorporation of dissolved organic substrates by microorganisms in marine sediments. *Marine Ecology Progress Series* 31: 143–149.

Meyer-Reil, L.A. 1987. Seasonal and spatial distribution of extracellular enzymatic activities and microbial incorporation of dissolved organic substrates in marine sediments. *Applied and Environmental Microbiology* 53: 1748–1755.

Mitchell, J.G., Okubo, A. and J.A. Fuhrman. 1985. Microzones surrounding phytoplankton form the basis for a stratified marine microbial ecosystem. *Nature* 316: 58–59.

Mow-Robinson, J.M. and G. Rheinheimer. 1985. Comparison of bacterial populations from the Kiel Fjord in relation to the presence or absence of benthic vegetation. *Botanica Marina* 28: 29–39.

Münster, U., Einö, P. and J. Nurminen. 1989. Evaluation of the measurements of extracellular enzyme activities in a polyhumic lake by means of studies with 4-methylumbelliferyl-substrates. *Archiv für Hydrobiologie* 115: 321–337.

O'Brien, M. and R.R. Colwell. 1987. A rapid test for chitinase activity that uses 4-methylumbelliferyl-N-acetyl-α-D-glucosaminide. *Applied and Environmental Microbiology* 53: 1718–1720.

Pancholy, S.K. and J.Q. Lynd. 1972. Quantitative fluorescence analysis of soil lipase activity. *Soil Biology and Biochemistry* 4: 257–259.

Pedersen, K. 1982. Factors regulating microbial biofilm development in a system with slowly flowing seawater. *Applied and Environmental Microbiology* 44: 1196–1204.

Perry, M.J. 1972. Alkaline phosphatase activity in subtropical Central North Pacific waters using a sensitive fluorometric method. *Marine Biology* 15: 113–119.

Petterson, K. and M. Jansson. 1978. Determination of phosphatase activity in lake water, a study of methods. *Verhandlungen der Internationalen Vereinigung für Theoretische und Angewandte Limnologie* 20: 1226–1230.

Priest, F.G. 1984. *Extracellular Enzymes.* Van Nostrand Reinhold (UK) Co. Ltd., Workingham. 79 pp.

Rheinheimer, G., Gocke, K. and H.G. Hoppe. 1989. Vertical distribution of microbiological and hydrographic-chemical parameters in different areas of the Baltic Sea. *Marine Ecology Progress Series* 52: 55–70.

Rosso, A.L. and F. Azam. 1987. Proteolytic activity in coastal oceanic waters: depth distribution and relationship to bacterial populations. *Marine Ecology Progress Series* 41: 231–240.

Somville, M. 1984. Measurement and study of substrate specificity of exoglucosidase in natural water. *Applied and Environmental Microbiology* 48: 1181–1185.

Somville, M. and G. Billen. 1983. A method for determining exoproteolytic activity in natural waters. *Limnology and Oceanography* 28: 190–193.

Snyder, A.P., Wang, T.T. and D.B. Greenberg. 1986. Pattern recognition analysis of in vivo enzyme-substrate fluorescence velocities in microorganisms detection and identification. *Applied and Environmental Microbiology* 51: 969–977.

Vives-Rego, J.V., Billen, G., Fontigny, A. and M. Somville. 1985. Free and attached proteolytic activity in water environments. *Marine Ecology Progress Series* 21: 245–249.

Zimmerman, R. 1977. Estimation of bacterial number and biomass by epifluorescence microscopy and scanning electron microscopy. pp. 103–120 in Rheinheimer, G. (editor), *Microbial Ecology of a Brackish Water Environment*. Springer Verlag, Berlin.

5

Ecological Aspects of Enzymatic Activity in Marine Sediments

Lutz-Arend Meyer-Reil

5.1 Introduction

The supply of organic material is a key factor determining the structure and activity of benthic microbial communities. The overwhelming portion of the organic matter entering the sediment via sedimentation is particulate organic carbon, which has to be extracellularly decomposed by enzymes prior to incorporation into microbial cells. The enzymatic hydrolysis of higher-molecular-weight material is considered to be the rate-limiting step in the process of organic matter oxidation in sediments (Billen, 1982; Meyer-Reil, 1987b). The decomposition processes are controlled by microbial (exclusively bacterial?) enzymes which degrade polymeric compounds extracellularly, and their oligomeric products become available substrates for uptake by microbial cells. If the hydrolysis products are not taken up directly, they are fed into the pool of dissolved substrates in sediments. Through microbial attack on particles, refractory organic carbon becomes more easily assimilable for higher trophic levels (Meyer-Reil, 1983).

The direct measurement of enzymatic activity by chemical analysis of the degradation products in natural ecosystems is usually limited by a number of methodological problems (e.g., unknown spectrum and composition of natural substrates and decomposition products, incorporation of products of enzymatic catalysis into microbial biomass). Therefore, radioactive or dye-labelled model substrates represent useful tools in the study of enzymatic activity in sediments. In this chapter, only dye-labelled substrates will be considered.

Extracellular enzymes, as opposed to intracellular enzymes, are defined as those acting outside the cell, however, they may be localized on the outer cell membrane. The term exoenzyme, as opposed to endoenzyme, is used to characterize the location of enzymatic hydrolysis. Exoenzymes attack polymers at terminal locations, liberating oligomeric or monomeric substrate subunits. Endoenzymes cleave the central structure of polymers, liberating relatively large units. According to the definitions given above, an enzyme that is bound to the outer cell membrane and releases small subunits from biopolymers has to be characterized as an extracellular exoenzyme in the strict sense. However, as will become evident from the discussion below, our limited knowledge of enzymes acting in natural sediments does not allow their unequivocal characterization according to their localization and mode of attack.

This article summarizes the ecological aspects of enzymatic activity in marine sediments. Beside the methodological approach, observations on the localization of enzymes, their spatial distribution, and the regulation of activity will be discussed. From these presentations it will become obvious that our present knowledge of the functioning of enzymes in natural sediments comprises a narrow range of casual observations and measurements which are far away from a detailed understanding of the enzymatic processes controlling organic matter degradation.

5.2 Methodological Approach

Although dye-labelled particulate enzyme substrates have been available for some time, they have not been extensively used in ecological studies (e.g., Kim and ZoBell, 1974; Little et al., 1979; Meyer-Reil, 1981; 1983). Recently, the dye-labelling of structural biopolymers with "reactive" covalent-bound dyes has been proposed by Reichardt (1986; 1988) as a simple method to label natural substrates for measuring their enzymatic solubilization. The decomposition processes can be easily followed as an increase of absorbance resulting from the release of soluble, dye-containing, hydrolytic products. The advantage of this method is that a broad spectrum of both chemically-defined and complex natural substrates can be labeled with reactive dyes. However, the assays have to be carefully standardized with regard to concentration and particle size of the substrate, extraction of enzymes from the sediment, and adsorption of dissolved reaction products to sediment particles (Reichardt, 1988). My own investigations using commercially available particulate dye-labelled substrates (Amylopectin Azure, Hide Powder Azure) have shown that microbial colonization was necessary for hydrolytic attack in natural sediments. This observation provides a possible explanation for the dependence of the hydrolysis rates on substrate-particle size reported by Reichardt (1988).

Beside the above mentioned dye-labelled insoluble particulate substrates, soluble fluorogenic analogs of organic substrates have been applied to natural sediment samples (King, 1986; Meyer-Reil, 1986; 1987a). Among these substrates, fluorescein diacetate (FDA) and methylumbelliferyl (MUF) derivatives of various organic compounds are probably most commonly used. Whereas the dye derivative is relatively nonfluorescent, the hydrolytic degradation product is highly fluorescent, thus facilitating the sensitive determination of enzymatic hydrolysis rates after relatively short incubation times.

Fluorescein diacetate (FDA) is hydrolyzed by nonspecific esterases (e.g., phosphatase, lipase, carbohydrate and protein-degrading enzymes) resulting in fluorescein release. The fluorogenic ester has been used to estimate microbial biomass on coniferous needle surfaces (Swisher and Carroll, 1980) as well as to follow total microbial activity in soil and litter (Schnürer and Rosswall, 1982). The mode of penetration of FDA into microbial cells and its intracellular hydrolysis is still under discussion (e.g., Lundgren, 1981; Chrzanowski et al., 1984). As pointed out by the latter authors, the outer membrane of heterotrophic Gram-negative bacteria reveals a low permeability for the fluorogen, thus restricting the use of FDA as a vital stain for microorganisms to environments dominated by eukaryotes and Gram-positive bacteria. In my own investigations, the hydrolysis of FDA has been proved as a useful assay to very quickly assess the hydrolytic potential of unknown sediments, especially those with relatively low enzymatic activity (cf. below). However, as mentioned above, it can not be used to distinguish between an extra- and intracellular hydrolysis. Parallel analysis of the sediments by epifluorescence microscopy revealed a low percentage (usually far less than 10% of the acridine orange counts) of weakly greenish fluorescent bodies (bacteria?). This observation would indirectly support the findings of Chrzanowski et al. (1984).

Methylumbelliferyl (MUF) substrates are probably hydrolyzed extracellularly. For a detailed discussion of the composition and application of MUF-substrates, see Hoppe (1983), Somville and Billen (1983), Somville (1984) and Rego et al. (1985), Chapters 3, 4, and 17. The specificity of the MUF-substrates needs further attention. By using commercially available enzymes, it could be shown that MUF-substrates are generally less specifically decomposed then was previously thought (Meyer-Reil, unpublished observations).

Since the major portion of the input of organic material into sediments are particles, soluble model substrates may not adequately reflect the enzymatic hydrolysis processes in sediments (compare additional remarks on the access of soluble substrates to the organic matrix discussed below). On the other hand, particulate dye derivatives may best simulate the particulate nature of the organic material settling to the sea floor. However, problems arise from the distribution of the particulate substrates in intact sediments as well as from the relatively long incubation time required (hours to days).

Special attention has to be paid to the concentration of the model substrates used in ecological studies. The use of relatively low substrate concentrations appears to be preferable from the ecological point of view. However, in enzyme investigation, the concentration and the spectrum of natural substrate analogs are mostly unknown, and therefore it is difficult to evaluate in advance what concentration of the added substrate is actually low. Furthermore, limiting substrate concentration results in first order kinetics, i.e., enzyme activity is dependent upon the substrate concentration. This makes a comparison of enzymatic activity in different sediments extremely difficult. Therefore, whenever possible, the substrate concentration should be at the saturation level, so that the kinetics approach zero order and are no longer dependent upon substrate concentrations. For a detailed discussion of the kinetic aspects of enzyme assays, the reader is referred to the excellent article concerning phosphatase activity in soil by Malcolm (1983).

The application of enzyme substrates to sediments still presents problems. Generally, the dissection and suspension of sediment horizons results in a considerable increase in microbial activity (including enzymatic hydrolysis) as compared to "undisturbed" sediments (Meyer-Reil, 1986; and literature cited therein). In sediment slurries, microhabitats are disturbed, chemical and biological gradients are diminished, and the substrate is evenly distributed. Based on the observed increase in activity measured in sediment slurries, considerable potential activity has to be attributed to sediments. Natural "disturbance" processes such as bioturbation or resuspension of sediments by currents may also play an important role in the stimulation of microbial activity.

For the assessment of enzymatic activity in undisturbed sediments, soluble enzyme substrates may be directly injected with a minimum of disturbance into intact sediment cores ("core-injection" technique; Meyer-Reil, 1986). The application of particulate enzyme substrates to individual horizons, however, requires the preparation of sediment slurries. If the investigation of particulate organic matter in intact sediments is desired, dye-labelled particles should be added to the water overlying the sediment core. After the appropriate incubation time, the enzymatic hydrolysis may be followed by dissecting the sediment core and analyzing the released dye in the individual sediment horizons. Using this approach, beside the enzymatic degradation, the transport of particles may be followed. However, since dye released through enzymatic hydrolysis undergoes diffusion in the interstitial water, decomposition rates related to the individual horizons have to be interpreted with care.

The problem of proper controls to account for "nonbiological" enzymatic activity is especially critical in sediments. Treatment with chemicals (acetone, mercuric chloride, glutaraldehyde, formaldehyde) only partly inhibits enzymatic activities (Meyer-Reil, 1981; 1986). This observation is of specific interest for the fixation of organic matter in sediment traps, which frequently are exposed in the water column for months. It can be expected, that despite fixation, enzymatic decomposition processes can continue, causing considerable changes in the composition of the organic matter sampled. Fixation of sediments with heat results in a complete termination of enzymatic activity. High temperature, however, also destroys the structure of sediments and alters their physical and chemical properties. Heat-fixed sediment "controls" should not be used to measure the enzymatic activity in intact sediments caused by nonbiological "decomposition". If proper controls cannot be obtained, enzyme assays should be run in time-course experiments from which abiotic activity may be estimated by extrapolation of the activity curve to a zero time intercept (Meyer-Reil, 1986).

The artificial substances used for enzymatic studies represent model substrates which are the analogs for a variety of naturally occurring substrates of unknown composition and concentration. Because of this, the measured rates of microbial enzymatic hydrolysis in samples describe potential enzyme activities. However, the measured activities reflect the pool of naturally occurring enzymes, which is the result of variations in concentration and composition of natural substrates. At the present, the calculation of natural decomposition rates is difficult due to the limited knowledge of concentration and spectrum of natural substrates, which may react quite differently with regard to their individual kinetic characteristics.

5.3 Localization of Extracellular Enzymes

In prokaryotes, extracellular enzymes are synthesized in cytoplasmic membrane-bound ribosomes, and are transported through the membrane by various mechanisms which allow the transport of charged proteins through the hydrophobic lipid layer (Burns, 1978; Priest, 1978). Dead and decaying organisms (algae, meio- and macrofauna) certainly contribute to the pool of extracellular enzymes in sediments as well. Their contribution to the pool of extracellular enzymes, however, is difficult to quantify.

Extracellular enzymes may be distributed in two different compartments of the sediment: free "dissolved" in the interstitial water, or bound to particles or cell surfaces. The activity of free extracellular enzymes in the interstitial water is usually very low. This may be due to the fact that enzymes excreted by microbial cells into the surrounding water are not of great use for the parent cells: these enzymes may diffuse away or undergo rapid denaturation or decomposition. Most of the extracellular enzymes in sediments are bound to particles or cell surfaces. By adsorption to inorganic and organic particles, enzymes can be physically and chemically immobilized (Burns, 1978; Ladd, 1978). As an adsorption site, the microbial slime layer (glycocalyx) is probably most important (Costerton et al., 1978). Enzymes adsorbed to particles may be more stable and may also be more resist to microbial attack.

Burns (1980) describes the adsorption of enzymes to humic acid complexes, which are stabilized by binding to clay colloids. Based on a model, the author discusses the possible importance of the enzyme-humic complex for the decomposition of organic matter in soil. A substrate may be hydrolyzed by persisting exoenzymes. The product of the enzymatic hydrolysis may become available to bacteria by diffusion or the bacteria may react by positive chemotaxis. The product, after uptake by bacteria, may stimulate the synthesis and secretion of the appropriate enzyme. According to this hypothesis, enzymes persisting in soil would act as "starter" enzymes; this "obviates the need for a microbe to continuously and wastefully produce exoenzymes" (Burns, 1980).

It is difficult to imagine that enzymes which become trapped during the genesis of the organic matrix contribute to the degradation of higher molecular weight material due to the limited access of this material to the matrix. Investigations of the decomposition of organic material by epilithic microorganisms in streams have shown that material of higher molecular weight became preferentially attached to the matrix. When this material was decomposed, then low-molecular-weight substrates were adsorbed to the matrix (Ford and Lock, 1987). Dispersion of sediments from the Kiel Bight by sonication resulted in an increase in hydrolytic activity (Kähler, 1985; Meyer-Reil, 1987b). Interestingly enough, various groups of enzymes were stimulated differentially. Whereas the activity of carbohydrate-decomposing enzymes was enhanced by 0–30%, the activity of proteolytic enzymes was increased by 50–70%. From these observations it may be concluded that a high percentage of proteolytic enzymes was trapped within the matrix with limited access to dissolved organic substrates. Carbohydrate-decomposing enzymes, however, seem to be preferentially located on the surface of the matrix, where they have access to dissolved carbohydrates. From these observations, consequences may be derived for the application of dissolved compounds as model substrates for measuring enzymatic activities in sediments (see above).

It can be expected that most of the organic material entering the sediments via sedimentation is preferentially decomposed by microbial cell-bound enzymes which gain contact with the substrate following microbial colonization (see diagram by Meyer-Reil, 1990). However, nothing is known about the contribution of free-living bacteria in interstitial water versus bacteria attached to particles to the decomposition of particles in sediments.

Rates of enzymatic hydrolysis in sediments are only partly inhibited by acetone, toluene, formaldehyde, glutaraldehyde or mercuric chloride, but completely inhibited by boiling the sediment samples (King, 1986; Meyer-Reil, 1990). Temperature optimum of hydrolytic enzymes (as determined with FDA as a model substrate) in natural sediments was between 30 and 40°C, even for sediments from the permanently cold Norwegian-Greenland Sea. Freezing of the sediment did not significantly influence enzymatic activity. Hydrolytic enzymes were active over a broad range of salinities (0–35‰, Meyer-Reil and Köster, unpublished observations). Based on these properties, the nature of the enzymatic response in sediments may be described as a "conservative" one.

5.4 Spatial Distribution of Enzymatic Activity in Sediments

Most recently, fine-scale investigations in sediments have demonstrated that interfaces are characterized by high enzymatic activity. These interfaces (e.g., sediment/water boundary, tubes and burrows of macrofauna, redox potential discontinuity layer) are generally known as zones of high microbial abundance and metabolism (Craven et al., 1986).

Sediment/water interface Since sediments in deep waters are exclusively dependent upon the nutrient supply from the water column via sedimentation, the major transformations of organic material occur at the sediment/water boundary. This becomes especially obvious in deep sea sediments which are characterized by a rather episodic supply of organic material (Bender and Heggie, 1984; Tsunogai and Noriki, 1987). In these sediments enzymatic degradation of particulate organic material is the key step in the decomposition process.

Surprisingly high hydrolytic activity rates were observed at a number of stations located in the Jan Mayen Fracture Zone (Norwegian-Greenland Sea) (see Chapter 19). The importance of the uppermost sediment layer for the decomposition of organic material becomes especially obvious from fine-scale studies. In the surface 0- to 2.5-mm horizon (the direct contact zone with the water), the enzymatic hydrolysis rates were enhanced by more than two orders of magnitude over horizons only a few millimeters deeper.

This extremely steep gradient in enzymatic activity is obviously closely related to a mass abundance of agglutinated foraminiferans which densely colonized the sediment surface. The measurement of hydrolytic activity of individual foraminiferans confirmed this hypothesis. This may explain the unexpected magnitude and the steep gradient in hydrolytic activity which has never been observed before even in shallow water coastal sediments (Meyer-Reil, 1987a). Parallel to the steep gradient in enzymatic potential, a pronounced fine-scale gradient in bacterial biomass was observed. Bacterial biomass was highest in the 0- to 2.5-mm horizon and decreased gradually with depth. Size fractionation of the bacteria by epifluorescence microscopy revealed that the larger volume of rod-shaped cells was mainly responsible for the high bacterial biomass observed in the uppermost sediment horizon (Meyer-Reil and Köster, unpublished observations). From these observations it becomes obvious that under certain ecological situations, other groups of microorganisms (in this case foraminiferans) may contribute significantly to the pool of hydrolytic enzymes in sediments. Nevertheless, as demonstrated by the increase of bacterial biomass in the uppermost sediment horizon, bacteria may also derive benefit from the decomposition of particles by foraminiferans.

Other examples from shallow water coastal sediments confirm that the sediment/water boundary is characterized by high enzymatic activity. This is especially the case when sedimentation events have caused a nutrient enrichment of the sediment surface (Meyer-Reil et al., 1987a). In shallow water sediments, however, the organic material entering the seafloor may be transported relatively quickly into deeper sediment horizons caused by physical and biological (bioturbation) processes (Meyer-Reil, 1987b). In this case, high enzymatic activity may occur in subsurface sediment horizons.

Biogenic structures Recent studies have documented that the tubes and burrow walls of macrofauna are zones of intensive microbial metabolism. This applies to the bacterial autotrophic fixation of carbon dioxide, mineralization of dissolved organic substrates, biomass production (thymidine incorporation) and enzymatic activity. The elevated microbial activity of the burrow in connection to burrow walls of invertebrates could be attributed to physical disturbance of the sediment (Eckman, 1985; Findlay et al., 1985) as well as to infaunal metabolism (Yingst and Rhoads, 1980; Alongi and Hanson, 1985).

Reichardt (1988) investigated the activity of protease, phosphatase, and sulfatase at burrow walls of the lungworm in intertidal sediments of the North Sea. The author showed that phosphatase and sulfatase had maximum activity in the burrow walls as compared to surface or subsurface sediments. Protease, however, revealed maximum activity in the fecal casts deposited by the worm at the sediment surface. It is difficult to decide where the hydrolytic enzymes are originated from. They may have been derived or even excreted by the worm. This is most likely for the protease, whose activity was three- to fivefold higher in the fecal casts as compared to the burrow walls. Reichardt (1988) speculated that the excretion of protease through the gut of the worm could be a mechanism by which an essential biocatalyst is transported from subsurface to surface sediments which are the major enrichment sites for organic material. Köster et al. (see Chapter 19) analyzed the hydrolytic activity associated with macrofauna organisms in sediments of the Norwegian Sea. The authors demonstrated that major decomposition processes of organic material in these nutrient limited sediments occurred in connection with biogenic structures.

Redox potential discontinuity layer The redox potential discontinuity layer, which is characterized by the change from oxic to anoxic conditions, is very important for the early diagenesis of organic material in shallow water coastal sediments. In this layer the use of different electron acceptors for the oxidation of organic matter and the chemoautotrophic oxidation of reduced compounds of anaerobic processes cause diverse microbial metabolic pathways. Fine-scale studies of the redox potential discontinuity layer revealed high hydrolytic activity at certain seasons of the year which may even exceed the enzymatic activity observed at the sediment surface (Köster, unpublished observations). Parallel measurements of the protein concentrations showed an enrichment of organic material in this layer. One possible reason for the subsurface enrichment of hydrolytic activity may be that organic material can be transported by bioturbation into deeper sediment layer, or that storms can cause a mixing of the sediment surface, thus depositing organic matter in subsurface horizons. Enhanced microbial metabolism, however, directly influences the physicochemical conditions (e.g., redox potential) of the sediments.

The influence of anaerobic conditions on the activity of hydrolytic enzymes is still under discussion. According to King (1986), β-glucosidase activity in intertidal marine sediments was insensitive to the presence or absence of oxygen. My own investigations of sediments of the Kiel Bight have shown that proteolytic enzymes may be inhibited during periods of anoxia in summer (Meyer-Reil, 1983).

From the foregoing discussion of their fine-scale distribution, it becomes clear that microbial biomass and activity as well as enzymatic hydrolysis of polymers are rather unevenly distributed in sediments. The enrichment of enzymatic activity at interfaces may be easily overlooked if arbitrary depth intervals are analyzed.

5.5 Regulation of Enzymatic Activity in Sediments

Very little is known about the regulation of enzymatic activity and its impact on the stimulation of microbial metabolism (uptake of hydrolysis products, biomass production, cell division) in natural marine sediments, although these questions are of fundamental importance in understanding the cycle of organic matter. Investigations have shown that the input of organic material into sediments of the Kiel Bight following the breakdown of phytoplankton blooms in spring and autumn caused an immediate response in enzymatic activity. Subsequently, microbial uptake of dissolved organic substrates, biomass production and eventually cell division were stimulated (Meyer-Reil, 1983; 1987a, 1987b; Meyer-Reil et al., 1987).

Aspects of the relationships between the different manifestations of microbial activity were followed during a nutrient enrichment experiment in sediments from the Norwegian Sea (Köster and Meyer-Reil, unpublished observations). Deep sea sediments are best suited for nutrient enrichment experiments, because these sediments are generally nutrient limited and only occasionally receive an input of organic material. Sediment cores were kept on board ship under simulated in situ conditions with and without addition of natural aged detritus. The input of organic material could be demonstrated by following protein concentration, which started to increase in the top 1-centimeter layer of the sediment within one day after nutrient enrichment. The increase in protein concentration in deeper sediment horizons, however, was delayed by at least three days. Protein enrichment comprised the top 3 centimeters of the sediment core. With a time lag of less than three days enzymatic activity, as analyzed by the hydrolysis of fluorescein diacetate, was stimulated. This was especially obvious in the top 1-centimeter layer of the sediment and decreased gradually with depth. Only three days later, however, enzymatic activity had already decreased and with prolonged time course of the experiment, enzymatic activity leveled off. Similar curves demonstrating enzyme induction were obtained by analysis of the activity of protein-decomposing enzymes using leucine-methylcoumarinylamide as a model substrate.

Parallel measurements of microbial incorporation of tritiated leucine into macromolecular cellular compounds (a measure of protein synthesis) revealed a slight increase within three days after nutrient enrichment, which comprised only the top centimeter of the sediment. The response in deeper horizons, however, was delayed by at least three days. The main stimulation of microbial biomass production was observed within six days after nutrient enrichment and was measurable down to a sediment depth of 4 centimeters. At the time of highest microbial biomass production, enzymatic activity was already considerably reduced. Parallely, the development of bacterial numbers and biomass was followed by epifluorescence microscopy. Although the analysis is still in progress, preliminary results showed that during the first three days of nutrient enrichment, cell numbers did not increase significantly. A considerable increase, however, was observed in microbial biomass, which was dominated by the increase in cell volume of rod-shaped cells. Interestingly enough, microbial biomass production as measured by the incorporation of leucine into protein fraction correlates with the increase in cell volume as determined by epifluorescence microscopy. An even stronger correlation exists between the average biomass per cell and the percentage of dividing cells.

Although a number of questions may be raised with regard to the experiment described, the results clearly demonstrate that the enzymatic response was inducible within a few days following nutrient enrichment (sediment temperature 1°C). The induction of enzymatic activity turned out to be the key process for the stimulation of microbial uptake of dissolved hydrolysis products. Subsequently, microbial biomass production and cell division were stimulated (Köster and Meyer-Reil, unpublished observations).

5.6 Conclusions

The enzymatic decomposition of organic material is the initial and rate-limiting step in organic carbon oxidation in sediments. However, our present knowledge of the enzymatic degradation process is far away from a detailed understanding of the relevant processes.

For the measurement of enzymatic hydrolysis, dye-labelled particulate and soluble artificial as well as natural substrates are available. However, before their use in sediments, their nature, concentration, and mode of application have to be considered. Dye-labelled particles may simulate the particulate nature of the organic material entering the sediments via sedimentation. Enzyme assays, however, have to be carefully standardized with regard to particle size and concentration of the substrate, as well as with regard to adsorption of the dye-containing hydrolytic product to sediment particles.

Sediments are complex environments characterized by strong gradients in chemical and microbiological parameters. The accumulation of enzymatic activity at interfaces, such as sediment/water interface, biogenic structures, and the redox potential discontinuity layer, may be easily overlooked if arbitrary depth intervals are analyzed.

Very little is known about the location and regulation of enzymatic activity in natural sediments. Most of the enzymes are bound to cell surfaces or adsorbed to particles. It can be expected that most of the particulate organic material entering the sediment via sedimentation is hydrolyzed by cell-bound enzymes which contact the substrate following microbial colonization. Experiments with nutrient-limited deep-sea sediments have demonstrated that the enzymatic response is inducible within short periods of time. The induction of enzymatic activity turned out to be the key process in initiating microbial biomass production and eventually cell division.

At present, the extrapolation of enzymatic activity measured in vitro to natural activity in situ is difficult due to our limited knowledge of the concentration and spectrum of natural substrates. They compete for enzymes which may react differently dependent upon their kinetic characteristics.

Microbiologists have only recently become aware of the dynamics of enzymatic decomposition of organic materials in sediments. In future research a number of problems have to approached. Among them, investigations on the origin, location and regulation of enzymatic activity are probably most urgently needed.

Acknowledgments I am grateful to my students O. Charfreitag, H. Held, and M. Köster for sharing and discussing their data with me. This work was supported by Deutsche Forschungsgemeinschaft, publication No. 86 of the Joint Research Program at Kiel University (Sonderforschungsbereich 313).

References

Alongi, D.M. and R.B. Hanson. 1985. Effect of detritus supply on trophic relationships within experimental benthic food webs. II. Microbial responses, fate and composition of decomposing detritus. *Journal of Experimental Marine Biology and Ecology* 88: 167–182.

Bender, M.L. and D.T. Heggie. 1984. Fate of organic carbon reaching the deep sea floor: a status report. *Geochimica et Cosmochimica Acta* 48: 977–986.

Billen, G. 1982. Modelling the processes of organic matter degradation and nutrient recycling in sedimentary system. pp. 15–52 in Nedwell, D.B. and Brown, C.M. (editors), *Sediment Microbiology*, Academic Press, London.

Burns, R.G. 1978. Enzyme activity in soil: some theoretical and practical considerations. pp. 295–340 in Burns, R.G. (editor), *Soil Enzymes*, Academic Press, London.

Burns, R.G. 1980. Microbial adhesion to soil surfaces: consequences for growth and enzyme activities. pp. 249–262 in Berkeley, R.C.W., Lynch, J.M., Melling, J., Rutter, P.R. and Vincent, B. (editors), *Microbial Adhesion to Surfaces*, Ellis Horwood Limited, Chichester.

Chrzanowski, T.H., Crotty, R.D., Hubbard, J.G. and R.P.Welch. 1984. Applicability of the fluorescein diacetate method of detecting active bacteria in freshwater. *Microbial Ecology* 10: 179–185.

Costerton, J.W., Geesey, G.G. and K.J. Cheng. 1978. How bacteria stick. *Scientific American* 238: 86–95.

Craven, D.B., Jahnke, R.A. and A.F. Carlucci. 1986. Fine-scale distributions of microbial biomass and activity in California Borderland sediments. *Deep-Sea Research* 33: 379–390.

Eckman, J.A. 1985. Flow disruption by an animal-tube mimic effects sediment bacterial colonization. *Journal of Marine Research* 43: 419–435.

Findlay, R.H., Pollard, P.C., Moriarty, D.J.W. and D.C. White. 1985. Quantitative determination of microbial activity and community nutritional status in estuarine sediments: evidence for a disturbance artifact. *Canadian Journal of Microbiology* 31: 493–498.

Ford, T.E. and M.A. Lock. 1987. Epilithic metabolism of dissolved organic carbon in boreal forest rivers. *Federation of European Microbiological Societies Microbiology Ecology* 45: 89–97.

Hoppe, H.-G. 1983. Significance of exoenzymatic activities in the ecology of brackish water: measurements by means of methylumbelliferyl-substrates. *Marine Ecology Progress Series* 11: 299–308.

Kähler, P. 1985. Mikrobiologische Untersuchungen an Sedimentprofilen der Ostsee in der Kieler Bucht (bei Boknis Eck). *Diplomarbeit, Universität Kiel.*

Kim, J. and C.E. ZoBell. 1974. Occurrence and activities of cell-free enzymes in oceanic environments. pp. 368–385 in Colwell, R.R. and Morita, R.Y. (editors), *Effect of the Ocean Environment on Microbial Activities*, University Park Press, Baltimore.

King, G.M. 1986. Characterization of β-glucosidase activity in intertidal marine sediments. *Applied and Environmental Microbiology* 51: 373–380.

Ladd, J.N. 1978. Origin and range of enzymes in soil. pp. 51–96 in Burns, R.G. (editor), *Soil Enzymes*, Academic Press, London.

Little, J.E., Sjogren, R.E. and G.R. Carson. 1979. Measurement of proteolysis in natural waters. *Applied and Environmental Microbiology* 37: 900–908.

Lundgren, B. 1981. Fluorescein diacetate as a stain of metabolically active bacteria in soil. *Oikos* 36: 17–22.

Malcolm, R.E. 1983. Assessment of phosphatase activity in soils. *Soil Biology and Biochemistry* 15: 403–408.

Meyer-Reil, L.-A. 1981. Enzymatic decomposition of proteins and carbohydrates in marine sediments: methodology and field observations during spring. *Kieler Meeresforschung Sonderheft* 5: 311–317.

Meyer-Reil, L.-A. 1983. Benthic response to sedimentation events during autumn to spring at a shallow water station in the Western Kiel Bight. II. Analysis of benthic bacterial populations. *Marine Biology* 77: 247–256.

Meyer-Reil, L.-A. 1986. Measurement of hydrolytic activity and incorporation of dissolved organic substrates by microorganisms in marine sediments. *Marine Ecology Progress Series* 31: 143–149.

Meyer-Reil, L.-A. 1987a. Seasonal and spatial distribution of extracellular enzymatic activities and microbial incorporation of dissolved organic substrates in marine sediments. *Applied and Environmental Microbiology* 53: 1748–1755.

Meyer-Reil, L.-A. 1987b. Bakterien in Sedimenten der Kieler Bucht: Zahl, Biomasse und Abbau von organischem Material. *Habilitationsschrift, Universität Kiel.*

Meyer-Reil, L.-A. 1990. Microorganisms in marine sediments: considerations concerning activity measurements. *Archiv für Hydrobiologie Beihefte Ergebnisse Limnologie* 34: 1–6.

Meyer-Reil, L.-A., Faubel, A., Graf, G. and H. Thiel. 1987. Aspects of benthic community structure and metabolism. pp. 69–110 in Rumohr, J., Walger, E. and Zeitzschel, B. (editors), *Seawater-Sediment Interactions in Coastal Waters*, Springer Verlag, Berlin.

Priest, F.G. 1984. *Extracellular Enzymes*. Van Nostrand Reinhold (UK) Co. Ltd., Wokingham. 79 pp.

Rego, J.V., Billen, G., Fontigny, A. and M. Somville. 1985. Free and attached proteolytic activity in water environments. *Marine Ecology Progress Series* 21: 245–249.

Reichardt, W. 1986. Polychaete tube walls as zonated microhabitats for marine bacteria. *IFREMER Actes de Colloques* 3: 415–425.

Reichardt, W. (1988): Impact of bioturbation by *Arenicola marina* on microbiological parameters in intertidal sediments. *Marine Ecology Progress Series* 44: 149–158.

Schnürer, J. and T. Rosswall. 1982. Fluorescein diacetate hydrolysis as a measure of total microbial activity in soil and litter. *Applied and Environmental Microbiology* 43: 1256–1261.

Somville, M. 1984. Measurement and study of substrate specificity of exoglucosidase activity in eutrophic water. *Applied and Environmental Microbiology* 48: 1181–1185.

Somville, M. and G. Billen. 1983. A method for determining exoproteolytic activity in natural waters. *Limnology and Oceanography* 28: 190–193.

Swisher, R. and C.C. Carroll. 1980. Fluorescein diacetate hydrolysis as an estimator of microbial biomass on coniferous needle surfaces. *Microbial Ecology* 6: 217–226.

Tsunogai, S. and S. Noriki. 1987. Organic matter fluxes and the sites of oxygen consumption in deep water. *Deep-Sea Research* 34: 755–767.

Yingst, J.Y. and D.C. Rhoads. 1980. The role of bioturbation in the enhancement of bacterial growth rates in marine sediments. pp. 407–421 in Tenore K.R. and Coull, B.C. (editors), *Marine Benthic Dynamics*. University of South Carolina Press, Columbia.

6

Extracellular Enzyme Activity in Eutrophic and Polyhumic Lakes

Uwe Münster

6.1 Introduction

For many years, the synthesis, secretion, and activity of extracellular enzymes of different organisms have been intensively studied by workers in many different disciplines, such as applied microbiology, biotechnology, biochemistry, and medicine (Pollock, 1962; Priest, 1977; 1984; Kreutzberg et al., 1986; Chaloupka and Krumphanzl, 1987). In aquatic sciences, the first reports on extracellular enzymes were written in middle of the 1960s (Overbeck and Babenzien, 1964; Reichardt et al., 1967; Kim and ZoBell, 1974). However, in the last decade, interest has increased in the role of microbial extracellular enzymes in natural aquatic environments.

When reviewing the current literature on extracellular enzymes in aquatic environments, most publications describe the activity and role of hydrolases such as phosphatases; a few report data on proteases, glucanases, lipases, and sulfatases; and a very few deal with ligninases and polyphenoloxidases. Many of these hydrolases participate in the cleavage of natural polymers. Microheterotrophs, particularly bacteria, are the most effective organisms producing the extracellular enzymes that cleave natural polymers to smaller molecules which can subsequently penetrate microbial cell membranes and be utilized. This high catabolic flexibility to changing substrate composition and availability offers a significant advantage to microbes living in natural habitats with fluctuating environmental conditions like temperature, oxygen, nutrients, and dissolved organic matter (DOM).

The total amount of organic matter (OM) in aquatic environments consists of 80 to 90% of DOM, and only a small fraction of DOM can penetrate cell membranes and be utilized directly by heterotrophic microorganisms. The amount of this easily utilizable dissolved organic matter (UDOM), however, is highly variable (Münster, 1984; Münster and Chróst, 1990) and is not always sufficient to support microbial nutrition and growth requirements (Jannasch, 1970; Sieburth, 1979). Roughly 80 to 90% of the DOM exists as polymeric substrate, is partially recalcitrant, and has to be depolymerized before microbial utilization. Therefore, microbial extracellular-substrate depolymerization can be regarded as an important tool and a rate-limiting step in the nutrition of microheterotrophs. Further, comparisons of the composition of pools of dissolved free amino acids (DFAA), dissolved free carbohydrates (DFCHO), dissolved combined amino acids (DCAA), and dissolved combined carbohydrates (DCCHO) revealed that DCCHO pools have higher variability and diversity than DFAA and DFCHO pools (Riley and Segar, 1970; Münster, 1984; 1985; Sigleo et al., 1983; Coffin 1989). According to Hollibaugh and Azam (1983) protein derived amino acids were utilized preferentially compared to DFAA and do not mix completely with DFAA in the bulk phase. Amano et al. (1982) could show, that marine bacteria with proteolytic activity used DCAA more rapidly and selectively than DFAA. Therefore DCAA or DCCHO pools may accomplish better metabolic requirements of aquatic microheterotrophs than DFAA or DFCHO pools alone (Amano et al., 1982; Hollibaugh and Azam, 1983; Münster, 1985; Münster et al., 1989; Coffin, 1989). Operating with an enzymatic "switch on-switch off" mechanism (Chróst, 1989), microheterotrophs may utilize different substrate pools of the DOM depending on their energetic demands, metabolic flexibility, and UDOM availability.

In aquatic ecosystems, heterotrophic microorganisms and particularly bacteria (Overbeck, 1968) are considered as the priority utilizers and modifiers of detritus (DOM and POM). They are not only responsible for recycling of nutrients in general, but they also represent an exceptional trophic link (through microbial loops) between the detritus and the classical grazing food chain. In these loops, nutrients, carbon, and energy are efficiently transferred from the lower microbial to the higher trophic food chain levels (Pomeroy, 1974; Sorokin, 1977; Azam et al., 1983; Pomeroy and Wiebe, 1989).

Studies on microflora in aquatic environments have a long tradition (ZoBell, 1946; Kuznetsow, 1959; 1968; Brock, 1966; Ohle, 1962; 1968; Overbeck, 1965; 1968; Morita, 1966). The role of microheterotrophs in former studies was mostly restricted to nutrient regeneration. However, measurements of microbial activities have rapidly developed in recent years and have been focused in most cases on radiolabelled-substrate-uptake studies, such as glucose and acetate (Wright and Hobbie, 1965; 1966; Overbeck, 1972; 1975; 1979) or amino acids (Crawford et al., 1974; Jørgensen, 1982; 1987; Simon, 1985), and bacterial production (Fuhrman and Azam, 1980; 1982; Riemann et al., 1982; Riemann and Søndergard, 1984; 1986). Many of these publications have clarified and sharpened the role of microbes in aquatic food chain processes (Sorokin, 1977; Williams, 1981; Witzel et al., 1982; Azam et al., 1983; Cho and Azam, 1988; Søndergaard et al., 1985; 1988; Scavia and Laird, 1987; Sherr et al., 1989; Wright, 1984; 1988). Less work has been done, however, on studies of microbial activities in relation to polymeric substrate utilization and the regulation of the synthesis and activities of the corresponding extracellular enzymes (Chróst, 1990).

From earlier studies on DOM composition and concentration in Plußsee (Münster, 1984; 1985), it could be shown that the UDOM pool and the pool of excretory products of phytoplankton (RDOM) are highly variable (Münster and Chróst, 1990). Depending on the light/nutrient regime of phytoplankton and bacterioplankton activities, UDOM substrates (DF-glucose, glycolic acid, etc.) fluctuate significantly in concentration and composition on the diurnal scale (Münster, 1984; Münster and Chróst, 1990). Release and uptake of UDOM substrates is highly dynamic and well controlled by metabolic processes (Søndergard et al., 1985; Fuhrman, 1987). Two key biochemical processes are of significant importance: extracellular release of UDOM substrates and substrate depolymerization with subsequent uptake and utilization.

Recently, Chróst et al. (1989) have shown that during spring phytoplankton bloom there was a significant contribution of β-glucosidase and β-galactosidase activity to substrate depolymerization in the euphotic zone of an eutrophic lake. Furthermore, significant relationships were found between β-glucosidase activities and [^{14}C]-glucose uptake and bacterial production, leading to the conclusion that bacterioplankton responds significantly to polymeric substrate supply during the decline of phytoplankton bloom development (Chróst, 1989; 1990; Chróst et al., 1989). Based on these results, Chróst (1989) developed a hypothetical scheme for enzymatic synthesis, substrate depolymerization, and utilization by bacterioplankton in aquatic ecosystems. Similar studies have also been done by Somville (1984) for β-glucosidase activities in brackish water, for proteases (Billen, 1984; Hoppe, 1983; Hollibaugh and Azam, 1983; Hoppe et al., 1988), for chitinases in marine brackish waters (Smucker and Kim, 1987), and in sediments and mudflats for proteases and glucosidases (King, 1986; Meyer-Reil, 1986; 1987; Cunningham and Wetzel, 1989; Mayer, 1989). However, in most investigations artificial substrates, such as Hide Powder Azur (HPA; Little et al., 1979; Chróst et al., 1986; Halemejko and Chróst, 1986), radiolabelled cellulose (Bengtsson, 1988), bovine serum albumin (Bengtsson, 1988), and radiolabelled hemoglobin (Hollibaugh and Azam, 1983) were used to follow the depolymerization activities. Chróst et al. (1989) however, have shown that there is a reasonably close relation between natural polymeric substrate supply via phytoplankton production and subsequent extracellular substrate modification and utilization by bacterioplankton.

Similar relationships were found by Coffin (1989) for commercial (^{14}C)-labelled algal protein and labelled amino acid uptake and utilization by bacterioplankton in the Delaware (U.S.A.) estuary. These results support the hypothesis that monomeric substrate pools like DFAA and DFCHO may not completely fulfil the microbial substrate requirements for growth and energy (Coffin, 1989). They may rather function as activation potential for the corresponding transport and catabolic enzyme systems (Jannasch, 1970; Sieburth, 1979). Further studies on the composition, concentrations, dynamics, and biochemical processing of DOM under varying environmental conditions regarding nutrient contents, species composition, and acidity are urgently needed.

In this chapter the results of the relationship between extracellular aminopeptidases and glucanases activities and the concentration and composition of natural substrates in a eutrophic (Plußsee, F.R.G.) and a polyhumic (Mekkojärvi, Finland) lake are presented. The two lakes differ in nutrient status, species composition, environmental conditions, and DOM content.

Extracellular enzymes activities were measured fluorometrically with 4-methylumbelliferyl-(4-MUF)-hexoses and L-leucine-4-methyl-7-coumarinylamid (Leu-MCA) as substrates, as described by Hoppe (1983), Chróst et al. (1989), Chróst (1989), and Münster et al. (1989). Carbohydrate composition and concentration was measured by capillary gas liquid chromatography (Münster, 1985) and amino acids by high performance liquid chromatography (HPLC) according to Lindroth and Mopper (1979).

6.2 Environmental Characteristics of the Studied Lakes

The basic parameters for both lakes are summarized in Tables 6.1 and 6.2. The lakes differ markedly in chemical composition of lake water (e.g., inorganic nutrients and DOC) and in primary production, however, smaller differences are observed in their heterotrophic activity and biomass. Plußsee had about 90% higher P_{tot} values compared to Mekkojärvi and about 70% higher dissolved inorganic carbon (DIC), but NO_3^- was only 50% higher in Plußsee than in Mekkojärvi. Primary production in Plußsee was 4.3 times higher than in Mekkojärvi because of its higher nutrient content and availability. However, bacterial numbers, measured as acridine orange direct counts (AODC), and acriflavine direct counts (AFDC), bacterioplankton biomass and heterotrophic activity ([^{14}C]-glucose uptake) displayed similar ranges in both lakes, although the more eutrophic Plußsee had much lower DOC concentration (35%) and its water was much less colored compared to the polyhumic Mekkojärvi water. The contents of UDOM solutes (DFAA, DFCHO) and the polymeric substrates (DCAA and DCCHO) in Plußsee was about 10- to 25- times higher than in Mekkojärvi. Available substrate pool sizes (monomers and polymers) probably differ in both lakes. Microbial DOM processing is therefore a key function in understanding metabolism and food chain processes in both lakes. According to Salonen (1981) and Salonen et al. (1983), Mekkojärvi has an exceptionally high predominance of heterotrophy compared to clear-water and classical temperate lakes (e.g., Plußsee). Heterotrophic production exceeds that of primary production (Salonen, 1981). Recalcitrant DOM and allochthonous DOC may serve in polyhumic lakes as additional nutrient and energy sources (Salonen, 1981; Tranvik, 1988; Hessen, 1989). According to Ilmavirta (1988) and Arvola (1986) most lakes in Finland with such high dissolved humic matter (DHM) content have predominately phytoflagellata species with partially mixotrophic nutrition (Salonen and Hammar, 1986). Microbial loop interactions are therefore expected to be the

Table 6.1 Basic limnological data for Plußsee (0- to 1-m depth)

Parameter	Range	Average
Surface area (ha)	—	13.5
Maximum depth (m)	—	29.2
Mean depth (m)	—	9.4
Secchi visibility (m)	0.9–3.4	2.2
Conductivity ($\mu S\ cm^{-2}$)	290–240	266
pH	7.3–9.4	8.3
P_{tot} ($\mu g\ P\ l^{-1}$)	42–385	204
$P\text{-}PO_4^{3-}$ ($\mu g\ P\ l^{-1}$)	5–190	90
$N\text{-}NO_3^-$ ($\mu g\ N\ l^{-1}$)	14–450	101
$N\text{-}NH_4^+$ ($\mu g\ N\ l^{-1}$)	13–431	124
DIC (mg $C\ l^{-1}$)	10.2–19.5	17.4
DOC (mg $C\ l^{-1}$)	7.0–9.0	8.8
DFAA ($\mu g\ C\ l^{-1}$)	9–109	59
DFAA (% C of DOC)	0.1–0.5	0.3
DCAA ($\mu g\ C\ l^{-1}$)	15–234	124
DCAA (% C of DOC)	0.2–2.6	1.4
DFCHO ($\mu g\ C\ l^{-1}$)	15–800	200
DFCHO (% C of DOC)	0.2–8	2.2
DCCHO ($\mu g\ C\ l^{-1}$)	120–1,600	650
DCCHO (% C of DOC)	1.7–18.0	7.4
$Chlorophyll_a$ ($\mu g\ l^{-1}$)[1]	1.2–128	65.0
Primary production (mg C $m^2\ d^{-1}$)[1]	17–1,493	512
Bacteria (AODC; 10^6 cells ml^{-1})[2]	3–5	4
Bacterial biomass ($\mu g\ C\ l^{-1}$)[2]	30–50	40
Glucose uptake ($\mu g\ l^{-1}\ h^{-1}$)[2]	0.2–1.2	0.5

[1]$Chlorophyll_a$ and primary production from Meffert and Overbeck (1985).
[2]Bacterial numbers and biomass, and glucose uptake from Overbeck (1979).

prevailing food web processes (Salonen et al., 1983). Because of such highly specialized food chain interactions, nutrition, and energy relationships, fluctuations in DOM (and the substrate pools included) seem to play an important role in polyhumic lakes. This may require alterations in the traditional models of aquatic food web structures (Salonen 1981; Tranvik, 1988; 1989; Hessen and Schartau, 1988; Hessen, 1989).

6.3 DOM Composition and Dynamics

It has been emphasized that detrital organic matter (DOM, POM) plays important role in whole lake metabolism (Wetzel, 1983; Mann, 1988). Knowledge about the composition, dynamics, and utilization of detritus is therefore of great importance. During studies on the concentration and composition of DOM in Plußsee in 1976/77 (sampled weekly), it was found that dissolved free carbohydrates (DFCHO) in the DOM pool varied less than dissolved combined carbohydrates (DCHO), which displayed much higher variations in concentrations

Table 6.2 Basic limnological data for Mekkojärvi (0– to 1–m depth)[1]

Parameter	Range	Average
Surface area (ha)	—	0.35
Maximum depth (m)	—	5.0
Mean depth (m)	—	2.2
Secchi visibility (m)	0.3–0.5	0.4
Mixing depth (m)	0.5–0.9	0.7
Water color (mg Pt l^{-1})	300–600	450
Conductivity (μS cm^{-2})	2–10	6
pH	4.2–6.6	5.4
P_{tot} (μg P l^{-1})	10–25	17
P-PO_4^{3-} (μg P l^{-1})	2–12	7
N_{tot} (μg N l^{-1})	500–1,500	1,000
N-NO_3^- (μg N l^{-1})	20–80	50
N-NH_4^+ (μg N l^{-1})	10–70	40
DIC (mg C l^{-1})	0.1–0.5	0.3
DOC (mg C l^{-1})	15–35	25
DFAA (μg C l^{-1})	3–110	56
DFAA (% C of DOC)	0.02–0.60	0.3
DCAA (μg C l^{-1})	160–850	505
DCAA (% C of DOC)	0.85–3.20	2.20
DFCHO (μg C l^{-1})	1–50	25
DFCHO (% C of DOC)	0.01–0.10	0.05
DCCHO (μg C l^{-1})	0.5–100	50
DCCHO (% C of DOC)	0.02–0.5	0.25
$Chlorophyll_a$ (μg l^{-1})	5–30	18
Primary production (mg C m^2 d^{-1})	—	120
Phytoplankton biomass (μg C l^{-1})	—	3
Bacteria (AODC; 10^6 cells ml^{-1})	1–6	3
Bacterial biomass (μg C l^{-1})	10–40	25
Glucose uptake (μg glucose l^{-1} h^{-1})	0.05–0.3	0.16

[1]Modified from Münster et al. (1989).

and composition (Münster, 1985). Shorter sampling intervals during a diurnal study on DOM in Plußsee in 1981 showed, however, that UDOM substrates such as dissolved free glucose (DF-glucose) and dissolved free glycine (DF-glycine) varied significantly with the light and dark periods (Figure 6.1). In short-term experiments, UDOM substrates displayed a highly dynamic character dependent on the intensity of DOM release by autotrophs and subsequent uptake by heterotrophs. In comparison to UDOM substrates, as shown in Figure 6.1, polymeric substrates do not necessarily follow the same distribution pattern, but displayed different and more complex distribution, as shown in Figure 6.2 for dissolved combined glucose (DC-glucose). There was an inverse relationship between DF-and DC-glucose pools, which may reflect important microbial biochemical processes, like substrate uptake and substrate depolymerization, which modify and regulate the pool sizes of both compounds. Good correlation was found between DF-glucose and [^{14}C]-glucose uptake by bacterioplankton

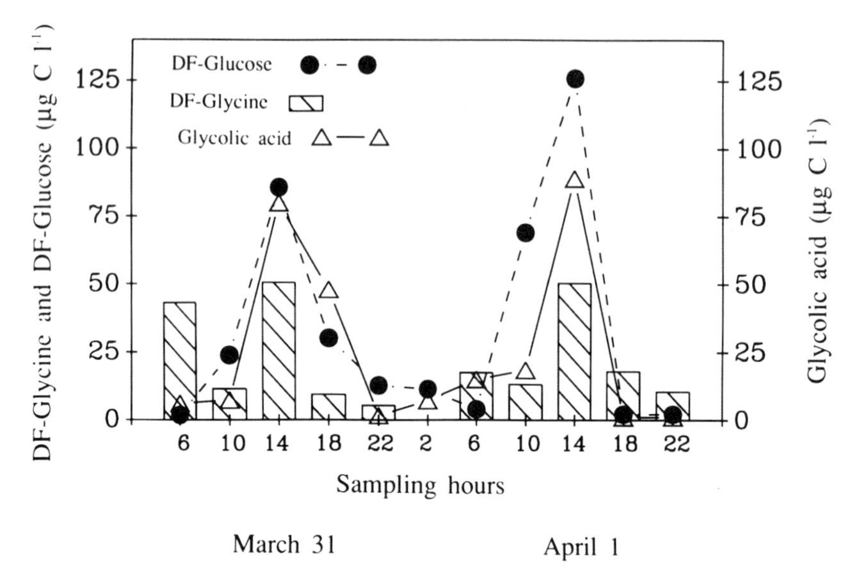

Figure 6.1 Diurnal variation of DF-glucose, DF-glycine, and glycolic acid in the euphotic zone of Plußsee.

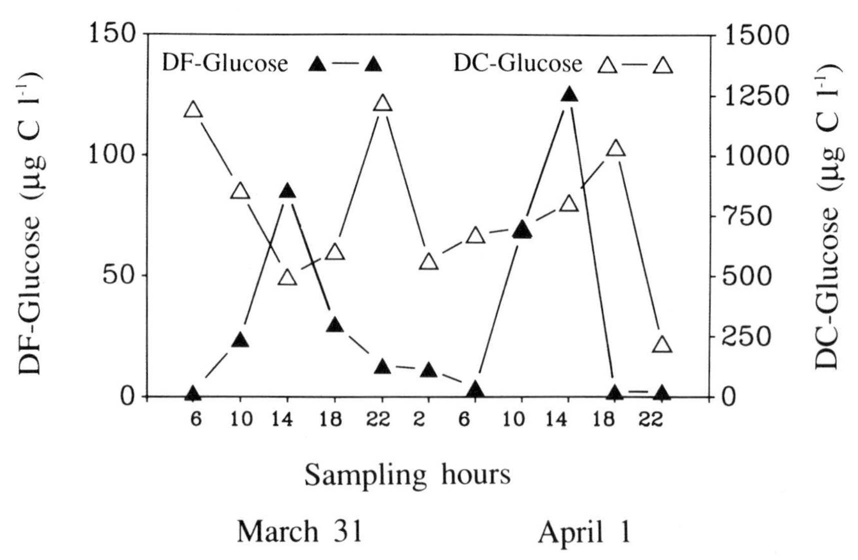

Figure 6.2 Diurnal distribution of DF-glucose and DC-glucose in the euphotic zone of Plußsee.

(Figure 6.3). The DF-glucose pool seemed therefore to be predominantly regulated by the uptake-transport systems of bacterioplankton. Kinetic studies on uptake systems of bacteria in Plußsee have shown that bacteria have high-affinity-uptake systems (Overbeck, 1972; 1975; 1979), with partially multiphasic kinetic patterns (Overbeck, 1975; 1979), which enable them rapidly adapt to fluctuating UDOM substrate availability and pool size. In deeper lake water layers, these relatively clear substrate-distribution pattern became more complex (Overbeck, 1979; Münster, 1984; Münster and Chróst 1990), as DFCHO compounds did not strictly follow the light-dark periods (Münster, 1984; Münster and Chróst, 1990). More complex substrate supply and utilization processes were found in deeper water layers (Overbeck, 1979; Witzel et al., 1982; Krambeck, 1984; Münster, 1985; Overbeck and Sako, 1988).

Studies on substrate composition of the DOM in Mekkojärvi confirmed that UDOM was much lower than in Plußsee and did not exceed 0.1 to 1% of DOC (Table 6.2). Heterotrophic nutrition of bacterioplankton depended upon the availability of UDOM substrates (e.g., DFAA and DFCHO compounds) and, to a in much higher extent, on the cleavage of polymeric substrates compared to Plußsee, where UDOM substrates may represent 1–5% of DOM (Table 6.1). Because the DOM content displayed higher amounts of combined polymeric compounds (e.g., DCAA, DCCHO, lignins, DHM), extracellular substrate depolymerization may be the significant rate-limiting step of substrate utilization in Mekkojärvi.

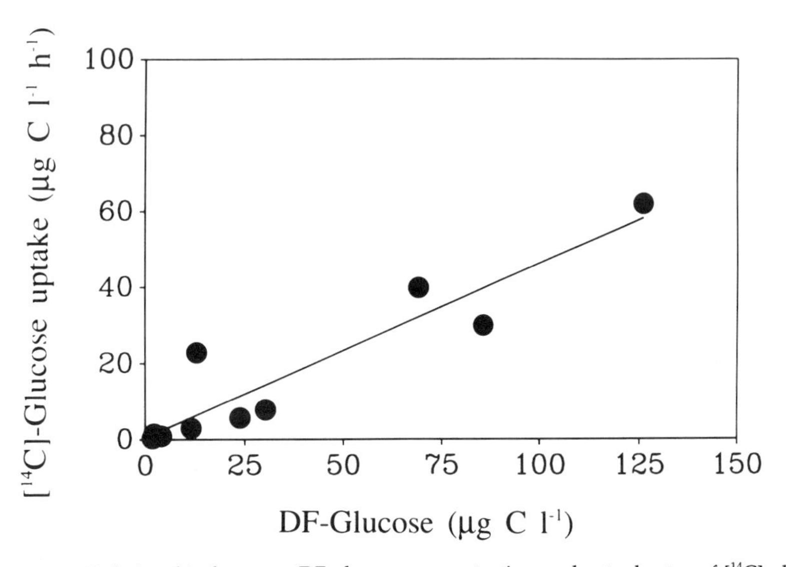

Figure 6.3 Relationship between DF-glucose concentration and actual rates of [^{14}C]-glucose uptake in Plußsee surface water. Data from Münster and Chróst (1990).

6.4 DOM-Enzyme Interactions

Because of substantial differences in the composition of DOM and the levels of availability of UDOM solutes for microheterotrophs in the two lakes, DOM-extracellular enzyme interactions were studied. DOM studies during the phytoplankton spring bloom in Plußsee in 1984 partially explained the dynamics and distribution of DOM substrates on the seasonal or monthly scale. Distribution patterns of DC-glucose and of the DOC released by phytoplankton (PhDOC) covaried well with the increase of their amounts at the end of spring phytoplankton bloom (Figure 6.4). DF-glucose also showed maxima at the end of phytoplankton bloom but distribution pattern differed from that of DC-glucose (Figure 6.4). A slightly positive relationship was observed between the DC-glucose concentration and PhDOC release (Figure 6.5). This suggests that, besides DF-glucose, polymeric-bound carbohydrates were also released at the end of the phytoplankton bloom. According to Chróst et al. (1989) and Chróst (1989), extracellular enzyme activity (e.g., β-glucosidase and β-galactosidase) in Plußsee correlated well with glucose uptake and bacterial biomass. This suggests that polymeric substrates of the DOM pools may be effectively utilized at the highest rates during phytoplankton bloom breakdown. PhDOC and DC-glucose concentrations were approximately parallel to the V_{max} of β-glucosidase (Figure 6.6). A positive relationship was found for DC-glucose and PhDOC and β-glucosidase activity (Figure 6.7). A similar relationship was found between DC-galactose and β-galactosidase activity (data not shown).

Monitoring of nine different extracellular enzymes in Mekkojärvi over a

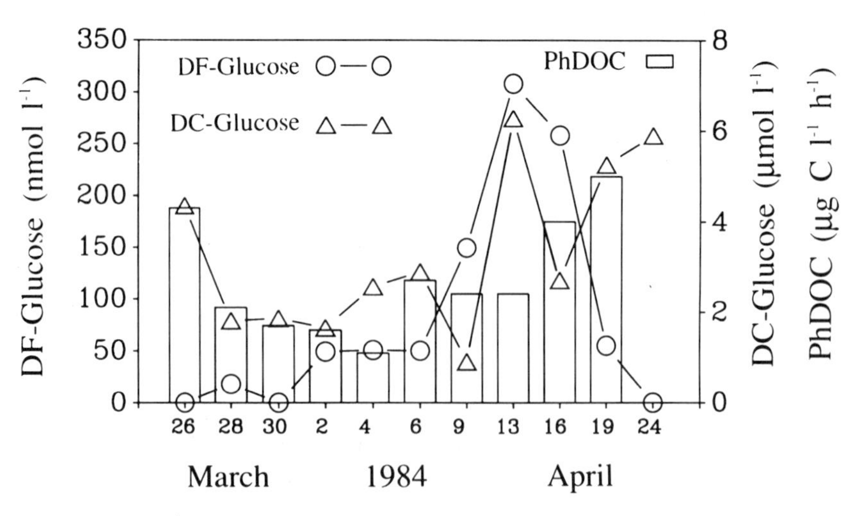

Figure 6.4 Concentrations of DF-glucose, DC-glucose, and PhDOC in the euphotic zone (0–3 m) of Plußsee during the course of a spring phytoplankton bloom period. PhDOC data from Chróst et al. (1989).

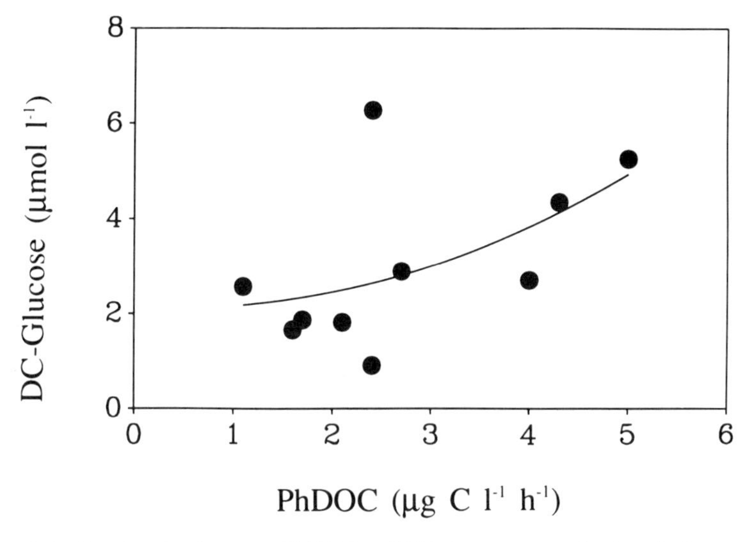

Figure 6.5 Relationship between PhDOC and DC-glucose concentrations in the euphotic zone (0–3 m) of Plußsee during the spring phytoplankton bloom. PhDOC data from Chróst et al. (1989).

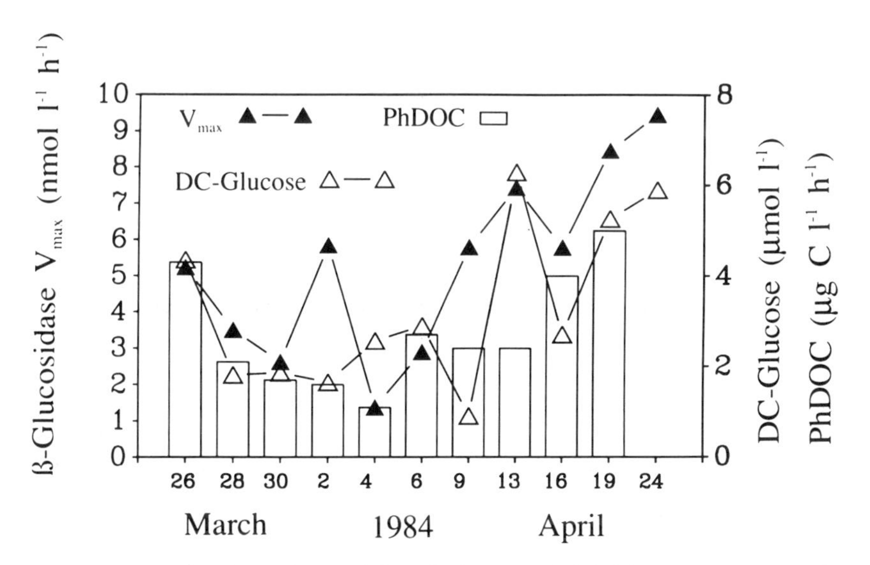

Figure 6.6 V_{max} of β-glucosidase and the DC-glucose and PhDOC concentrations in the euphotic zone of Plußsee during the course of the spring phytoplankton bloom. β-Glucosidase and PhDOC data from Chróst et al. (1989).

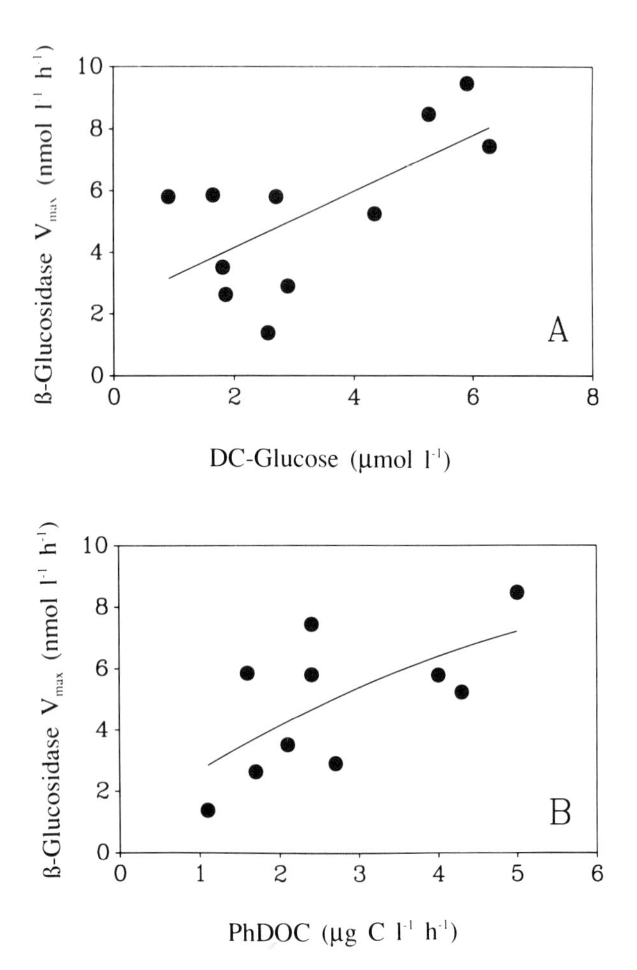

Figure 6.7 Relationship between V_{max} of β-glucosidase and (A) DC-glucose and (B) PhDOC concentration in the euphotic zone of Plußsee. β-Glucosidase and PhDOC data from Chróst et al. (1989).

three-month period in 1986 revealed that three groups of enzymes displayed significantly high activities. Phosphatases, aminopeptidases and α- and β-glucosidases showed the highest activities among all the extracellular enzymes measured (Figure 6.8). One reason for the high level of phosphatase activity may be explained by the relatively low $P\text{-}PO_4^{3-}$ level in Mekkojärvi water; phosphorus was therefore recycling rapidly in lake water. β-Glycosidase activity was higher than the α-glycosidase activity (Table 6.3). This result is somewhat surprising and difficult to explain. But it might be possible that available DOM contains relatively high amounts of β-linked or -bound carbohydrates, as can be expected from lignocellulosic or hemicellulosic material. The DOM of Mekkojärvi certainly contains such recalcitrant organic matter. Therefore, a similar

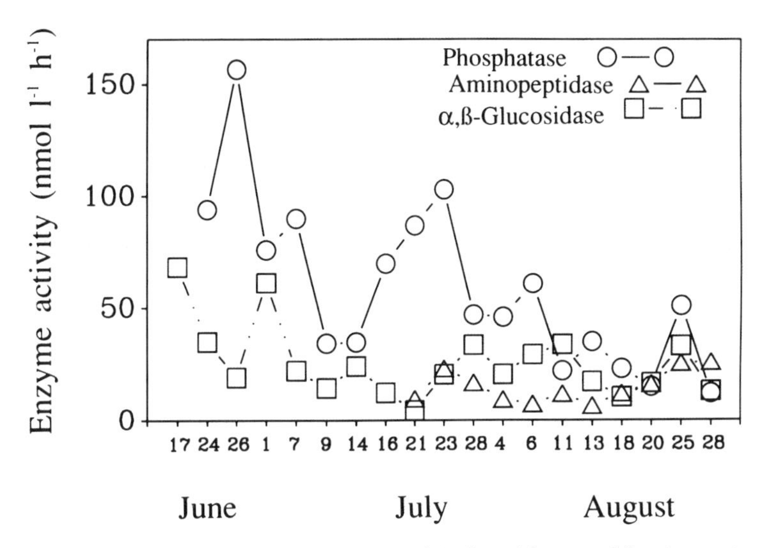

Figure 6.8 Distribution of phosphatase, α- and β-glucosidase, and leucine aminopeptidase activity in Mekkojärvi (0–1 m in depth) during one summer.

Table 6.3 The α:β ratio of glycosidase activities in Mekkojärvi (0- to 1-m depth) for three different enzymes

	α-activity ÷ β-activity		
Sampling date 1986	Glucosidase	Galactosidase	Mannosidase
1 July	0.31	0.14	3.00
7 July	0.01	0.01	0.01
9 July	0.13	0.01	0.01
21 July	0.57	0.17	0.10
28 July	0.10	—	—
4 August	0.20	—	—
6 August	0.32	0.08	—
11 August	0.18	0.71	—
13 August	0.54	—	1.71
18 August	0.79	1.77	0.35
20 August	0.79	0.95	2.95
25 August	0.62	1.83	0.01
28 August	0.47	0.83	1.32

ratio of α:β may be expected in substrate (carbohydrates) composition. Analytical results of carbohydrate pools (DFCHO and DCCHO) gave some information about structure of the substrate. Ten to 40% of identified carbohydrate subunits (e.g., glucose, galactose, and mannose) had a higher amount of β-form than of the α-form; that is similar α:β ratios to those found for the enzyme activities. However, the α-form predominated in hexoses analyzed.

6.5 β-Glucosidase Studies

β-Glucosidase kinetics and inhibition studies with various added substrates displayed the maximum inhibition for cellobiose (Table 6.4) and to some minor part for β-glucose (data not shown). According to enzyme kinetic theory, the addition of cellobiose creates a classical competition mechanism. The V_{max} was relatively unchanged but a significant increase of K_m values was observed indicating a decrease of affinity of enzyme active centers for 4-MUF-substrate. Cellobiose and 4-MUF-β-glucose competed for the reaction center of the enzyme. Addition of β-glucose, the reaction product of the enzyme reaction, also showed a competitive inhibition mechanism. In contrast, α-glucose, α,β-galactose, α,β-mannose, maltose, and lactose supplementation revealed noncompetitive inhibition patterns. However, such enzyme inhibition patterns were not always present during whole experimental season. In many cases, no inhibition was observed. Questions regarding the control and regulation of microbial enzyme activities in natural habitats should therefore be of fundamental importance in future studies (Chróst, 1989; 1990).

Glucosidase activity seems to be positively correlated to DC-glucose concentration in Mekkojärvi (Figure 6.9A). An inverse relation was found between DF-glucose and glucosidase activity (Figure 6.9B). This example shows that there are close interactions between UDOM availability and substrate utilization processes of microheterotrophs. As shown above, there was a significant higher β-activity than α-activity in all measured glycosidases. Assuming a specific cleaving mechanisms of β-glucosidase for β-linked glucose subunits (and for α-glucosidases) the clear relationship should be expected for a mixed substrate pool of α- and β-glucose. This relationship between glucosidases and DF- and DC-glucose (α- and β-anomer) may involve additional mechanisms (e.g., anomerization), which change the configuration in the C-1 position (Pigman and Anet, 1972; Campbell and Bentley, 1973). This can be done by enzymatic (aldose-epimerase) or physical-chemical (mutarotation) reactions (Campbell and Bentley, 1973). We do not know which reaction is essential for microbial substrate processing and utilization in Mekkojärvi. There is some evidence, that mutarotation and the enzyme mutarotase are essential during sugar transport through cell membranes (Keston, 1954). The mutarotation, however, may take

Table 6.4 Effect of cellobiose on β-glucosidase kinetics in Mekkojärvi surface water (0- to 1-m depth)

Cellobiose added (nmol l^{-1})	V_{max} (nmol l^{-1} h^{-1})	K_m (μmol l^{-1})
0	2.5	9.9
100	2.5	60.6
500	3.0	192.5
1,000	3.5	432.5
1,500	4.6	741.7

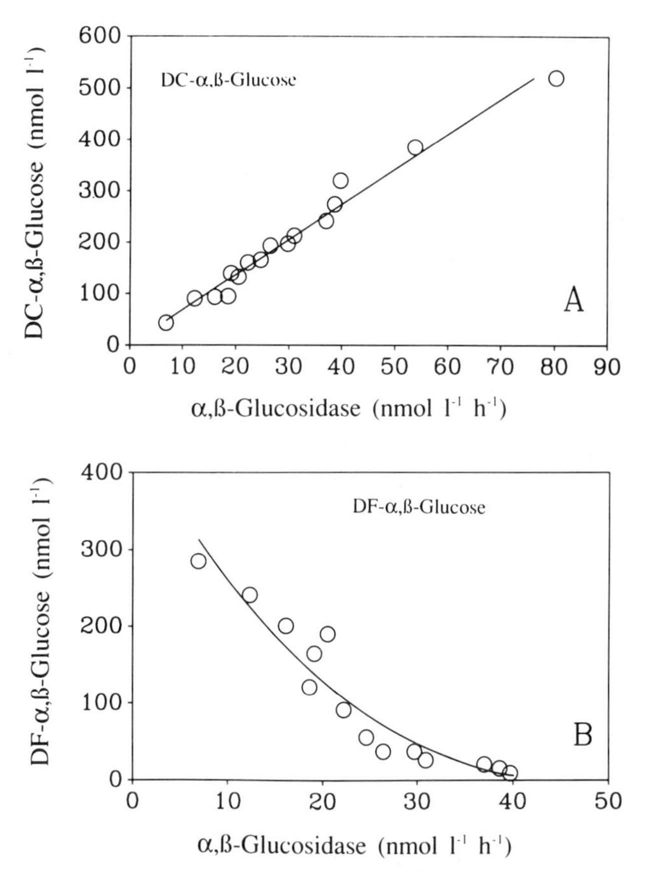

Figure 6.9 Relationship between α- and β-glucosidase activity and (A) DC-glucose and (B) DF-glucose in surface waters of Mekkojärvi.

place extracellularly or even during the chemical analysis of samples (hydrolysis and derivatization). Studies on carbohydrate-hydrolyzing enzymes and their relationship to natural substrate composition and microbial utilization have to consider such biochemical reactions. These results may demonstrate the importance of such biochemical approaches and studies in an ecological context.

6.6 Aminopeptidase Studies

Beside carbohydrates, dissolved amino acids (DAA) are another important group of substrates which are easily utilized by bacterioplankton. The concentration and dynamics of these important substrates have been studied in Plußsee and Mekkojärvi (Tables 6.1 and 6.2). An essential gap in our knowledge is, however, how efficiently microheterotrophs (e.g., bacteria) can utilize different pools and sources of DFAA and DCAA substrates in lakes with changing DOM of recalcitrant character. The DCAA pool seems to be more utilized than DFAA, because of its higher dynamics and more complex composition. Knowledge about the mechanisms of DCAA utilization and processing are therefore of fundamental interest in aquatic microbial ecology. In Mekkojärvi, leucine aminopeptidase was studied with special reference to comparing the enzyme activity and DFAA and DCAA composition and concentration. As shown in Tables 6.1 and 6.2, Mekkojärvi has a much lower content of DFAA than Plußsee. Consequently, the UDOM pool size in Mekkojärvi is smaller, but bacterioplankton biomass reaches similar levels to those in Plußsee. DOM composition and concentration is one of the regulating factors of activity and biomass in both lakes. Organic substrate availability may essentially be regulated by microbial extracellular enzymes, in the case of DAA, by aminopeptidases. Measured leucine aminopeptidase in Mekkojärvi gave similar levels for DF- and DC-leucine compared to β-glucosidases and DF-and DC-glucose in Plußsee (Figure 6.10). However, leucine aminopeptidase is not specific for leucyl peptides but also seems to hydrolyze many different peptides. As shown in Table 6.5, there is a sig-

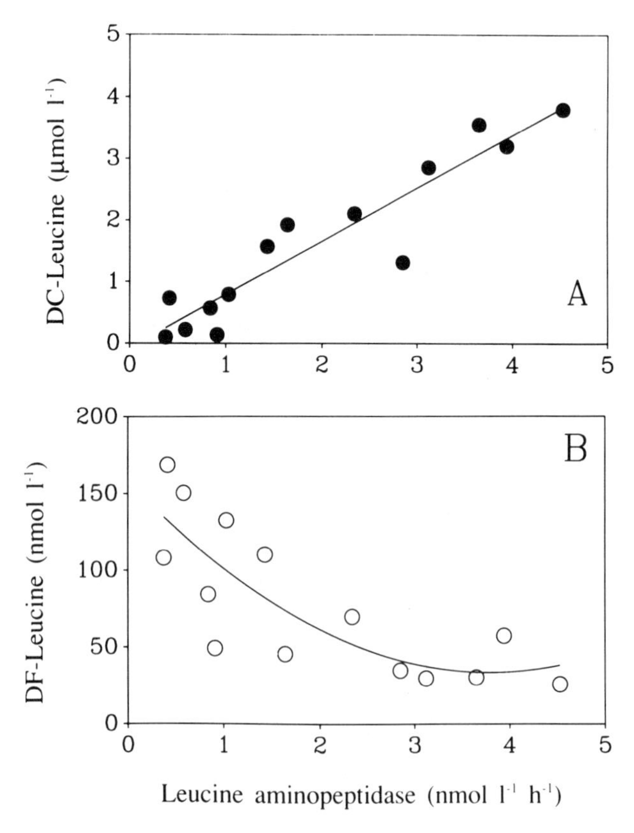

Figure 6.10 Relationship between leucine aminopeptidase activity and (A) DC-leucine and (B) DF-leucine concentration in Mekkojärvi surface waters.

Table 6.5 Correlation coefficients of leucine aminopeptidase activity and some dissolved amino acids (DAA) in Mekkojärvi (0- to 1-m depth)

	Correlation coefficient $(r^2)^1$	
DAA	DFAA	DCAA
Aspartic acid	−0.696	+0.870
Glutamine	−0.557	+0.868
Asparagine	−0.504	+0.874
Serine	−0.587	+0.814
Glycine	−0.790	+0.784
Threonine	−0.781	+0.871
Arginine	−0.625	+0.769
Tyrosine	−0.772	+0.851
Valine	−0.773	+0.649
Methionine	−0.530	+0.612
Phenylalanine	−0.610	+0.814
Isoleucine	−0.732	+0.778
Leucine	−0.740	+0.926

[1]$p < 0.01$.

nificant positive correlation between the enzyme and all measured DCAA compounds and a negative correlation with DFAA components. This non-specific cleaving mechanisms is not surprising for aminopeptidases in such a complex environment. It was previously reported that aminopeptidases are not very specific in their cleaving mechanisms (Hagihara, 1960). Hoppe (1983) and Hoppe et al. (1988) have found similar results in enzyme studies in marine environments. For a better understanding of these enzyme reaction mechanisms, more work must be done on substrate analysis, enzyme synthesis, and the regulation of activity and cleaving mechanisms.

These data show that UDOM substrates are rapidly oscillating on the diurnal scale because of their simultaneous release and utilization, whereas polymeric substrates vary over longer time intervals and tend to balance the pool of UDOM. Consistently, extracellular substrate depolymerization seem to be an essential tool used by aquatic microbes to compensate for short-term variations in external UDOM supply. From the data on extracellular enzyme activities and polymeric substrate composition, about 5–15% of polymeric DOC (DCAA and DCCHO) may be processed and utilized by bacterioplankton. This aspect of substrate function and utilization is essential during low PhDOC release, although it was not well known for bacterioplankton nutrition in the past.

6.7 Enzyme Endo- and Exo-Cleaving Mechanisms and Synergistic Action

Because I used only simple fluorogenic substrates (4-MUF-hexoses and Leu-MCA) during all enzyme assays, predominantly exo-cleaving mechanisms were measured. Because of lack of suitable fluorogenic substrates for endo-cleaving mechanisms studies, e.g., for studies on carbohydrate-cleaving mechanisms, 4-MUF-cellobiose was used and the activity of an exo-cellobiohydrolases was measured (Figure 6.11). From these studies, it became clear, that exo-cellobiohydrolases was measured, but its activity was generally lower than that of β-glucosidase. During sampling periods in Mekkojärvi in 1987, the ratio of β-glucosidase activity to exo-cellobiohydrolase activity varied between 5:1 and 10:1. This ratio is difficult to explain, because exo-cleaving activity is mostly accompanied by endo-cleaving activity (Heptinstall et al., 1986). According to Tilbeurgh et al. (1985) cellulolytic enzyme systems may involve two different 1,4-β-glucan cellobiohydrolases (CBH I and II), which hydrolyze cellobiose, cellotriose, cellotetraose and cellopentaose (e.g., $[\text{glucose}]_n$, n = 2–5) subunits from large polymers, which are subsequently cleaved by β-glucosidase to glucose monomers (Tilbeurgh et al., 1985). This experiment does not completely describe the actual cleaving mechanisms and processes in the natural habitats, but may be a model for synergistic enzyme action during DOM processing, a "cooperating endo-exo-cleaving mechanism." From many studies of microbial cellulose degradation and utilization (Ljungdahl and Eriksson, 1985) it is known that many polymeric carbohydrates are cleaved by a synergistic enzyme system (Wood and McCrea, 1979; Wood, 1985). These results indicate that similar mechanisms may occur in the natural environments of Plußsee and Mekkojärvi during substrate depolymerization and utilization processes. Because of the high amount of allochthonous organic matter entering Mekkojärvi during snow-melting period and heavy rainfalls (Salonen, unpublished observations.), DOM may contain heteropolymers such as lignocellulosic material or hemicelluloses of ter-

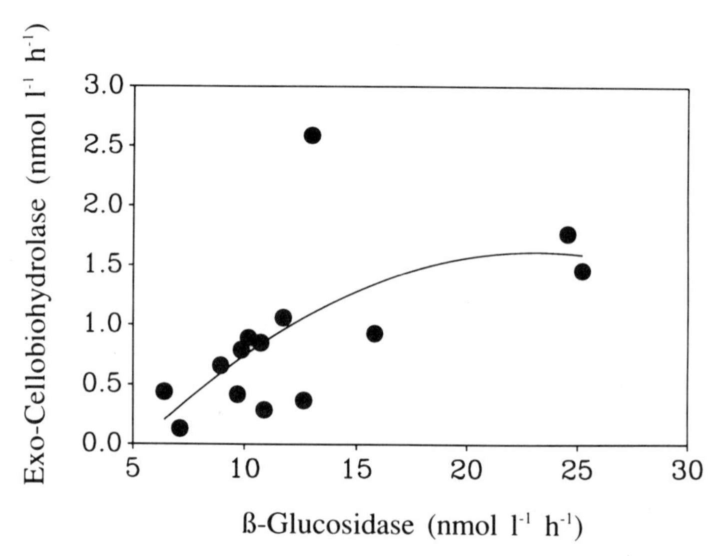

Figure 6.11 Relationship between exo-cellobiohydrolase and β-glucosidase activity in surface waters of Mekkojärvi.

restrial origin. Also additional glucanases, ligninases, and polyphenoloxidases may be involved in Mekkojärvi DOM depolymerization and processing. Recent publications (Zeikus, 1981; Kirk and Farrell, 1987; Tien, 1987; Viikari et al., 1989; Haider, 1988) suggest that lignin degradation by fungi and bacteria may occur in a synergistic pattern involving the action of several different glucanases, peroxidases, and ligninases. In many other aquatic environments, similar enzymatic DOM processing mechanisms may also exist.

6.8 Environmental and Regulatory Aspects of Microbial Extracellular Enzyme Activity

Our current knowledge about the activities and regulation of microbial enzymes in natural habitats is rather limited. Most information on this topic was obtained from chemostat studies (Harder and Dijkhuizen, 1982). However, in most of these experiments, microbial substrate utilization was studied at high substrate concentrations, that is at close to substrate saturations of the corresponding enzyme systems. In these experiments, sequential utilization of substrates and diauxic growth were often observed (Matin, 1979; Harder and Dijkhuizen, 1982). Substrate saturation however, is an exceptional situation, whereas substrate limitation is more regularly found in natural aquatic environments (Morita, 1984; Williams, 1986; Azam and Cho, 1987). Under substrate-limited conditions, chemostat experiments demonstrate a simultaneous uptake and utilization of different substrates (Harder and Dijkhuizen, 1982; Kuenen and Harder, 1982). It is assumed that microheterotrophs (e.g., bacterioplankton) are adapted to such a low nutrient supply by appropriate metabolic mechanisms (Azam and Cho 1987; Overbeck, 1990). Most studies on microbial substrate utilization focused on uptake of simple monomeric compounds (Overbeck, 1975; 1979; Hobbie and Williams, 1984; Wright, 1984). In some cases, first-order reaction kinetics were observed (Overbeck, 1972; 1975; 1979), but on many occasions, multiphasic reaction kinetics were measured (Overbeck, 1975; Azam and Hodson, 1981; Overbeck, 1990). Such variable transport systems were considered to be due to the flexibility of natural processes in microbes to rapidly changing microenvironments (Azam and Cho, 1987; Overbeck, 1990).

Flexibility in the synthesis rate and regulation of enzyme systems is therefore a prerequisite for the fitness of survival and growth of microbes in natural environments (Morita, 1984). This also includes a versatile genetic and metabolic control of enzyme activities. Studies on the genetic and metabolic control of polymeric substrate utilization by natural microheterotrophs have been neglected for many years. A close genetic and metabolic coupling of substrate depolymerization and utilization may be of essential advantage during substrate limitations. Only a few papers have been focused concerning these aspects in aquatic microbial ecology (Billen, 1984; Hoppe, 1989; Chróst, 1989; 1990). But our knowledge about the genetics and biochemistry of extracellular hydrolytic enzymes in natural habitats is rather limited. Essential questions regarding the regulation and inhibition mechanisms of extracellular enzyme actions have still to be answered. The DOM of natural environments is highly variable and it rapidly changes concentration and composition. A large number of different solutes with activating or inhibiting effects on extracellular enzyme actions are always present. Enzyme actions in natural habitats can be significantly modified by cooccurring substrates, reaction products of other enzymes, or potential inhibitors such as heavy metals or polyphenols. Microbes may develop certain strategies and mechanisms to circumvent and outcompete potential inhibitory compounds and solutes present in the dissolved phase or can create favorable microenvironments where enzyme reactions occur. Especially in extreme environments such as Mekkojärvi, with high DHM content, low pH values, and rapidly changing oxygen conditions, microbes are concerned with such environmental effects.

In Mekkojärvi, pH can change rapidly from pH 4.2 to 6.5. Therefore, extracellular enzyme activities were measured at different pHs to evaluate the effect of acidity on their activity. Results of these experiments are summarized in Figure 6.12. Of the five enzymes measured only aminopeptidase seems not to be adapted to the natural pH of Mekkojärvi water. Maximum of activity was found at pH 5.4 for phosphatase, β-glucosidase, and exo-cellobiohydrolase, whereas leucine-aminopeptidase displayed highest activity near neutral pH (pH 7.8). Therefore, most tested enzymes have their maximum of activity at the

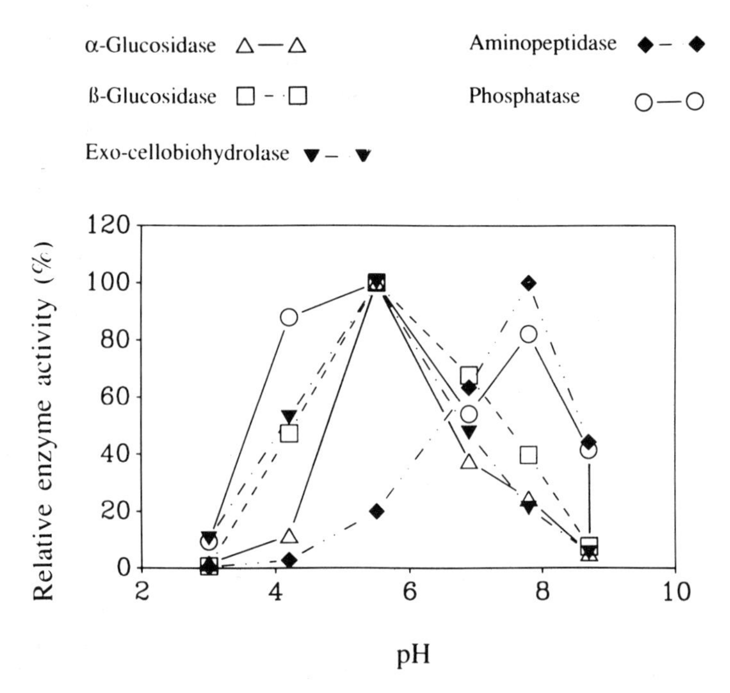

Figure 6.12 Response of α- and β-glucosidase, exo-cellobiohydrolase, phosphatase, and aminopeptidase to varying pHs of water samples from Mekkojärvi.

natural pH of Mekkojärvi (pH 5.4). In the case of phosphatases, predominantly acid phosphatases were measured. Thus it appears that most enzymes in natural habitats are adapted to the pH of their environments (Jansson et al., 1981; King, 1986; Chróst, 1989; Olsson, 1990). Temperature experiments revealed, however, that all tested enzymes were far below their temperature optimum (35–45°C).

Another important aspect of understanding enzyme activity in natural habitats is related to enzyme inhibition by secondary metabolites having high binding affinity to proteins and polysaccharides. According to Haslam (1974), McManus et al. (1985), and Spencer et al. (1988), many naturally occurring solutes such as gallus acid derivatives, tannins, and polyphenolic DHM may bind to the substrates and to enzymes as well and inhibit enzymatic reactions (see also Chapter 2). The origin of these secondary metabolites are rather complex; they may be derived from allochthonous or autochthonous processes. Higher plants, eukaryotes, or even prokaryotes can also be producers of such compounds. According to Cannell et al. (1987; 1988a, 1988b) and Nishizawa et al. (1985) many freshwater algae produce such enzyme inhibitors. The occurrence and dynamic of polyphenolic compounds and their significance for the whole lake metabolism have already been emphasized for Plußsee (Münster, 1985) and in general for lacustrine environments by Steinberg and Münster (1985) and Münster and Chróst (1990). In Plußsee, polyphenols having the OH-group in meta- and para-ring position predominate and constitute 10–20% of the DOC. This ratio is certainly higher in Mekkojärvi, approximately 50–80%. Nishizawa et al. (1985) and Cannell et al. (1988b) have identified pentagalloyl-glucose as a significant metabolite of freshwater green algae. This compound may inhibit glucanase and protease enzyme reactions. The regulatory mechanisms for inhibitor production and enzyme inhibition are still unclear.

6.9 Position and Importance of Extracellular Enzymes in Detritus Processing

There is increasing evidence that the majority of energy and nutrients flow through detritus food chains (Wetzel, 1984; Mann, 1988). Detritus processing and utilization play a key function in understanding the metabolism of aquatic ecosystems. Phagotrophic and osmotrophic microorganisms are probably the most efficient utilizer of detritus. As described in chapter 2 aquatic ecosystems primarily consist of immobilized enzymes. Such enzyme systems have been shown to be efficient modifiers of detritus. From the data given for the two lakes, I have tried to construct a model for detritus processing including the role and position of microbial extracellular enzymes (Figure 6.13). It can be seen that extracellular enzymes operate at several different levels of detritus processing and they are tightly coupled metabolically to substrate uptake systems. Synthesis, activity and regulation of extracellular enzymes are assumed to be under environmental, metabolic, and genetic control (see chapter 3). I hope

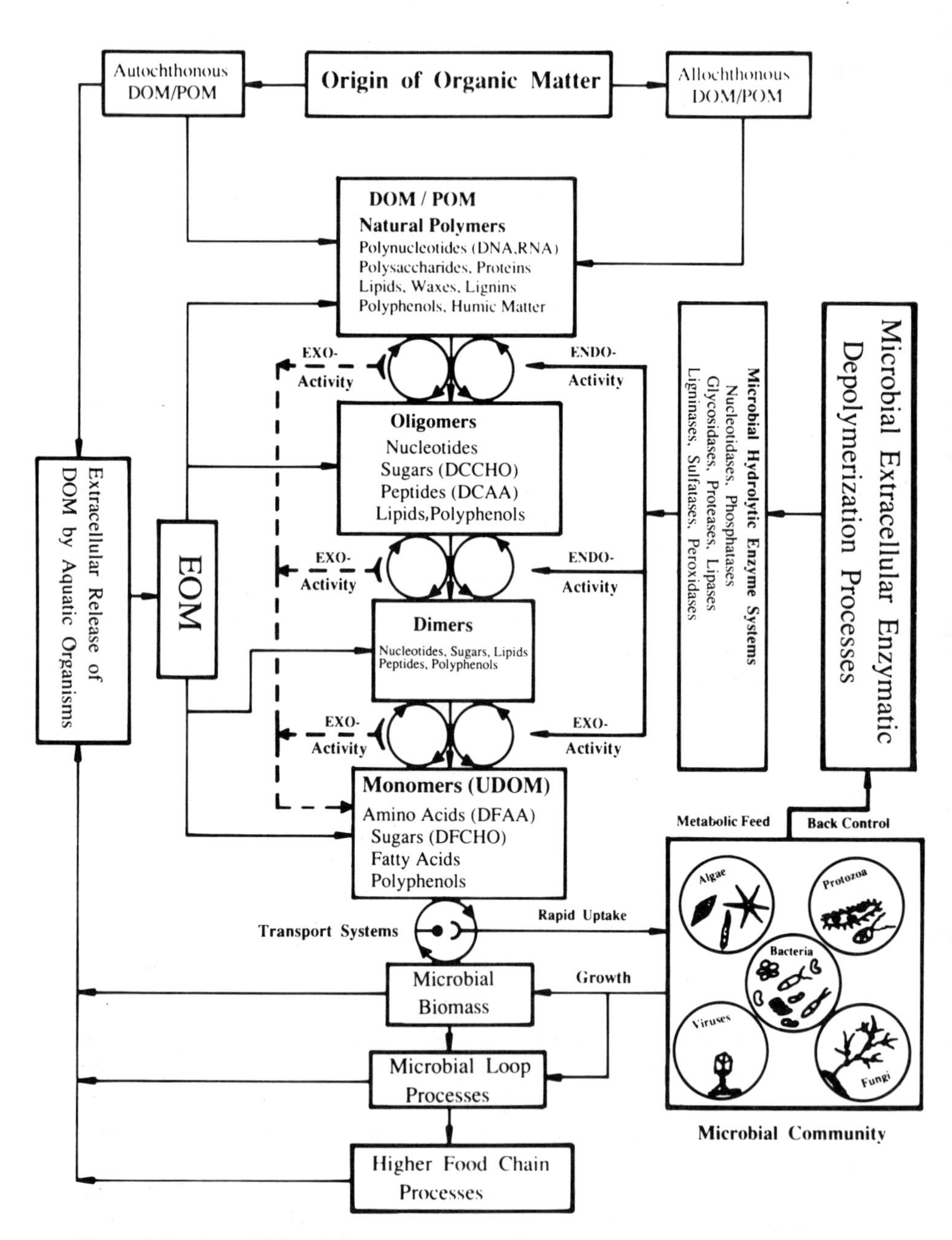

Figure 6.13 A model for detritus processing in lacustrine environments, including the location, action, and regulation of extracellular enzymes.

that the data I have present may exemplify and encourage studies in aquatic microbial ecology using modern biochemical approaches to provide a deeper insight into microbial DOM interactions and processes, food web interactions, and ecological functions.

Acknowledgments I dedicate this report to Professor. Dr. Jürgen Overbeck on the occasion of his retirement and 68th birthday. This work was supported by the Maj and Tor Nessling Foundation, Finland and the Max-Planck-Gesellschaft, F.R.G. This work also benefitted from stimulating discussions with Prof. Dr. J. Overbeck, Dr. K. Salonen, and Dr. L. Arvola. Dr. R.J. Chróst contributed some data on PhDOC and β-glucosidase activity and reviewed this manuscript. I am also grateful to P. Einiö and J. Nurminen, who helped me carry to the experimental work at Lammi Biological Station in Finland, and B. Albrecht at the MPI for Limnology in Plön for photographic assistance.

References

Amano, M., Hara, S. and Taga, N. 1982. Utilization of dissolved amino acids in seawater by marine bacteria. *Marine Biology* 68: 31–36.

Arvola, L. 1986. Spring phytoplankton of 54 small lakes in Southern Finland. *Hydrobiologia* 137: 125–134.

Azam, F. and R.E. Hodson. 1981. Multiphasic kinetics for D-glucose uptake by assemblages of natural marine bacteria. *Marine Ecology Progress Series* 6: 213–222.

Azam, F. and B.C. Cho. 1987. Bacterial utilization of organic matter in the sea. pp. 260–281 in Fletcher, M., Gray, T.R.G. and Jones, J.G. (editors), *Ecology of Microbial Communities*, Cambridge University Press, Cambridge.

Azam, F., Fenchel, T., Field, J.G., Gray, J.S., Meyer-Reil, L.A. and F. Thingstad. 1983. The ecological role of water column microbes in the sea. *Marine Ecology Progress Series* 10: 257–263.

Bengtsson, G. 1988. The impact of dissolved amino acids on protein and cellulose degradation in stream waters. *Hydrobiologia* 164: 97–102.

Billen, G. 1984. Heterotrophic utilization and regeneration of nitrogen. pp. 313–355 in Hobbie, J.E. and Williams, P.J.LeB. (editors), *Heterotrophic Activity In The Sea*, Plenum Press, New York.

Brock, T.D. 1966. *Fundamentals of Microbial Ecology*. Prentince Hall, Englewood Cliffs. 306 pp.

Campbell, I.M. and R. Bentley. 1973. Analytical methods for the study of equilibria. *Advances in Chemistry Series* 117: 1–19.

Cannell, R.J.P., Kellam, S.J., Owsianka, A.M. and J.M. Walker. 1987. Microalgae and Cyanobacteria as a source of glycosidase inhibitors. *Journal of General Microbiology* 133: 1701–1705.

Cannell, R.J.P., Kellam, S.J., Owsianka, A.M. and J.M. Walker. 1988a. Results of a large scale screen of microalgae for the production of protease inhibitors. *Planta Medica* 54: 10–14.

Cannell, R.J.P., Farmer, P. and J.M. Walker. 1988b. Purification and characterization of pentagalloylglucose, an α-glucosidase inhibitor/antibiotic from the freshwater green alga *Spirogyra varians*. *Biochemical Journal* 255: 937–941.

Chaloupka, J. and V. Krumphanzl. 1987. *Extracellular Enzymes of Microorganisms*. Plenum Press, New York. 215 pp.

Cho, B.C. and F. Azam. 1988. Major role of bacteria in biogeochemical fluxes in the ocean's interior. *Nature* 332: 441–443.

Chróst, R.J. 1989. Characterization and significance of β-glucosidase activity in lake water. *Limnology and Oceanography* 34: 660–672.
Chróst, R.J. 1990. Microbial ectoenzymes in aquatic environments. pp. 47–78 in Overbeck, J. and Chróst, R.J. (editors), *Aquatic Microbial Ecology: Biochemical and Molecular Approaches*. Springer Verlag, New York. 190 pp.
Chróst, R.J., Münster, U., Rai, H., Albrecht, D., Witzel, P.K. and J. Overbeck. 1989. Photosynthetic production and exoenzymatic degradation of organic matter in euphotic zone of an eutrophic lake. *Journal of Plankton Research* 11: 223–242.
Coffin, R.B. 1989. Bacterial uptake of dissolved free and combined amino acids in estuarine waters. *Limnology and Oceanography* 34: 531–542.
Crawford, C.C., Hobbie, J.E. and K.L. Webb. 1974. The utilization of dissolved free amino acids by estuarine microorganisms. *Ecology* 55: 551–563.
Cunningham, H.W. and R.G. Wetzel. 1989. Kinetic analysis of protein degradation by freshwater wetland sediment community. *Applied and Environmental Microbiology* 55: 1963–1967.
Fuhrman, J. 1987. Close coupling between release and uptake of dissolved free amino acids in seawater studied by an isotope dilution approach. *Marine Ecology Progress Series* 37: 45–52.
Fuhrman, J. and F. Azam. 1980. Bacterioplankton secondary production estimates for coastal waters of British Columbia, Antarctica, and California. *Applied and Environmental Microbiology* 39: 1085–1095.
Fuhrman, J. and F. Azam. 1982. Thymidine incorporation as a measure of heterotrophic bacterioplankton production in marine surface waters: evaluation and field results. *Marine Biology* 66: 109–120.
Hagihara, B. 1960. Bacterial and mold proteases. pp. 193–213 in Boyer, P.D., Lardy, H. and Myrbäck, K. (editors), *The Enzymes*, vol. 4. Plenum Press, New York.
Haider, K. 1988. The microbial degradation of lignin and its role in the carbon cycle. *Forum Mikrobiologie* 11: 477–489.
Hałemejko, G.Z. and R.J. Chróst. 1986. Enzymatic hydrolysis of proteinaceous particulate and dissolved material in an eutrophic lake. *Archiv für Hydrobiologie* 107: 1–21.
Harder, W. and L. Dijkhuizen. 1982. Strategies of mixed substrate utilization in microorganisms. *Philosophical Transactions of Royal Society London* 297: 459–479.
Haslam, E. 1974. Polyphenol-protein interactions. *Biochemical Journal* 139: 285–288.
Heptinstall, J., Stewart, J.C. and M. Seras. 1986. Fluorometric estimation of exo-cellobiohydrolase and β-glucosidase activities in cellulase from *Aspergillus fumigatus* Fresenius. *Enzyme Microbiology Technology* 8: 70–74.
Hessen, D.O. 1989. Factors determining the nutritive status and production of zooplankton in a humic lake. *Journal of Plankton Research* 11: 649–664.
Hessen, D.O. and A.K. Schartau. 1988. Seasonal and spatial overlap between cladocerans in humic lakes. *Internationale Revue der gesamten Hydrobiologie* 73: 379–405.
Hobbie, J.E. and LeB.P.J. Williams. 1984. *Heterotrophic Activity In The Sea*. Plenum Press, New York. 560 pp.
Hollibaugh, J.T. and F. Azam. 1983. Microbial degradation of dissolved proteins in seawater. *Limnology and Oceanography* 28: 1104–1116.
Hoppe, H.G. 1983. Significance of exoenzymatic activities in the ecology of brackish water: measurements by means of methylumbelliferyl-substrates. *Marine Ecology Progress Series* 11: 299–308.
Hoppe, H.G., Kim, S.J. and K. Gocke. 1988. Microbial decomposition in aquatic environments: combined processes of extracellular enzyme activity and substrate uptake. *Applied and Environmental Microbiology* 54: 784–790.
Ilmavirta, V. 1988. Phytoflagellates and their ecology in Finnish brown-water lakes. *Hydrobiologia* 161: 255–270.
Jannasch, H. 1970. Threshold concentration of carbon sources limiting bacterial growth

in seawater. pp. 321–330 in Hood, D.W. (editor), *Symposium On Organic Matter In Natural Waters.* University of Alaska Press, Fairbanks.

Jansson, M., Olsson, H., and Broberg, O. 1981. Characterization of acid phosphatase in the acidified lake Gårdsjön, Sweden. *Archiv für Hydrobiologie* 92: 377–395.

Jørgensen, N.O.G. 1982. Heterotrophic assimilation and occurrence of dissolved free amino acids in a shallow estuary. *Marine Ecology* 8: 145–159.

Jørgensen, N.O.G. 1987. Free amino acids in lakes: concentrations and assimilation rates in relation to phytoplankton and bacterial production. *Limnology and Oceanography* 32: 97–111.

Keston, A.S. 1954. Occurrence of mutarotase in animals: its proposed relation to transport and reabsorption of sugars and insulin. *Science* 143: 698–699.

Kim, J. and C.E. ZoBell. 1974. Occurrence and activities of cell-free enzymes in oceanic environments. pp. 367–385 in Colwell, R.R. and Morita, R.Y. (editors), *Effect Of The Ocean Environment On Microbial Activities*, University Park Press, College Park.

King, G.M. 1986. Characterization of β-glucosidase activity in intertidal marine sediments. *Applied and Environmental Microbiology* 51: 373–380.

Kirk, T.K. and R.L. Farrell. 1987. Enzymatic "combustion": the microbial degradation of lignin. *Annual Revue Microbiology* 41: 465–505.

Krambeck, C. 1984. Diurnal response of microbial activity and biomass in aquatic ecosystems. pp. 502–508 in Klug, M.J. and Reddy, C.A. (editors), *Current Perspectives in Microbial Ecology*, American Society for Microbiology, Washington DC.

Kreutzberg, G.W., Reddington, M. and H. Zimmermann. 1986. *Cellular Biology of Ectoenzymes*. Springer Verlag, Berlin. 313 pp.

Kuenen, G. and W. Harder. 1982. Microbial competition in continuous culture. pp. 342–367 in Burns, R.G. and Slater, J.H. (editors), *Experimental Microbial Ecology*. Blackwell Scientific Publication, London.

Kuznetsov, S.I. 1959. *Die Rolle Der Mikroorganismen Im Soffkreislauf Der Seen*. VEB Deutscher Verlag der Wissenschaften, Berlin. 225 pp.

Kuznetsov, S.I. 1968. Recent studies on the role of microorganisms in the cycling of substances in lakes. *Limnology and Oceanography* 13: 211–224.

Lindroth, P. and K. Mopper. 1979. High performance liquid chromatographic determination of subpicomole amounts of amino acids by precolumn fluorescence derivatization with o-phthaldehyde. *Analytical Chemistry* 51: 1667–1674.

Little, J.E., Sjøgren, R.E. and G.R. Carson. 1979. Measurement of proteolysis in natural waters. *Applied and Environmental Microbiology* 37: 900–908.

Ljungdahl, L.G. and K.E. Eriksson. 1985. Ecology of microbial cellulose degradation. pp. 237–299 in Marshall, K.C. (editor), *Advances in Microbial Ecology*, vol. 8, Plenum Press, New York.

Mann, K.H. 1988. Production and use of detritus in various freshwater, estuarine, and coastal marine ecosystems. *Limnology and Oceanography* 33: 910–930.

Matin, A. 1979. Microbial regulatory mechanisms at low nutrient concentrations as studied in chemostat. pp. 323–339 in Shilo, M. (editor), *Strategies of Microbial Life in Extreme Environments*, Verlag Chemie, Weinheim.

Mayer, L.M. 1989. Extracellular proteolytic enzyme activity in sediments of an intertidal mudflat. *Limnology and Oceanography* 34: 973–981.

McManus, J.P., Davis, K.G., Beart, J.E., Gaffney, S.H., Lilley, T.H. and Haslam, E. 1985. Polyphenol interactions. Part 1. Introduction; Some observations on the reversible complexation of polyphenols with proteins and polysaccharides. *Journal of Chemical Society Perkin Transactions* 2: 1419–1438.

Meffert, M.E. and J. Overbeck. 1985. Dynamics of chlorophyll and photosynthesis in natural phytoplankton associations. *Archiv für Hydrobiologie* 104: 219–234.

Meyer-Reil, L.-A. 1986. Measurements of hydrolytic activity and incorporation of dissolved organic substrates by microorganisms in marine sediments. *Marine Ecology Progress Series* 31: 143–149.

Meyer-Reil, L.-A. 1987. Seasonal and spatial distribution of extracellular enzymatic activities and microbial incorporation of dissolved organic substrates in marine sediments. *Applied and Environmental Microbiology* 53: 1748–1755.
Morita, R.Y. 1966. Marine psychrophilic bacteria. *Oceanography and Marine Biology Annual Review* 4: 187–203.
Morita, R.Y. 1984. Substrate capture by marine heterotrophic bacteria in low nutrient waters. pp. 83–100 in Hobbie, J.E. and Williams, LeB.P.J. (editors), *Heterotrophic Activity in the Sea*, Plenum Press, New York.
Münster, U. 1984. Distribution, dynamic and structure of free dissolved carbohydrates in the Plußsee, a North German eutrophic lake. *Verhandlungen der Internationalen Vereinigung für Theoretische und Angewandte Limnologie* 22: 929–935.
Münster, U. 1985. Investigations about structure, distribution and dynamics of different organic substrates in the DOM of lake Plußsee. *Archiv für Hydrobiologie Supplement* 70: 429–480.
Münster, U. and R.J. Chróst. 1990. Origin, composition and microbial utilization of dissolved organic matter. pp. 8–46 in Overbeck, J. and Chróst, R.J. (editors), *Aquatic Microbial Ecology: Biochemical and Molecular Approaches*, Springer Verlag, New York. 190 pp.
Münster, U., Einiö, P. and J. Nurminen. 1989. Evaluation of the measurements of extracellular enzyme activities in a polyhumic lake by means of studies with 4-methylumbelliferyl-substrates. *Archiv für Hydrobiologie* 115: 321–337.
Nishizawa, M., Yamagishi, T., Nonaka, G.I., Nishioka, I. and M.A. Ragan. 1985. Gallotannins of the freshwater green alga *Spirogyra* sp. *Phytochemistry* 24: 2411–2413.
Ohle, W. 1962. Der Stoffhaushalt der Seen als Grundlage einer allgemeinen Stoffwechseldynamik der Gewässer. *Kieler Meeresforschung* 18: 107–120.
Ohle, W. 1968. Chemische und mikrobiologische Aspekte des biogenen Stoffhaushaltes der Binnengewässer. *Mittellungen der Internationalen Vereinigung für Theoretische und Angewandte Limnologie* 14: 122–133.
Olsson, H. 1990. Phosphatase activity in relation to phytoplankton composition and pH in Swedish lakes. *Freshwater Biology* 23: 353–362.
Overbeck, J. and H. Babenzien. 1964. Über den Nachweis von freien Enzymen in Gewässer. *Archiv für Hydrobiologie* 60: 107–114.
Overbeck, J. 1965. Primärproduktion und Gewässerbakterien. *Naturwissenschaften* 51: 145–153.
Overbeck, J. 1968. Prinzipielles zum Vorkommen der Bakterien im See. *Mitteilungen der Internationalen Vereinigung für Theoretische und Angewandte Limnologie* 14: 134–144.
Overbeck, J. 1972. A computer analysis of the distribution pattern of phytoplankton and bacteria, measurement of rate of microbial decomposition of organic matter by means of kinetic parameters and remarks on the bacterial production in a stratified lake. pp. 227–237 in Hilbricht-Ilkowska, H. and Kajak, Z. (editors), *Productivity Problems of Freshwaters*. Polskie Wydawnictwo Naukowe, Warszawa.
Overbeck, J. 1975. Distribution pattern of uptake kinetic responses in stratified eutrophic lake. *Verhandlungen der Internationalen Vereinigung für Theoretische und Angewandte Limnologie* 19: 2600–2615.
Overbeck, J. 1979. Studies on the heterotrophic function and glucose metabolism of microplankton in Plußsee. *Archiv für Hydrobiologie Beihefte Ergebnisse Limnologie* 13: 56–76.
Overbeck, J. 1990. Aspects of aquatic microbial carbon metabolism: regulation of phosphoenolpyruvate carboxylase. pp. 79–95 in Overbeck, J. and Chróst, R.J. (editors), *Aquatic Microbial Ecology: Biochemical and Molecular Approaches*. Springer Verlag, New York. 190 pp.
Overbeck, J. and Y. Sako. 1988. Heterotrophic bacteria-how do they adapt to limited substrates in aquatic ecosystems? Studies on regulatory mechanisms. *Verhandlungen der Internationalen Vereinigung für Theoretische und Angewandte Limnologie* 23: 1815–1820.

Pigman, W. and E.F.L.J. Anet. 1972. Mutarotations and actions of acids and bases. pp. 165–193 in Pigman, W. and Horton, D. (editors), *The Carbohydrates, Chemistry and Biochemistry*. Academic Press, New York.

Pollock, M.R. 1962. Exoenzymes. pp. 121–178 in Gunsalus, I.C. and Stanier, R.Y. (editors), *The Bacteria*, vol. 4. Academic Press, New York.

Pomeroy, L.R. 1974. The ocean's food web, a changing paradigm. *Bioscience* 24: 499–504.

Pomeroy, L.R. and W.J. Wiebe. 1988. Energetics of microbial food webs. *Hydrobiologia* 156: 7–18.

Priest, F.G. 1977. Extracellular enzyme synthesis in the genus *Bacillus*. *Bacteriological Review* 41: 711–753.

Priest, F.G. 1984. *Extracellular Enzymes*. Van Nostrand Reinhold (UK), Wokingham. 79 pp.

Reichardt, W., Overbeck, J. and L. Steubing. 1967. Free dissolved enzymes in lake water. *Nature* 216: 1345–1347.

Riemann, B., Fuhrman, J. and F. Azam. 1982. Bacterial secondary production in freshwater measured by ^{3}H-thymidine incorporation method. *Microbial Ecology* 8: 101–114.

Riemann, B. and M. Søndergaard. 1984. Bacterial growth in relation to phytoplankton primary production and extracellular release of organic carbon. pp. 233–248 in Hobbie, E.J. and Williams, P.J.LeB. (editors), *Heterotrophic Activity in the Sea*. Plenum Press, New York.

Riemann, B. and M. Søndergaard. 1986. Regulation of bacterial secondary production in two eutrophic lakes and in experimental enclosures. *Journal of Plankton Research* 8: 519–536.

Riley, J.P. and D.A. Segar. 1970. Seasonal variation of the free and combined dissolved amino acids in the Irish Sea. *Journal of Marine Biological Association United Kingdom* 50: 713–720.

Salonen, K. 1981. The ecosystem of the oligotrophic lake Pääjärvi. 2. Bacterioplankton. *Verhandlungen der Internationalen Vereinigung für Theoretische und Angewandte Limnologie* 21: 448–453.

Salonen, K. and T. Hammar. 1986. On the importance of dissolved organic matter in the nutrition of zooplankton in some lake waters. *Oecologia* 68: 246–253.

Salonen, K., Kononen, K. and L. Arvola. 1983. Respiration of plankton in two small polyhumic lakes. *Hydrobiologia* 101: 65–70.

Scavia, D. and G.A. Laird. 1987. Bacterioplankton in Lake Michigan: dynamics, control, and significance to carbon flux. *Limnology and Oceanography* 32: 1017–1033.

Sherr, B.F., Sherr, E.B. and C.S. Hopkinson. 1989. Trophic interactions within pelagic microbial communities: Indications of feedback regulation of carbon flow. *Hydrobiologia* 159: 19–26.

Sieburth, McN.J. 1979. *Sea Microbes*. Oxford University Press, New York. 657 pp.

Sigleo, A.C., Hare, P.E. and G.R. Helz. 1983. The amino acid composition of estuarine colloidal material. *Estuarine and Coastal Shelf Sciences* 17: 87–96.

Simon, M. 1985. Specific uptake rates of amino acids by attached and free-living bacteria in a mesotrophic lake. *Applied and Environmental Microbiology* 49: 1254–1259.

Smucker, R.A. and C.K. Kim. 1987. Chitinase induction in an estuarine system. pp. 347–355 in Llevellyn, G.C. and O'Rear, C.O. (editors), *Biodeterioration Research*. Plenum Press, New York.

Somville, M. 1984. Measurement and study of substrate specificity of exoglucosidase activity in eutrophic water. *Applied and Environmental Microbiology* 48: 1181–1185.

Søndergaard, M., Riemann, B. and N.O.G. Jørgensen. 1985. Extracellular organic carbon (EOC) released by phytoplankton and bacterial production. *Oikos* 45: 323–332.

Søndergaard, M., Rieman, B., Møller-Jensen, L., Jørgensen, N.O.G., Bjørnsen, P.K., Olesen, M., Larsen, J.B., Geertz-Hensen, O., Hansen, J., Christoffersen, K., Jespersen, A.M., Andersen, F. and S. Bosselmann. 1988. Pelagic food web processes in an oligotrophic lake. *Hydrobiologia* 164: 271–286.

Sorokin, Y.I. 1977. The heterotrophic phase of plankton succession in the Japan Sea. *Marine Biology* 41: 107–117.

Spencer, C.M., Cai, Y., Martin, R., Gaffney, S.H., Goulding, P.N., Magnolato, D., Lilley, T.H. and E. Haslam. 1988. Polyphenol complexation—some thoughts and observations. *Phytochemistry* 27: 2397–2409.

Steinberg, C. and Münster, U., 1985. Geochemistry and ecological role of humic substances in lake water. pp. 105–145 in Aiken, G.R., McKnight, D.M., Wershaw, R.L., and McCarthy, P. (editors), *Humic Substances in Soil, Sediment and Water. Geochemistry, Isolation, and Characterization.* Wiley and Sons, New York.

Tien, M. 1987. Properties of ligninase from *Phanerochaete chrysosporium* and their possible applications. *CRC Critical Reviews in Microbiology* 15: 141–168.

Tilbeurgh, H., Pettersson, G., Bhikabhai, R., Boeck, H. and M. Claeyssens. 1985. Studies of the cellulolytic system of *Trichoderma reesei* QM 9414-reaction specificity and thermodynamics of interactions of small substrates and ligands with the 1,4-β-glucan cellobiohydrolase. *European Journal of Biochemistry* 148: 329–334.

Tranvik, L. 1988. Availability of dissolved organic carbon for planktonic bacteria in oligotrophic lakes of differing humic content. *Microbial Ecology* 16: 311–322.

Tranvik, L. 1989. Bacterioplankton in humic lakes—a link between allochthonous organic matter and pelagic food webs. *Ph.D. Thesis, University Lund, Sweden.* 104 pp.

Viikari, L., Keränen, S., Kantelinen, A. and M. Linko. 1989. Entsyymit puunjalostuksessa. *Kemia-Kemi* 16: 1139–1141.

Wetzel, R.G. 1983. *Limnology*. Saunders, Philadelphia. 755 pp.

Wetzel, R.G. 1984. Detrital dissolved and particulate organic carbon functions in aquatic ecosystems. *Bulletin of Marine Science* 35: 503–509.

Williams, P.J.LeB. 1981. Incorporation of microheterotrophic processes into the classical paradigm of the planktonic food web. *Kieler Meeresforschung Sonderheft* 5: 1–28.

Witzel, K.P., Overbeck, H.J. and K. Moaledj. 1982. Microbial communities in Lake Plußsee—An analysis with numerical taxonomy of isolates. *Archiv für Hydrobiologie* 94: 38–52.

Wood, T.M. 1985. Properties of the cellulolytic enzyme systems. *Biochemical Society Transactions* 13: 407–410.

Wood, T.M. and S.I. McCrea. 1979. Synergism between enzymes involved in the solubilization of native cellulose. *Advances in Chemistry Series* 181: 181–209.

Wright, R.T. 1984. Dynamics of pools of dissolved organic carbon. pp. 121–154 in Hobbie, J.E. and Williams, P.J.LeB. (editors), *Heterotrophic Activity in the Sea.* Plenum Press, New York.

Wright, R.T. 1988. A model for short-term control of the bacterioplankton by substrate and grazing. *Hydrobiologia* 159: 111–117.

Wright, R.T. and J.E. Hobbie. 1965. The uptake of organic solutes in lake water. *Limnology and Oceanography* 10: 22–28.

Wright, R.T. and J.E. Hobbie. 1966. Use of glucose and acetate by bacteria and algae in aquatic ecosystems. *Ecology* 47: 447–464.

Zeikus, J.G. 1981. Lignin metabolism and the carbon cycle. pp. 211–243 in Marshall, K.C. (editor), *Advances in Microbial Ecology*, vol. 8. Academic Press, London.

ZoBell, C. 1946. *Marine Microbiology*. Chronica Botanica, Waltham. 744 pp.

7

Protein Degradation in Aquatic Environments

Gilles Billen

7.1 Introduction

Organic matter degradation in aquatic environments has been studied for a long time by sanitary engineers and geologists concerned with predicting the rate of this process, either in polluted rivers, in sewage treatment plants or in sediment. Models have been established for this purpose, and most of them derive from the early Streeter and Phelps (1925) model where the rate of organic matter degradation is simply assumed to be proportional to the organic load. In order to take into account the differing susceptibilities to bacterial attack of the various classes of compounds making up the overall organic matter, Jørgensen (1978), Berner (1980) and Westrich and Berner (1984) suggested the use of "multi G's-first order kinetics," considering a number of organic matter fractions, each with its own first-order degradation constant.

Although these geochemical approaches often yielded reasonable predictions of the order of magnitude of the organic matter degradation rates, they are somewhat frustrating for the microbiologist, for they consider organic matter degradation as a chemical property of matter itself, without explicitly taking into account the activity of the microorganisms responsible. More recently, on the other hand, hydrobiologists have become aware of the fact that bacteria represent a significant component in the trophic structure and function of the aquatic ecosystem. Powerful methods have recently been developed for measuring in situ bacterial activity, such as uptake of individual organic substrates, bacterial growth rate, and grazing. The current trend however is to put more emphasis on the study of the trophic relationships between bacteria and higher trophic levels, than on a detailed study of the interaction between bacteria and organic matter. This has prevented aquatic microbiologists to propose an alternative operational approach to the simplified models currently used by engineers and geologists. In the long term, this might also be a severe handicap for understanding bacterioplankton dynamics in natural environments, because there is strong evidence that bacterial dynamics is controlled more strongly by resources (bottom-up) than by predators (top-down) control (Billen et al., 1990b).

There is therefore a need for studies on the supply and bacterial utilization of organic matter in aquatic environments. In particular, two main classes of processes must be distinguished: those supplying organic matter in the form of small molecules (monomers and oligomers) which can be directly taken up by bacteria (e.g., phytoplankton exudation), and those supplying high-molecular-weight polymers requiring extracellular hydrolysis (e.g., phytoplankton lysis, waste-water discharge, etc.). The dynamics of bacteria in response to these different types of organic matter appears to be very different (Münster and Chróst, 1990; Chróst, 1990; Chapter 3).

The main concern of this chapter is to reconcile the geochemical approach and the microbiological approach to the study of organic matter utilization. I will try to show: (1) that a realistic (although simplified) model of organic matter utilization by bacteria can be constructed on the basis of data from the application of new methods in microbial ecology, including those for measuring enzymatic activity; (2) that such a model is not only useful for applied and management studies, but can also be of significant help in understanding the dynamics of bacterioplankton and its relationships with phytoplankton.

The discussion in this chapter is based almost entirely on experimental data concerning protein and amino acid utilization, for the following reasons: first, proteins, peptides, and amino acids constitute simple models of organic carbon in general. Second, they are the backbone of organic matter circulation, since protides,—in contrast to glucides and lipids, which are largely used as storage compounds—make up the bulk of the structural and functional constituents of living organisms (Lancelot et al., 1986; Simon and Azam, 1989).

7.2 Experimental Studies of the Kinetics of Protein Degradation

Critical overview of the available methods Methods presently available for experimentally studying the kinetics of extracellular proteolytic enzymes involve either the use of fluorogenic substrates or of radiolabeled proteins. Several derivatives of amino acids which, give rise after hydrolysis to a fluorescent product have been proposed as model substrates for assaying extracellular proteolytic activity in water samples: L-leucyl-β-naphthylamide (LLβN) (Somville and Billen, 1983) and L-leucyl-4-methyl-7-coumarinylamide (Hoppe, 1983) for aminopeptidases; dansylglycyl-L-tryptophan (Hashimoto et al., 1985) for carboxypeptidases. These assays are cheap and straightforward. They usually require only short incubation times. The measurement can be performed at saturating concentration of artificial substrate, in which case they yield a potential rate of extracellular enzymatic activity. This can also be considered as a determination of enzyme concentration, which is useful for studying enzyme distribution and the regulation of enzyme synthesis. On the other hand, the use of several nonsaturating concentrations of the artificial substrates make it possible to investigate the kinetics and specificity of the extracellular enzymes. The rates of artificial fluorogenic-substrate hydrolysis was thus found to obey Michaelis-Menten kinetics (Somville and Billen, 1983; Hoppe, 1983). Using a procedure similar to that developed by Hobbie and Wright (1965) for studying the uptake of amino acids, it is therefore possible to determine the maximum rate of enzyme reaction (V_{max}) and the apparent half-saturation constant of substrate (K_m), i.e., the substrate concentration at which the rate of enzymatic reaction is equal to the 1/2 V_{max}. Moreover, extrapolation to zero concentration of the artificial substrate added allows one to estimate a mean turnover rate of the composite naturally occurring polymers affected by the enzymes being measured (Hoppe, 1983). A typical result is shown in Figure 7.1.

The simplicity of the assay with fluorogenic substrates has led to extensive

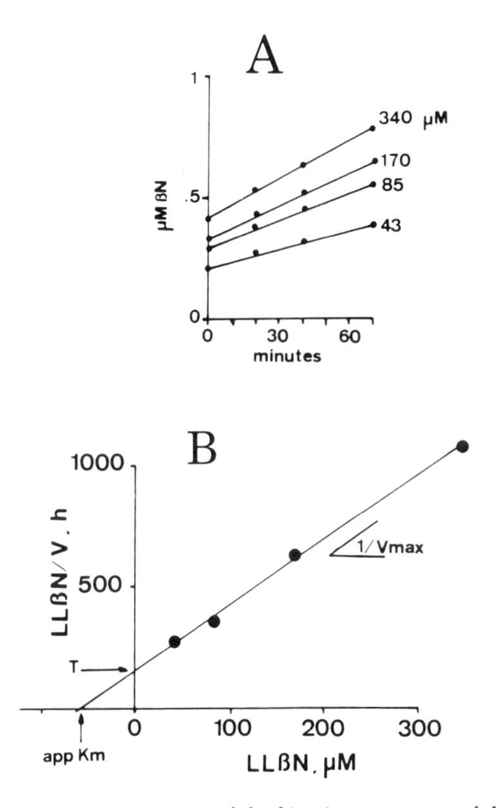

Figure 7.1 Example of the determination of the kinetic parameters of the hydrolysis of fluorogenic substrate by a sample from the Dutch coastal zone of the North Sea taken on May 1989. (A) Observed increase of fluorescence of β-naphtylamide as a function of time, for different added concentrations of LLβN. (B) Reciprocal plot of the relative rate of hydrolysis of LLβN as a function of LLβN concentrations, yielding V_{max} (reciprocal of the slope of the regression line), apparent K_m (x-intercept) and "turnover time" (T) of the natural polymers hydrolyzed by the enzymes acting on LLβN (y-intercept).

utilization of these methods in aquatic ecology (Chróst, 1990; Chapter 4). Few authors, however, seem to have considered the question of what can really be learned from the methodology concerning the kinetics of naturally occurring proteins. The K_m and V_{max} determined are related to the artificial substrate, not to naturally occurring proteins. The turnover time extrapolated to zero concentration of added substrate corresponds to that fraction of naturally occurring proteins which is the substrate of the enzyme responding to the artificial substrate used, provided it indeed reacts with a similar kinetics. Therefore, interpretation of the information gained from the assays with fluorogenic substrates in terms of real protein utilization requires careful investigation of the specificity of these substrates with respect to naturally occurring proteolytic enzymes.

Only a few authors have used radiolabeled proteins as a tool for studying bacterial proteolytic activity in waters, probably partly because of the high cost of these substrates (Chróst, 1990). Hollibaugh and Azam (1983) have developed a procedure involving the use of ^{125}I and ^{14}C labeled proteins. After adding the label to water samples, they followed the appearance of radioactivity in CO_2 and cold TCA-insoluble fraction. I have used a mixture of the ^{14}C-methylated proteins of different molecular weight (commercially supplied as molecular markers for gel electrophoresis; Amersham) to follow the disappearance of TCA-insoluble radioactivity in the <0.2-μm filtrate. Typical results are shown in Figure 7.2. This method does not suffer from the drawback of unknown specificity toward natural proteolytic enzymes and the results obtained are dependent on the size of the labeled proteins used (see Figure 7.2). Therefore the method does not allow one to deduce directly a mean turnover time of naturally occurring proteins, as was done by Hollibaugh and Azam (1983).

As shown, no one technique is straightforward for studying the kinetics of natural protein degradation. A combination of both approaches discussed above was used in the work reported here. The fluorogenic substrate L-leucyl-β-naphthylamide (LLβN), a highly specific substrate for aminopeptidases (N-terminal exoproteases), was used as a model substrate. The experiments reported below demonstrate that these enzymes play a major role in dissolved protein hydrolysis.

Half-saturation constant (K_m) for protein hydrolysis When proteins or peptides are added to samples of natural water, they act as competitive inhibitors for the hydrolysis of LLβN (Somville and Billen, 1983): they do not affect the maximum rate of hydrolysis (V_{max}) of the fluorogenic substrate but they do increase its apparent K_m for LLβN according to the relation:

$$\text{Apparent } K_m = K_m \, (1 + I/K_i)$$

where K_i is the inhibition constant of added proteins, which, in the case of competitive inhibition, also represents their K_m with respect to aminopeptidases, and I is the concentration of proteins in the assay. Using this equation,

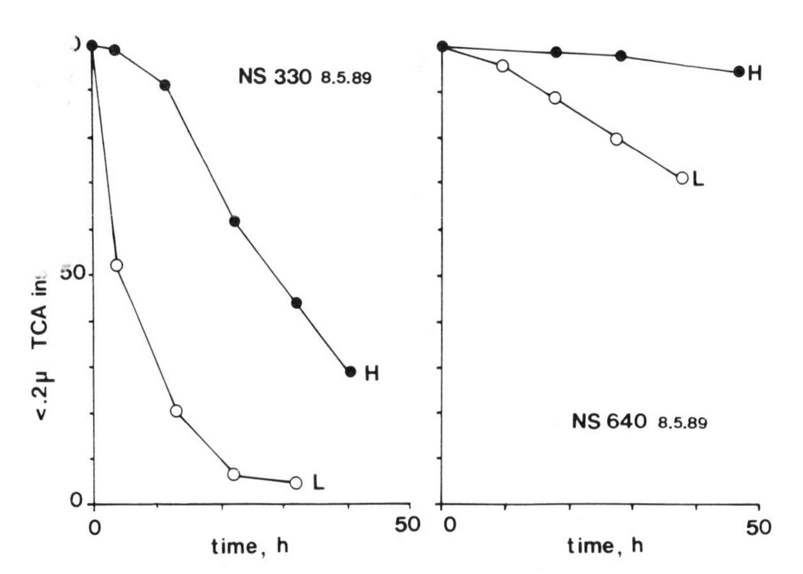

Figure 7.2 Example of results of [^{14}C]methylated protein hydrolysis, using two samples from the Belgian coastal zone of the North Sea, May 1989. The fraction of label remaining in the form of dissolved proteins (i.e., passing through an 0.2-μm membrane filter and precipitable by cold TCA) is plotted against time. Two different mixtures (L and H) of labeled proteins (Amersham) were used. Mixture L contains [^{14}C]methylated carbonic anhydrase (MW 30,000), [^{14}C]methylated trypsin inhibitor (MW 21,500), [^{14}C]methylated cytochrome c (MW 12,500), [^{14}C]methylated aprotinin (MW 6,500) and [^{14}C]methylated insulin (MW 5,740). Mixture H contains [^{14}C]methylated myosin (MW 200,000), [^{14}C]methylated phosphorylase b (MW 92,500), [^{14}C]methylated bovine serum albumin (69,000), [^{14}C]methylated ovalbumin (MW 46,000), [^{14}C]methylated carbonic anhydrase (MW 30,000) and [^{14}C]methylated lysozyme (MW 14,300).

it is indirectly possible to determine the K_m for a given added protein of natural aminopeptidases.

The same approach can be used to determine the affinity of extracellular proteolytic enzymes for the indigenous proteins present in seawater. The apparent K_m for LLβN was determined in series of assays with reconstituted samples containing different concentrations of indigenous proteins. These samples were obtained by mixing untreated seawater with ultrafiltered seawater (thus devoided of proteins) or with seawater of which the macromolecular fraction had been concentrated by ultrafiltration. Figure 7.3 shows two examples of apparent K_m determinations plotted against indigenous protein concentration of the assays, performed in an estuarine and a seawater environment. These results allow one to estimate the K_m for the hydrolysis of indigenous proteins by bacterial aminopeptidases. Table 7.1 summarizes the data obtained by this approach in three aquatic environments.

Maximum rate of aminopeptidase activity (V_{max}) and its significance in the overall process of protein degradation The rate (in particular the maximum rate, V_{max}) of degradation of an artificial substrate such as LLβN represents the

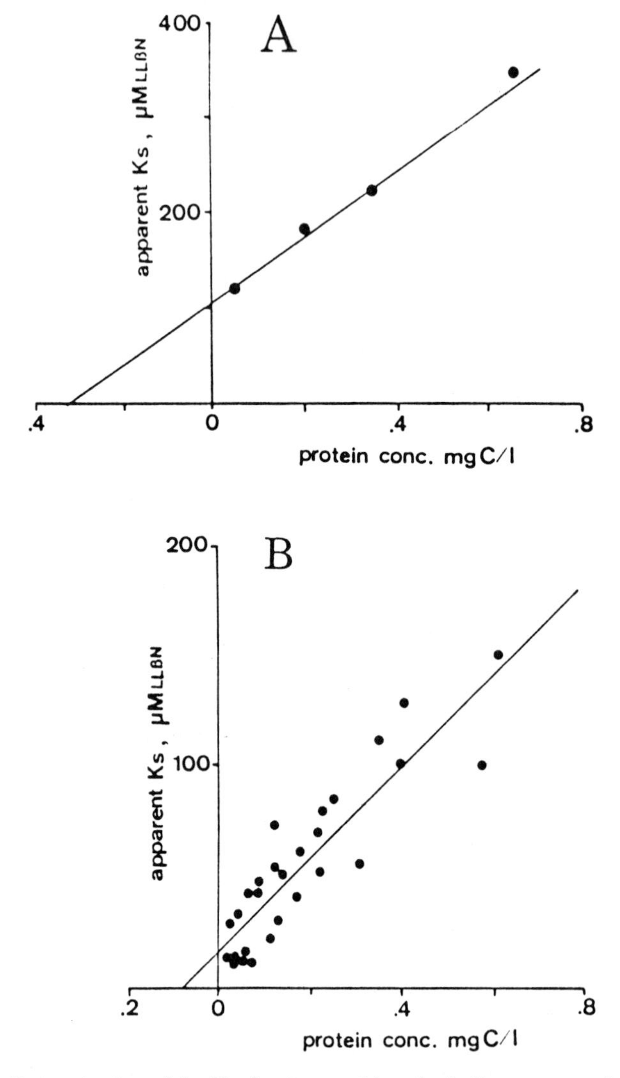

Figure 7.3 Determination of the K_s of aminopeptidase for indigenous proteins in aquatic environments. Relationship between apparent K_s and protein concentration. (A) Scheldt estuary, Belgium (autumn 1983), (B) Belgian coastal zone (spring 1984, redrawn from Fontigny et al., 1987).

rate of hydrolysis of a single peptide bond located in a N-terminal position. In order to assess the significance of this process in terms of natural protein degradation, parallel determination of the kinetics of LLβN utilization and of the degradation of added labeled proteins were made on several samples of sea and river water.

Interpretation of the labeled protein degradation experiments is complex,

Table 7.1 Experimental determination of the half saturation constant (K_m) of indigenous dissolved proteins by natural aminopeptidases in various aquatic environments

Environment	K_m ($\mu g\ C\ l^{-1}$)	Protein concentration range ($\mu g\ C\ l^{-1}$)
Belgian coastal zone	80	40–350
Meuse River	250	100–300
Scheldt Estuary	350	100–500

namely because of the fact that the results are strongly dependent on the molecular weight of the added labeled proteins (see Figure 7.2). Two extreme alternative models were elaborated in an attempt to interpreting the data. In the first model, it was assumed that all peptide bonds of the labeled substrate have equal probability to be hydrolyzed (hydrolysis predominantly by "ideal" endopeptidases). In the second model, on contrary, only N-terminal peptide bonds were assumed to be hydrolyzed (exopeptidase hydrolysis). In both cases, the Michaelis-Menten kinetics were assumed for the rate of labeled substrate hydrolysis, with an apparent K_m equal to that determined above for natural proteins plus the concentration of unlabeled proteins in the sample (which was assumed to be equal to the K_m in the cases where protein determination was lacking).

The time course of protein degradation, or more specifically, the rate of disappearance of proteins with molecular weight higher than 5,000 daltons (the limit for TCA precipitation) was calculated for each model. In the first model (endopeptidase hydrolysis), the distribution of each size class of peptides was calculated at every time interval as:

$$p(i,t + dt) = p(i,t) - V \times dt \left[\frac{p(i) \times i}{\sum_{j=2}^{\infty} p(j)/j} - \sum_{k=i+1}^{\infty} \frac{p(k) \times i \times 2/k}{\sum_{j=2}^{\infty} p(j) \times j} \right]$$

where: i is the size, in terms of the number of amino acids per peptide or protein; p(i,t) represents the concentration, expressed in number of amino acids in each size class i, at time t; V is the actual overall rate of proteolysis and is assumed to obey Michaelis-Menten kinetics with respect to labeled protein.

In the second model (N-terminal exopeptidase hydrolysis), the following relationship was used:

$$p(i,t + dt) = p(i,t) - V \times dt \left[\frac{p(i)}{\sum^{\infty} p(j)/i} - \frac{p(i + 1)/(i + 1)}{\sum^{\infty} p(j)/j} \times i \right]$$

The predictions of these two models using identical values of V_{max} and K_m are very different from each other. Figure 7.4 shows the rate of disappearance of radioactivity from the TCA-insoluble fraction calculated with both models for the two initial molecular weight distribution used in the experiments.

The general shape of the curves calculated by the N-terminal exopeptidase hydrolysis model and their strong dependence on the molecular weight distribution of the substrate are quite consistent with the experimental data. The curves calculated using the endopeptidase model are less consistent with the observations. Careful adjustment of calculated curves on the experimental results allow one to determine the value of V_{max} required for simulating the observed rate of protein hydrolysis in the scope of each model (Figure 7.5).

Surprisingly, the V_{max} values determined by this procedure with the exopeptidase hydrolysis model are quite similar to the V_{max} of LLβN hydrolysis determined on the same samples (Figure 7.6). The V_{max} determined with the endopeptidase hydrolysis model, however, are two orders of magnitude lower. This suggests that aminopeptidases as detected by LLβN do indeed have the dominant role in the hydrolysis of natural proteins in the environments studied.

Apparent turnover time of dissolved proteins The preceding discussion suggests that LLβN and natural dissolved proteins are hydrolyzed by the same enzymes, and with similar V_{max} value (but higher K_m for LLβN). Therefore, a kinetic analysis of LLβN hydrolysis similar to that described in Figure 7.1 should allow one to determine a meaningful turnover time of natural proteins. Results of parallel determination of turnover time of natural proteins and amino acids, obtained in the North Sea using LLβN and the standard method of Hobbie and Wright (1965), respectively, are plotted in Figure 7.7. This figure shows that the turnover time of proteins determined in this way is nearly one order of magnitude higher than that of amino acids. Also, the data available on the standing stock of dissolved proteins and amino acids for the same area (Table 7.2) show that the former is always about one order of magnitude higher and much more variable than the latter.

These results are consistent with the view that amino acid uptake is a rapid process that is highly coupled with the processes supplying the amino acids, so that their concentration is maintained at a low, steady level (Fuhrman, 1987), while the hydrolysis of protein is slower. The latter process constitutes the limiting step of bacterial growth when there is no direct supply of free amino acids.

Specific potential activity of proteolytic exoenzymes When determined at a reference temperature, the maximum rate of proteolytic activity (standard EPA) obtained by LLβN hydrolysis can be used as a measure of the amount of enzyme present. Over 60 such determinations were performed in temperate sea and river water throughout the annual cycle. A very good correlation was found with bacterial biomass (Figure 7.8A). The standard EPA/bacterial biomass ratio found (12 nmol h^{-1} μg^{-1} C bact) is just in the middle of the range observed in

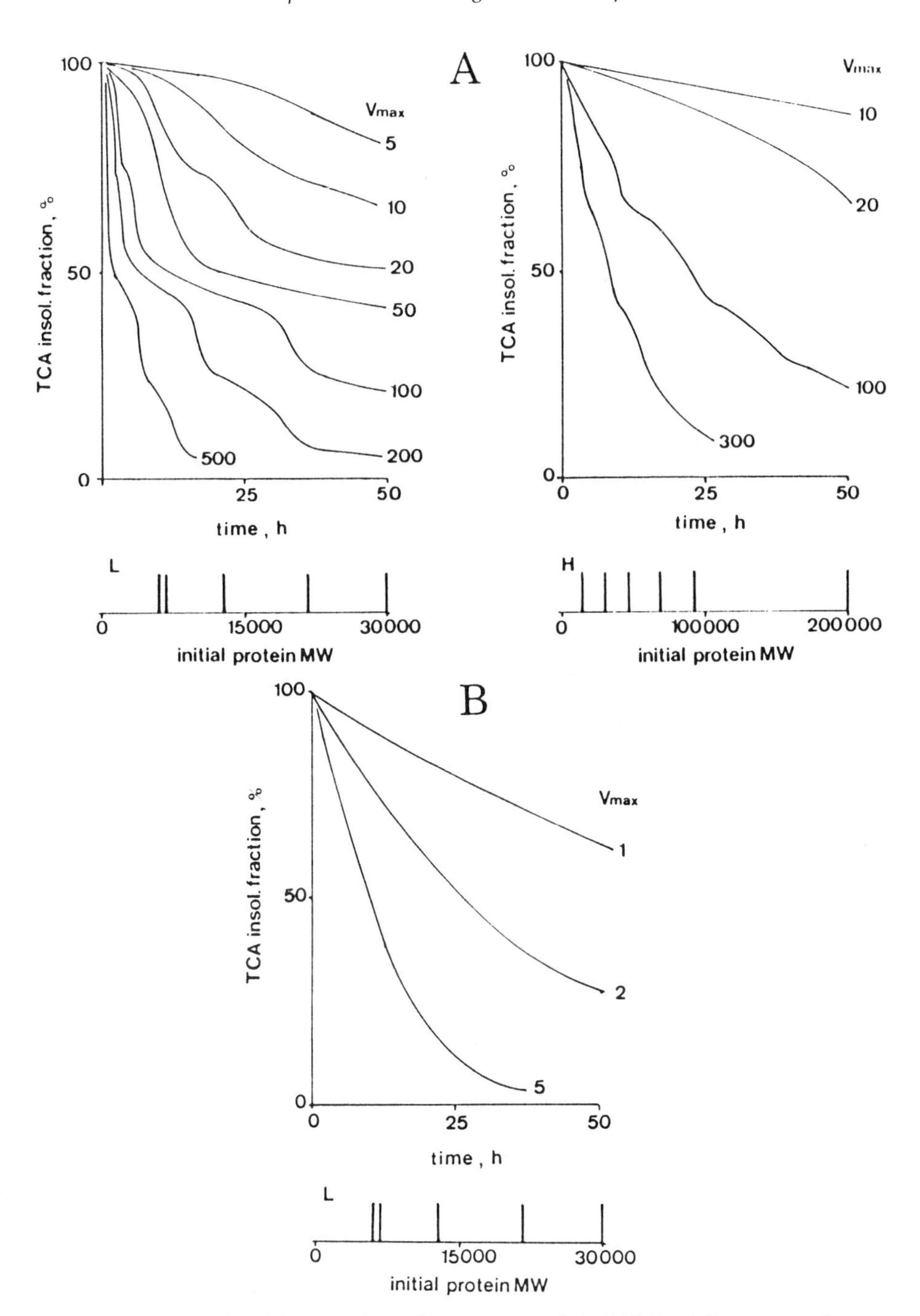

Figure 7.4 Calculated kinetics of the disappearance of the TCA-insoluble proteins for a number of values of the maximum rate of peptide bonds hydrolysis during: (A) N-terminal exoproteolytic hydrolysis (for different size distributions of labeled proteins); and (B) ideal endoproteolytic hydrolysis.

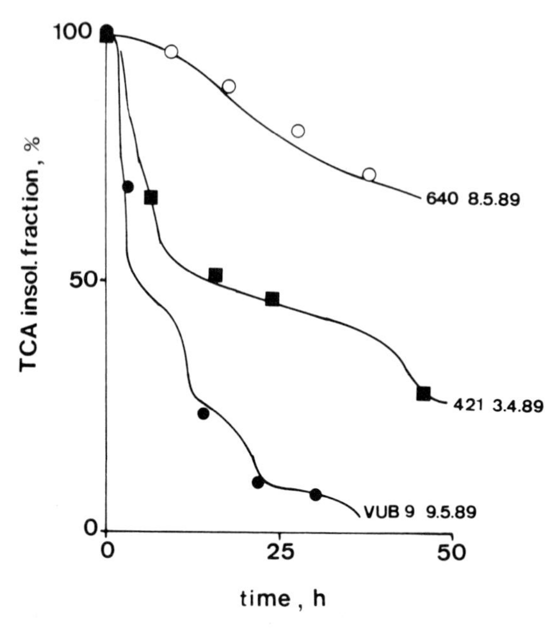

Figure 7.5 Example of the adjustment of the calculated hydrolysis kinetics of the TCA-insoluble labeled proteins with experimental data obtained from experiments with four different seawater samples from the Belgian and Dutch coastal zones. V_{max} values: sample 640 = 10 nmol l^{-1} h^{-1}; sample 421 = 75 nmol l^{-1} h^{-1}; sample VUB 9 = 300 nmol l^{-1} h^{-1}.

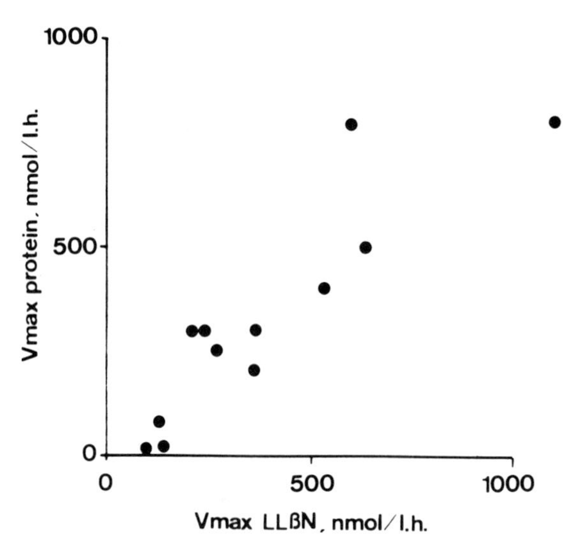

Figure 7.6 Relationship between the parallel determinations of the maximum rate of hydrolysis (V_{max}) of the peptide bond using labeled proteins and the fluorogenic substrate LLβN, respectively with samples from the Belgian coastal zone and from the river Seine.

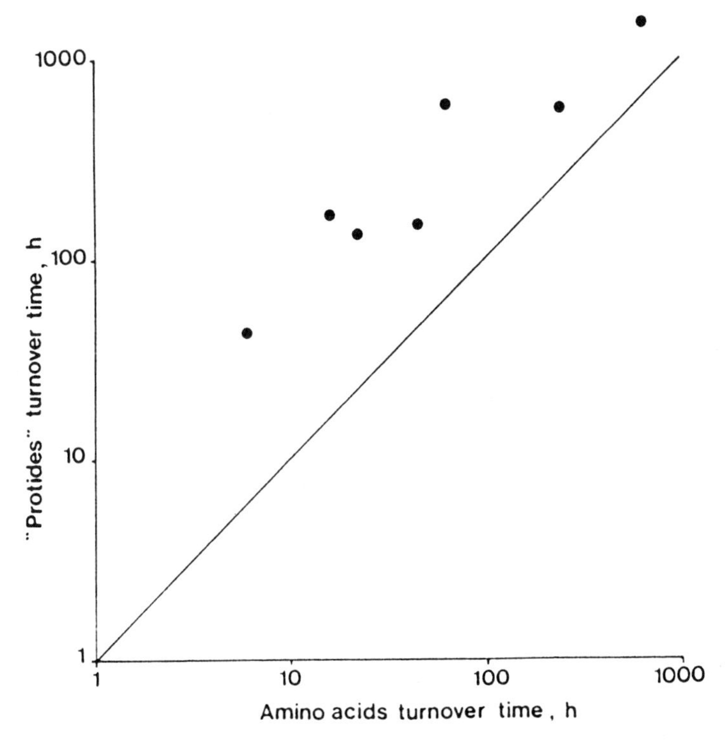

Figure 7.7 Parallel determination of the turnover time of proteins and amino acids in seawater from the Belgian and Dutch coastal zone. Note the log/log scale.

Table 7.2 Concentration of dissolved proteins and amino acids in various North Sea waters

Site	Concentration ($\mu g\ C\ l^{-1}$)	Reference
Proteins		
Belgian coast	50–350	Billen and Fontigny (1987)
Dutch coast	80–240	Billen (unpublished observations)
German Bight	180–360	Kattner et al. (1985)
Amino acids		
Ushant Front	8–17	Poulet et al. (1984)
English Channel	10–20	Andrews and Williams (1971)
German Bight	20–40	Kattner et al. (1985)
Belgian coast	10–45	Billen et al. (1980)

cultures enriched with proteins and amino acids respectively (Fontigny et al., 1987).

In a quite different environment, the Southern Ocean (Prydz Bay, February to March 1987), a similar relationship between standard EPA and bacterial bio-

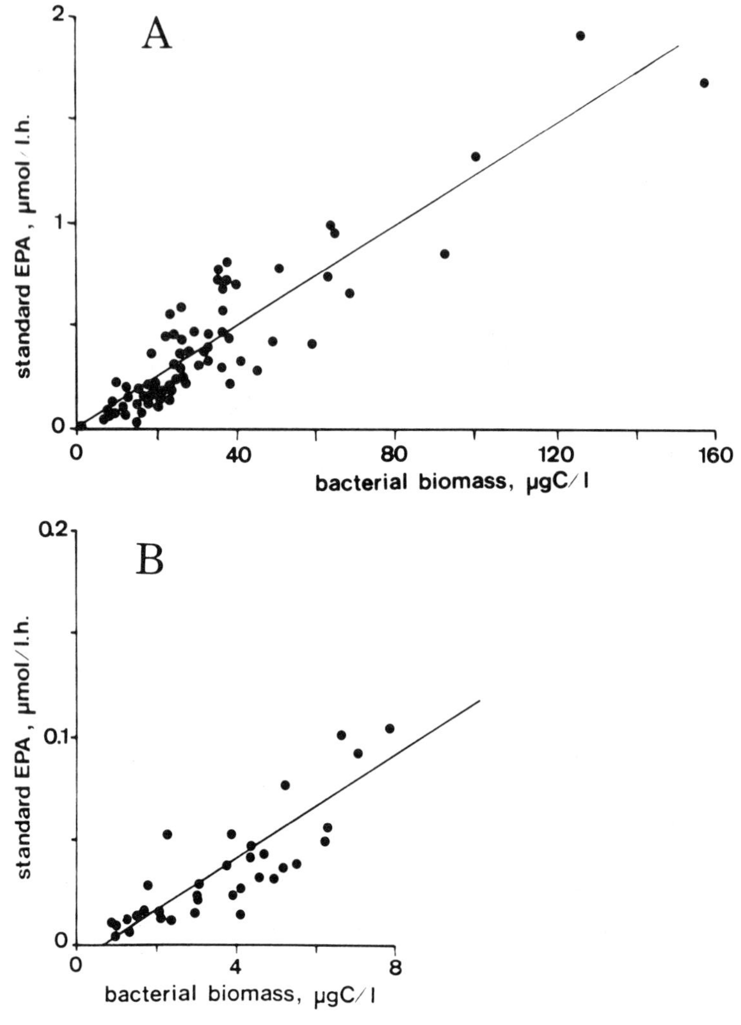

Figure 7.8 Relationship between standard extracellular proteolytic activity (EPA; V_{max} of LLβN hydrolysis at 20°C) and bacterial biomass (calculated from biovolumes and acridine orange cell counts) found in different aquatic environments. (A) temperate sea and fresh waters (coastal zones of the North Sea, pond water, river Seine) regression: EPA (μmol l^{-1} h^{-1}) = 12 . biomass (μgC l^{-1}) + 12; r = 0.90, n = 73. (B) Antarctic water (Prydz Bay, February-March 1987) regression: EPA μmol l^{-1} h^{-1}) = 11 . biomass (μgC l^{-1})—6, r = 0.82, n = 34.

mass was found again (Lancelot et al., 1989). Standard EPA was measured there at 13°C, the temperature found to be optimal for Antarctic bacterial growth. Aminopeptidase activity, however, did not follow the same temperature dependency as bacterial growth, showing an exponential relationship (with Q_{10} of 2.3) up to 20°C. Correcting measured EPA values to 20°C by using this relationship shows that exactly the same ratio between standard (20°C) EPA and bacterial biomass holds for Antarctic as for fresh waters (Figure 7.8B).

This consistency suggests that the regulatory processes affecting the synthesis of exoproteolytic enzymes, as described in the literature for several bacterial strains (McDonald and Chambers, 1966; Litchfield and Prescott, 1970; Priest, 1977; Holzer, 1980; Long et al., 1981), are not effective in the "normal" planktonic situation. At least, it indicates that such processes could be ignored for the purpose of modelling.

7.3 Modelling Bacterial Growth on Complex Organic Matter

Principle and mathematical formulation of the model As a guideline for our analysis of heterotrophic bacterial activity, Figure 7.9 present a diagram of the basic processes involved in organic matter utilization by bacteria in aquatic environments (Servais, 1986; Billen and Fontigny, 1987). Biodegradable organic matter in the sea is mostly supplied by phytoplankton, either through excretion or through lysis of cells. In the former process, small monomeric substrates are directly produced, while in the latter process, most of the organic matter is released under the form of macromolecular biopolymers (Billen, 1984).

Our preceding analysis of the kinetics of protein degradation, suggests that the utilization of macromolecules can be represented by a Michaelis-Menten relationship of the following form:

$$-e_1\text{max} \times \frac{H_1}{H_1 + KH_1} \times B$$

where: $e_1$max is the specific maximum rate of polymer hydrolysis per unit of bacterial biomass; H_1 is the concentration of polymer; B is the bacterial biomass; and KH_1 is the half-saturation constant of polymer hydrolysis.

For proteins, the following values can be assigned to the parameters $e_1$max and KH_1: $e_1$max = 12 nmol amino acid h^{-1} μg^{-1} bact C, or $e_1$max = 0.75 h^{-1}; KH_1 = 0.08–0.3 μg C l^{-1}. These values will be used for the general class of easily biodegradable polymers. A second class of slowly degradable com-

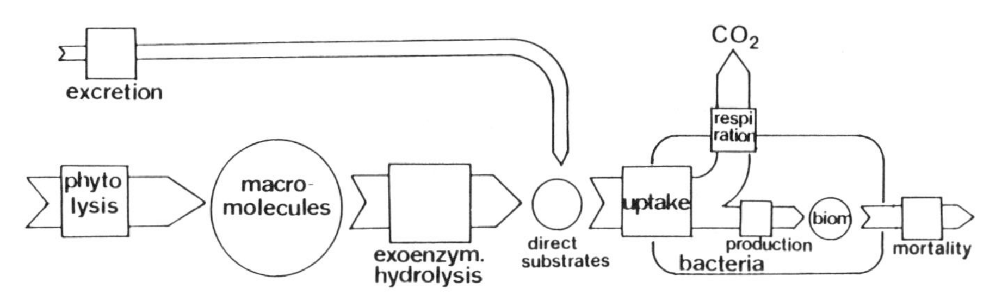

Figure 7.9 Schematic representation of the basic processes involved in the supply and bacterial utilization of dissolved organic matter in aquatic environments.

pounds will be considered, with e_2max and KH_2 values to be determined (see below).

The direct uptake of monomeric substrates (S) was also shown to obey an overall Michaelis-Menten kinetics (Parsons and Strickland, 1962; Wright and Hobbie, 1966). A constant fraction (Y) of the amount of substrates taken up is used for biomass production, the remaining part being respired (Servais, 1986). The process of bacterial mortality can be represented, as a first approximation, by the first order kinetics (Servais et al., 1986). The following equations can therefore be used to describe the dynamics of bacterial growth in aquatic environments:

$$\frac{dH_1}{dt} = -e_1max \times \frac{H_1}{H_1 + KH_1} \times B + pH_1 \tag{1}$$

$$\frac{dH_2}{dt} = -e_2max \times \frac{H_2}{H_2 + KH_2} \times B + pH_2 \tag{2}$$

$$\frac{dS}{dt} = e_1max \times \frac{H_1}{H_1 + KH_1} \times B + e_2max \times \frac{H_2}{H_2 + KH_2} \times B - bmax \frac{S}{S + K_s} B + pS \tag{3}$$

$$\frac{dB}{dt} = Y \times bmax \times \frac{S}{S + K_s} \times B - kd \times B \tag{4}$$

where: pH_1, and pH_2 are the rates of supply of polymers (e.g., through phytoplankton lysis); bmax is the maximum rate of substrate uptake by bacteria; K_s is the half-saturation constant of substrate uptake; Y is the growth yield, kd is the mortality constant, pS is the rate of supply of direct utilizable substrates (e.g., through phytoplankton exudation).

Y is generally close to 0.3 (Servais, 1986) except in situations of nutrient limitation where it can be as low as 0.1 (Lancelot and Billen, 1985; Billen et al., 1990a). kd varies little within the range 0.01–0.03 h^{-1} (Billen et al., 1987; Billen et al., 1990b). Ybmax (or μmax), the maximum specific growth rate of bacteria was determined in different aquatic environments by measuring either the increase in bacterial number or the incorporation of [^{3}H]thymidine (in samples prefiltered through 2.0 μm Nuclepore membrane filters to remove the grazers and to which a mixture of directly utilizable substrates at saturating concentrations were added). The results shown in Figure 7.10 have been fitted by a sigmoid relationship, as proposed by Lehman et al. (1975):

$$\mu max\ (h^{-1}) = 0.18 \exp - \left(\frac{T - T_{opt}}{T_i - T_{opt}}\right)^2$$

where: T_{opt} = 30°C and T_i = 12°C for temperate waters, and T_{opt} = 15°C and T_i = 5°C for Antarctic waters.

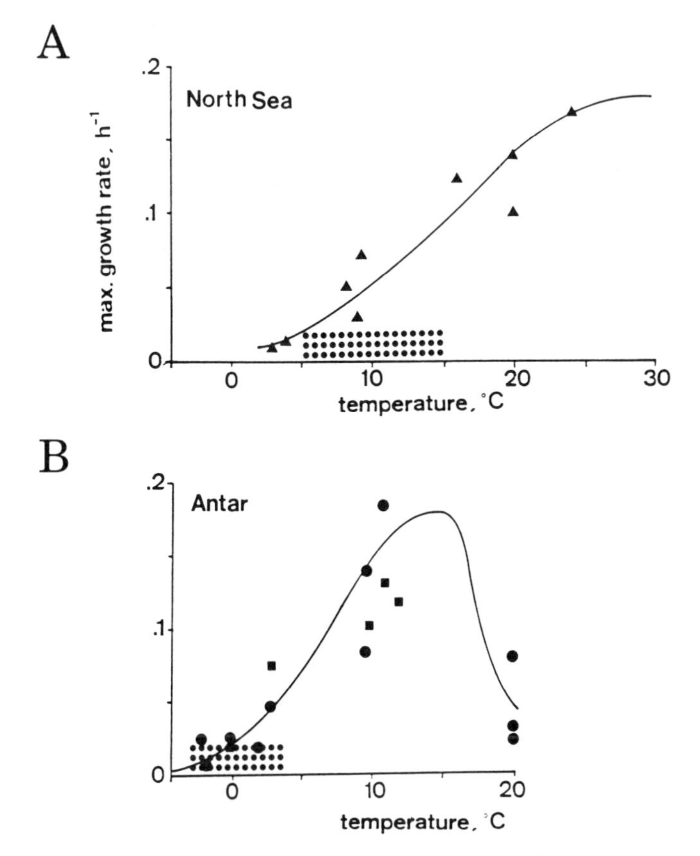

Figure 7.10 Relationship between maximum growth rate of bacteria (at saturating concentration of directly utilizable substrates) and temperature. (A) Temperate seawater (North Sea, coastal zone), and (B) Antarctic waters (Prydz Bay).

Bacterial utilization of phytoplankton lysis products and other complex material In order to characterize the kinetics of utilization of products of phytoplankton lysis by aquatic bacteria, the following experiment was carried out: an enriched culture of phytoplankton was sonicated (Branson Sonic S-75 adjusted at 4.2 A for 5 minutes) and filtered through 0.2-μm membrane filter. The dissolved organic material in the filtrate was considered as representative for phytoplankton lysis products. The culture was inoculated with a natural assemblage of aquatic bacteria filtered through 2-μm membrane filter in order to remove protozoa. The development of bacteria and the reduction in dissolved organic carbon was followed for about 15 days (Figure 7.11).

The growth pattern of the bacteria could be simulated according to the model presented above (equations (1) to (4)). The value of the parameters e_1max and KH_1 at 20°C were taken identical to those determined for the kinetics of protein hydrolysis by marine bacterial exoproteolytic enzymes (see above). The

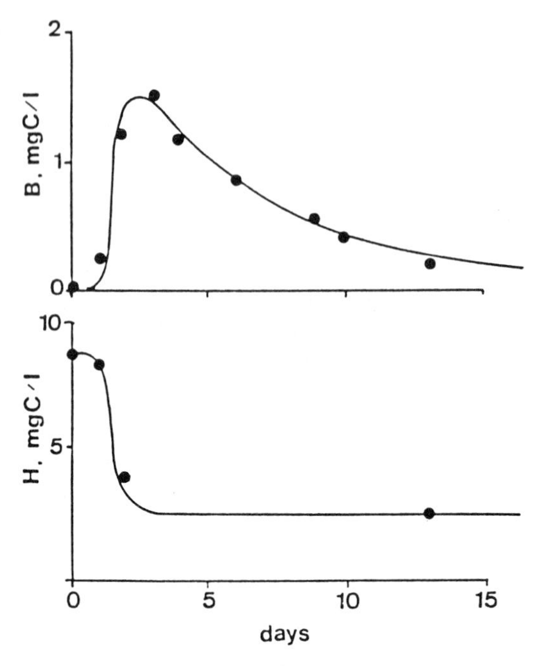

Figure 7.11 Time course of (top) bacterial biomass development and (bottom) dissolved organic carbon degradation in a batch experiment in which a sonicated and filtered phytoplankton culture was inoculated with a natural assemblage of aquatic bacteria. The curves represent the solution of the model described in the text for the following values of the parameters: $e_1max = 0.75\ h^{-1}$, $KH_1 = 0.1$ mgC l^{-1}; $e_2max = 0.25\ h^{-1}$, $KH_2 = 2.5$ mgC l^{-1}; a = 0.5.

parameters e_2max and KH_2, and the relative part (α and 1-α) of H_1 and H_2 fractions in H were determined by adjustment of the solution of the model to the experimental results (Figure 7.11).

For other similar experiments, we showed that bacterial growth on organic material of different origins, including soil-derived humic material, domestic sewage and cattle wastes, etc., could be simulated by the same equations using the same parameters for extracellular enzymatic hydrolysis of polymers, by simply varying the proportion (α:1-α) of rapidly (H_1) and slowly (H_2) degradable organic matter (Servais, 1986; Servais and Billen, unpublished observation).

Simulation of bacterial growth in response to phytoplankton blooms Two extreme paradigms have been proposed for describing the trophic relationship between phyto- and bacterioplankton. In the first paradigm (Larsson and Hagstrom, 1979; Wolter, 1982; Moller-Jensen, 1983), bacteria depend mainly on the exudation by phytoplankton of low molecular weight substrates with very short turnover times. In the second paradigm (Jassby and Goldman, 1974; Billen, 1984; Lancelot and Billen, 1985), lysis of aging phytoplankton cells or spillage of algae by zooplankton grazing causes the leakage of predominantly macromolecular organic material which constitutes the bulk of dissolved organic matter used by bacterioplankton after ectoenzymatic hydrolysis. In fact, both processes probably coexist (Chróst, 1989), with varying relative importance according to the season or the environment. Because of the predominantly different forms (either monomeric or macromolecular) under which organic matter is supplied by these two processes, they must have very different effects on the timing of bacterioplankton development.

Numerous observations have shown a distinct delay in the response of bacterioplankton to spring phytoplankton blooms (Chróst, 1989; Chapter 3). One example is the delay of about 10 days observed between the peak of the *Phaeocystis* bloom and the peak of heterotrophic bacteria (Figure 7.12) in the coastal zones of the North Sea (Billen and Fontigny, 1987; Billen et al., 1990a). Another example (Figure 7.13) is provided by observations carried out in Prydz Bay (Antarctica) which suggested that the bacterial peak is delayed by about one month with respect to the early phytoplankton bloom (Billen et al., 1988).

The model described above can be used to assess the relative effect of lysis and exudation in determining these delays. Equations 1 to 4 above will be used with the following expressions for pH_1, pH_2, and pS:

$$pS = k_{ex}\ PHY$$

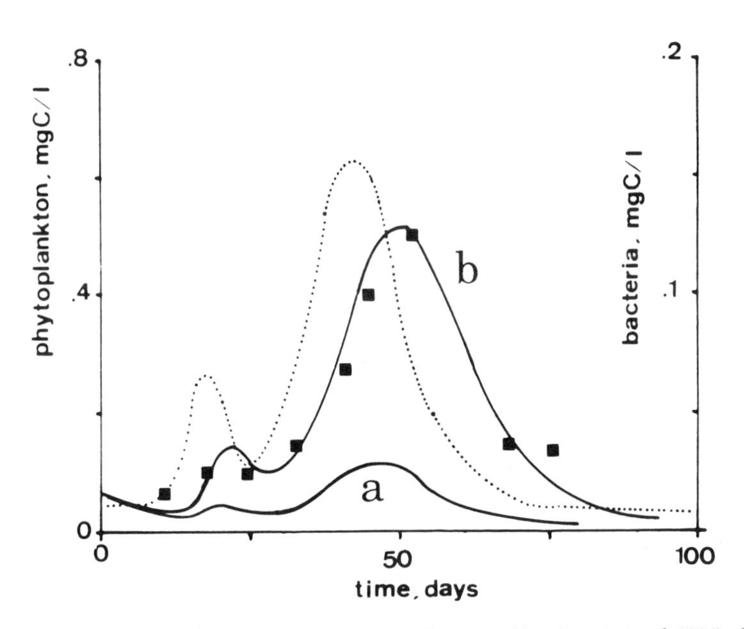

Figure 7.12 The spring bloom in Belgian coastal waters (day 0 = 1 April 1984). Simulation of bacterial growth in response to observed phytoplankton development (dotted curve) according to the model, for two different values of the rate of lysis (k_{lys}) and the rate of exudation (k_{ex}): (a) $k_{lys} = 0\ h^{-1}$, $k_{ex} = 0.002\ h^{-1}$; (b) $k_{lys} = 0.01\ h^{-1}$, $k_{ex} = 0.001\ h^{-1}$. Observed values of bacterial biomass are shown as black squares for comparison.

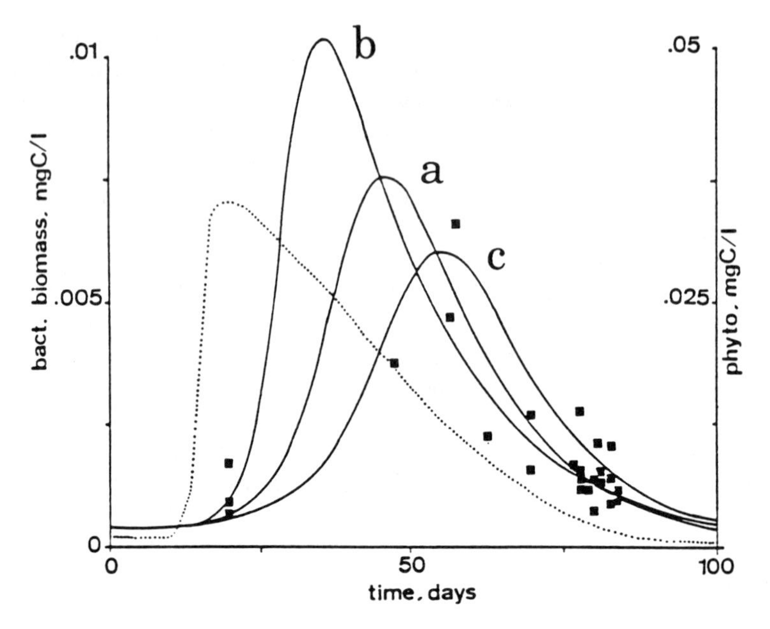

Figure 7.13 Phyto- and bacterioplankton development in Prydz Bay (Antarctica) after melting of the ice cover (day 0 = 1 January 1987). Simulation of bacterial growth (solid lines) in response to observed phytoplankton development (dotted line) according to the model, using three different values of the rate of lysis (k_{lys}) and the rate of exudation (k_{ex}): (a) $k_{lys} = 0.002\ h^{-1}$, $k_{ex} = 0.0005\ h^{-1}$; (b) $k_{lys} = 0.0005\ h^{-1}$, $k_{ex} = 0.002\ h^{-1}$; (c) $k_{lys} = 0.002\ h^{-1}$, $k_{ex} = 0.0001\ h^{-1}$. Observed values of bacterial biomass are shown as squares.

$$pH_1 = \alpha \times k_{lys}\ PHY$$

$$pH_2 = (1 - \alpha) \times k_{lys}\ PHY$$

Bjørnsen (1988) has argued that phytoplankton exudation should in fact be considered as a "property tax" instead of an "income tax" to the phytoplankton, being proportional to phytoplankton biomass and production. As an average, exudation of small molecules of substrates should represent about 2–5% of phytoplankton biomass per day (k_{ex} of about 0.001–0.002 h^{-1}).

Very few data are available, on the other hand, on the kinetics of phytoplankton lysis. A first order kinetics will be assumed as the first approximation. Lancelot and Mathot (1987) estimated the phytoplankton disappearance rate to be 0.006–0.012 h^{-1}. These rates, which include sedimentation, thus constitute an upper limit to the rate of lysis in this environment. The observed development of bacterioplankton in response to the phytoplankton bloom in the North Sea (Figure 7.12) could not be simulated by simply considering the process of exudation, even at the maximum value ($k_{ex} = 0.002\ h^{-1}$). On the contrary, assuming a phytoplankton lysis at a rate of 0.010 h^{-1} allows a good fit to both the size of the phytoplankton development and its timing. Similarly, bacterial development in Prydz Bay can be adequately simulated by the following values of the parameters:

$$k_{lys} = 0.002\ h^{-1}, \text{ and } k_{ex} = 0.0001\text{–}0.0005\ h^{-1}.$$

Higher values of the rate of exudation would lead to much shorter delay between phytoplankton and bacterioplankton development.

Application of the model therefore shows that in both environments considered here, lysis (and thus release of macromolecular substrates) instead of exudation (i.e., release of low-molecular-weight substrates) is the dominant process of dissolved organic matter supply from phytoplankton to bacteria. The requirement for an initial step of extracellular enzymatic hydrolysis of the macromolecular material so produced is responsible for a looser coupling between phyto- and bacterioplankton than would be the case if exudation was the dominant process of organic matter supply to aquatic bacteria.

Acknowledgments The author is Research Associate of the Fonds National de la Recherche Scientifique (Belgium). Part of the work reviewed here was funded by the Science Policy Office (Brussels, Belgium) under contracts ARC/Oceano, (ANTAR/05 and ENV/2).

References

Berner, R.A. 1980. A rate model for organic matter decomposition during bacterial sulfate reduction in marine sediments. pp. 35–45 in Daumas, R. (editor), *Biogeochimie de la Matiere Organique a L'interface Eau-Sediment Marin*. C.N.R.S., Paris.

Billen, G. 1984. Heterotrophic utilization and regeneration of nitrogen. pp. 313–355 in Hobbie, J.E. and Williams, P.J.LeB. (editors), *Heterotrophic Activity in the Sea*. Plenum, New York.

Billen, G. and A. Fontigny. 1987. Dynamics of *Phaeocystis*-dominated bloom in Belgian coastal waters. II. Bacterioplankton dynamics. *Marine Ecology Progress Series* 37: 249–257.

Billen, G., Lancelot, C. and S. Mathot. 1988. Ecophysiology of phyto- and bacterioplankton growth in the Prydz Bay area during the austral summer 1987. Part II. Bacterioplankton activity. *Proceedings of Belgian Colloqium on Antarctic Research*. Brussel. 146 pp.

Billen, G., Joiris, C., Meyer-Reil, L.A. and H. Lindeboom. 1990a. Ecological role of bacteria in the North Sea ecosystem. *Netherland Journal of Sea Research* (in press).

Billen, G., Servais, P. and S. Becquevort. 1990b. Dynamics of bacterioplankton in oligotrophic and eutrophic aquatic environments: Bottom-up or top-down control? Developments in Hydrobio In D.J. Bonin and H.L. Golterman (editors), *Fluxes Between Trophic Levels and Through the Water-Sediment Interface*. Developments in Hydrobiology. Elsevier (in press).

Bjørnsen, P.K. 1988. Phytoplankton release of organic matter: Why do healthy cells do it? *Limnology and Oceanography* 33: 151–159.

Chróst, R.J. 1990. Microbial ectoennzymes in aquatic environments. pp. 47–78 in J. Overbeck and R.J. Chróst (editors), *Aquatic Microbial Ecology: Biochemical and Molecular Approaches*. Springer-Verlag, New York.

Chróst, R.J. 1989. Characterization and significance of β-glucosidase activity in lake water. *Limnology and Oceanography* 34: 660–672.

Fontigny, A., Billen, G. and J. Vives-Rego. 1987. Kinetic characteristics of exoproteolytic activity in coastal seawater. *Est. Coastal Shelf Science* 25: 127–133.

Fuhrman, J.A. 1987. Close coupling between release and uptake of dissolved free amino acids in sea water studied by an isotope dilution approach. *Marine Ecology Progress Series* 37: 45–52.

Hashimoto, S., Fujiwara, K. and K. Fuwa. 1985. Distribution and characteristics of carboxypeptidase activity in pond, river and seawater in the vicinity of Tokyo. *Limnology and Oceanography* 30: 631–645.

Hobbie, J.E. and R.T. Wright. 1965. Competition between planktonic bacteria and algae for organic solutes. *Memorie Istituto Italiano di Idrobiologia, Supplement* 18: 175–185.

Hollibaugh, J.T. and F. Azam. 1983. Microbial degradation of dissolved proteins in sea water. *Limnology and Oceanography* 28: 1104–1116.

Holzer, H. 1980. Control of proteolysis. *Annual Reviews in Microbiology* 49: 63–91.

Hoppe, H.G. (1983): Significance of exoenzymatic activities in the ecology of brackish water: measurements by means of methylumbelliferyl-substrates. *Marine Ecology Progress Series* 11: 299–308.

Jassby, A.D. and C.R. Goldman. 1974. Loss rates from a lake phytoplankton community. *Limnology and Oceanography* 19: 618–627.

Jørgensen, B.B. 1978. A comparison of methods for the quantification of bacterial sulfate reduction in coastal marine sediments. II. Calculation of mathematical models. *Geomicrobiology Journal* 1: 29–47.

Lancelot, C. and G. Billen. 1985. Carbon-nitrogen relationships in nutrient metabolism of coastal marine ecosystems. *Advances in Aquatic Microbiology* 3: 2621.

Lancelot, C. and G. Billen. 1989. Ecophysiology of phyto- and bacterioplankton growth in the Southern Ocean. pp. 1–105 in Caschetto, S. (editor), *Plankton Ecology*, vol. 1. Science Policy Office, Brussels.

Lancelot, C. and S. Mathot. 1988. Dynamics of a *Phaeocystis*-dominated spring bloom in Belgian coastal waters. I. Phytoplanktonic activities and related parameters. *Marine Ecology Progress Series* 37: 239–248.

Lancelot, C., Mathot, S. and N.J.P. Owens. 1983. Modelling protein synthesis, a step to an accurate estimate of net primary production: The case of *Phaeocystis pouchetii* colonies in Belgian coastal waters. *Marine Ecology Progress Series* 32: 193–202.

Larsson, U. and A. Hagstrom. 1979. Phytoplankton exudate release as energy source for the growth of pelagic bacteria. *Marine Biology* 52: 199–206.

Lehman, J.T., Botkin, D.B. and G.E. Likens. 1975. The assumptions and rationales of a computer model of phytoplankton population dynamics. *Limnology and Oceanography* 20: 343–364.

Lichtfield, C.D. and J.M. Prescott. 1970. Regulation of proteolytic enzyme production by *Aeromonas proteolytica*. I. Extracellular endopeptidases. *Canadian Journal of Microbiology* 16: 17–22.

Long, S., Mothibeli, F.T., Robb, A. and D.R. Woods. 1981. Regulation of extracellular alkaline protease activity by histidine in a collagenolytic *Vibrio alginolyticus* strain. *Journal of General Microbiology* 127: 193–199.

McDonald, I.J. and A.K. Chambers. 1966. Regulation of protease formation in a species of *Micrococcus*. *Canadian Journal of Microbiology* 12: 1175–1185.

Münster, U. and R.J. Chróst. 1990. Origin, composition and microbial utilization of dissolved organic matter. pp. 8–46 in J. Overbeck and R.J. Chróst (editors), *Aquatic Microbial Ecology: Biochemical and Molecular Approaches*. Springer-Verlag, New York.

Parsons, T.R. and J.D.H. Strickland. 1962. On the production of particulate organic carbon by heterotrophic processes in sea water. *Deep-Sea Research* 8: 211–222.

Priest, F.G. 1977. Extracellular enzyme synthesis in the genus *Bacillus*. *Bacteriological Review* 41: 711–753.

Servais, P. 1986. Etude de la dégradation de matiére organique par les bactéries

hétérotrophes en riviére. Developpement d'une démarche méthodologique et application á la Meuse belge. Université Libre de Bruxelles, PhD. Thesis. 271 pp.
Servais, P., Billen, G. and J. Vives-Rego. 1985. Rate of bacterial mortality in aquatic environments. *Applied and Environmental Microbiology* 49: 1448–1454.
Simon, M. and F. Azam. 1989. Protein content and protein synthesis rates of planktonic marine bacteria. *Marine Ecology Progress Series* 51: 201–213.
Somville, M. and G. Billen. 1983. A method for determining exoproteolytic activity in natural waters. *Limnology and Oceanography* 28: 190–193.
Streeter, H.W. and E.B. Phelps. 1925. Study of the pollution and natural purification of the Ohio River. III. Factors concerned in the phenomena of oxidation and reaeration. *Bulletin of United States Public Health Services* No 146.
Westrich, J.T. and R.A. Berner. 1984. The role of sedimentary organic matter in bacterial sulfate reduction. The G model tested. *Limnology and Oceanography* 29: 236–249.
Wolter, K. 1982. Bacterial incorporation of organic substances released by natural phytoplankton populations. *Marine Ecology Progress Series* 7: 287–295.
Wright, R.T. and J.E. Hobbie. 1966. Use of glucose and acetate by bacteria and algae in aquatic ecosystems. *Ecology* 47: 447–453.

8

Peptidase Activity in River Biofilms by Product Analysis

Susan E. Jones and Maurice A. Lock

8.1 Introduction

There have been several reports of proteolysis resulting from the activities of pelagic marine microorganisms (Hollibaugh and Azam, 1983; Hoppe, 1983; Rosso and Azam, 1987; Hoppe et al., 1988). Exoproteolytic activities of intact microbial biofilms, other than in marine (Meyer-Reil, 1987) and freshwater sediments (Jones, 1979), have received less attention, despite the suggestion that attached bacteria have disproportionately higher protein-degrading activities than free-living bacteria (Hollibaugh and Azam, 1983). Substrate analogs have been used to investigate proteolytic activities in river sediments (Sinsabaugh and Linkins, 1988) and biofilms (Jones and Lock, 1989), but methods that rely on the addition of high concentrations of substrate identify potential, rather than actual, activities.

A parallel approach is to monitor putative enzyme activity through the appearance of enzymatic degradation products from naturally occurring substrates or added ones. The work described here incorporates the findings from traditional enzyme assays using substrate analogues and the direct determination of proteolytic activity by amino acid release monitored by high pressure liquid chromatography (HPLC).

8.2 Methods and Experimental Design

The site used in this study was the River Clywedog, a fourth-order mildly eutrophic river in North Wales, U.K. (Map Ref. OS SH 678 711).

Principle of the method Natural river biofilms were grown on 0.2-μm-pore-size polycarbonate membranes (Nuclepore). These were sandwiched between two chambers, thus forming the common wall of a double sided perfusion chamber. On the biofilm side of the membrane, the perfusion medium (river water or river water + casein) was pumped over the biofilm and discharged to waste. The other chamber, with the uncolonized membrane side, was perfused with organic-free (Millipore-Q water purification system) water in a small-volume, closed-loop circuit, the product loop (PL). A product loop was incorporated in the experimental design to allow for the retention, but spatial separation from the perfusing medium, of organic materials released from the biofilm. The presence of putative enzymatic degradation products were monitored in the product loop, and at the outlets of the perfusion chamber.

Colonization Membranes (47 mm, 0.2 μm; Nuclepore) were supported in a holder made from a 50-mm-wide plastic magnetic strip (Figure 8.1). The base magnetic strip was attached to an acrylic plastic plate. The membranes were then sandwiched between the base magnetic strip and a second magnetic strip placed on top. The second magnetic strip, from which a series of 37-mm holes had been cut, allowed the membranes to be held in place at their periphery. This arrangement allowed colonization of the membranes to take place in a largely unimpeded flow-regime.

Colonization of the membranes in the river was accomplished by placing the acrylic plastic plates (with magnetic membrane holder) into a 1.5 m long opaque pipe attached to the river bed. In this way primarily heterotrophic biofilms could be produced. The membranes were left to colonize for a minimum of 6 weeks and were returned to the laboratory in individual plastic petri dishes totally filled with river water.

Perfusion cell The perfusion cell, as shown in Figure 8.2 and Figure 8.3, consisted of two perspex holders, from which 37-mm-diameter central well (3 mm deep) had been cut. These holders were clamped together such that a 47

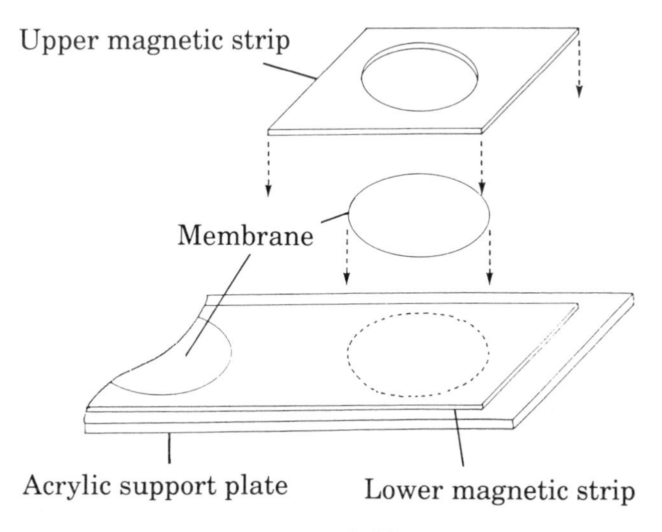

Figure 8.1 Diagram of a colonization support holder

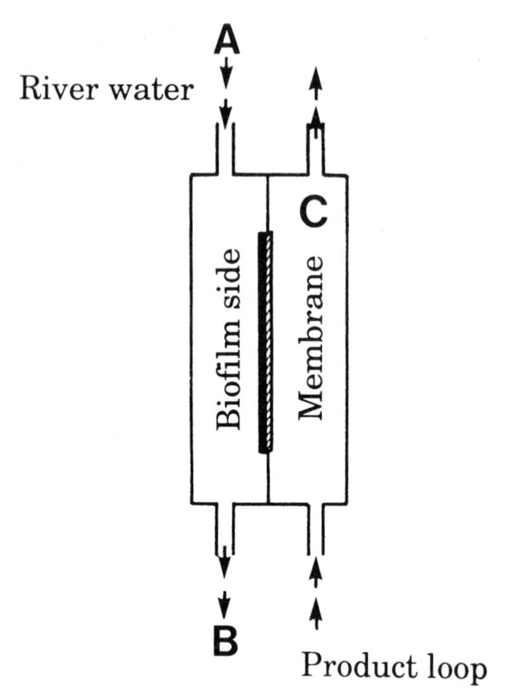

Figure 8.2 Diagram of a perfusion cell. A, B, and C denote sampling locations. A = perfusion inlet, B = perfusion outlet, C = product loop.

mm membrane placed between them formed two chambers. Each chamber had two inlets placed at right angles to one another to produce a circular, rather than laminar, flow path. An aerated perfusion medium (river water or river water + casein) was pumped at 1 ml min^{-1} across the chamber containing the biofilm colonized surface. The other side of the chamber, with the uncolonized membrane surface, was perfused with sterile, organic-free, deionized water (Millipore-Q) in a closed recirculation loop (7-ml volume), known as the product loop (PL). The product loop chambers were filled using a peristaltic pump and circulated across the membranes in the opposite direction to the perfusing medium at a flow rate of 2 ml min^{-1}. The perfusion cells were mounted onto an acrylic holder which could accommodate up to eight similar cells. Prior to use, the total perfusion assembly was sanitized by pumping through with H_2O_2 (0.2% v/v) for 12 h in the dark, followed by flushing with organic-free distilled water for 6 h.

Experiment 1 Colonized membranes were gently rinsed with river water to dislodge particulates. Six live biofilms and two fixed controls (2% v/v formaldehyde for 1 h) were placed in eight perfusion cells and the colonized side of the membranes perfused with river water (or river water + 0.2% formaldehyde for the controls) collected at the time of sampling. To reduce the risk of bacterial

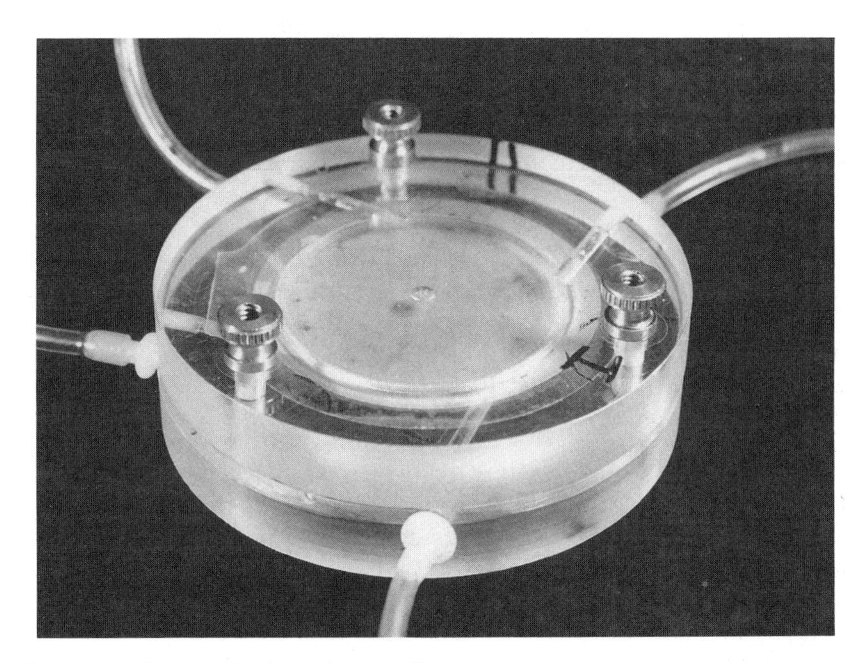

Figure 8.3 Photograph of a perfusion cell.

overgrowth in the river water reservoirs, the holding containers were autoclaved prior to use, and reservoirs and containers were changed every 6 h. The whole apparatus was maintained at the in situ temperature of the river at the time of sampling (10°C).

After 24 h, the product loop and perfusion medium lines ("in" and "waste") were sampled for amino acid analyses at the locations indicated in Figure 8.2. The perfusing medium was then changed to river water + 30 mg l^{-1} (final conc) casein enrichment (+ 0.2% formaldehyde for the killed controls). The product loop and perfusion media lines were sampled for amino acid analyses after 24 h and 48 h of perfusion with casein-enriched river water. In order to remove contaminating low-molecular-weight peptides and amino acids, the casein had been previously precipitated with 80% $(NH_4)_2SO_4$, followed by extensive dialysis against organic-free water. The amino acid concentrations from the live systems were corrected for abiotic amino acid genesis by subtraction of the equivalent amino acid concentrations of the fixed controls.

Experiment 2 This experiment was designed to test whether increases in amino acid concentrations in the product loop were related to concomitant increases in biofilm peptidase activity, assayed using fluorogenic substrates. Colonized Nuclepore filter membranes were removed from the River Clywedog as in Experiment 1. Prior to the perfusion procedure, the membranes were tested for peptidase activity (see below). Four colonized membranes (2 live, 2 formaldehyde-killed) were placed in the perfusion assembly and perfused for 72 h with either river water (+ 0.2% formaldehyde for killed controls) or river water + 30 mg l^{-1} casein (+ 0.2% formaldehyde for controls). The perfusion assembly was set up as in Experiment 1, except that each product loop was separated from the biofilm by a layer of polyethylene placed directly underneath the colonized side of the membrane to minimize contamination during the initial 72 h perfusion period. After 72 h of perfusion, the polyethylene separators were removed, and the product loop filled with organic-free water, and perfusion continued for another 24 h. The assembly was maintained at the in situ river temperature of 8°C. After a total of 96 h of perfusion, the product loop and perfusion inlet and outlets (locations A, B, and C, respectively in Figure 8.2) were sampled for amino acid analyses (HPLC) and total protein concentration (Coomassie Blue protein reagent, Biorad). The membranes were removed for peptidase activity analyses.

Peptidase assay Membranes (living and formaldehyde-treated) were cut in half using a sterile scalpel and placed into individual sterile petri dishes containing 3.5 ml of 300 μM fluorogenic substrate: methylumbelliferyl guanidinobenzoate (MUFG) or L-leucine-4-methyl-7-coumarinylamide hydrochloride (Leu-MCA), dissolved in 2 ml of 2-ethoxyethanol and made up to 100 ml with sterile distilled water. MUFG was used as a substrate for endopeptidase, and Leu-MCA was used as a substrate for exopeptidase. Previous work had shown that neither substrate was taken up whole by microorganisms (S.E. Jones, personal observation, and personal communication by H.G. Hoppe). The petri dishes were incubated in the dark at 8°C. After 1 h exactly, 1-ml aliquots were removed into darkened test tubes containing 1 ml of appropriate buffer (0.1 M Tris buffer, pH 7.6, for endopeptidase; 0.05 M glycine buffer, pH 10.4, for exopeptidase). Fluorescence was detected using a Perkin-Elmer LS3 Fluorescence Spectrophotometer at 455-nm emission under either 354-nm excitation (exopeptidase) or 365-nm excitation (endopeptidase) against standard ranges of methylumbelliferone (Sigma) or 7-amino-4-methylcoumarin (Aldrich) of 0 to 1 μM. Peptidase activities in the live biofilms were corrected for abiotic substrate degradation by subtraction of the equivalent fluorescence of the fixed controls.

After activity measurements had been made, the surface areas of the membranes were determined by weighing against known surface area-to-weight ratios (weight increase due to microbial biomass was considered negligible in this calculation).

Figure 8.4 (facing page) Amino acid concentrations in the perfusion medium inlet and outlets (Experiment 1). (A) product loop perfused with river water for 24 h; (B) perfusion medium (river water for 24 h); (C) product loop (perfused with river water + casein for 24 h); (D) perfusion medium (river water + casein for 24 h); (E) product loop (perfused with river water + casein for 48 h); (F) perfusion medium (river water + casein for 48 h). The bars denote standard errors (n = 6 for river water perfusion; n = 5 for river water + casein perfusion).

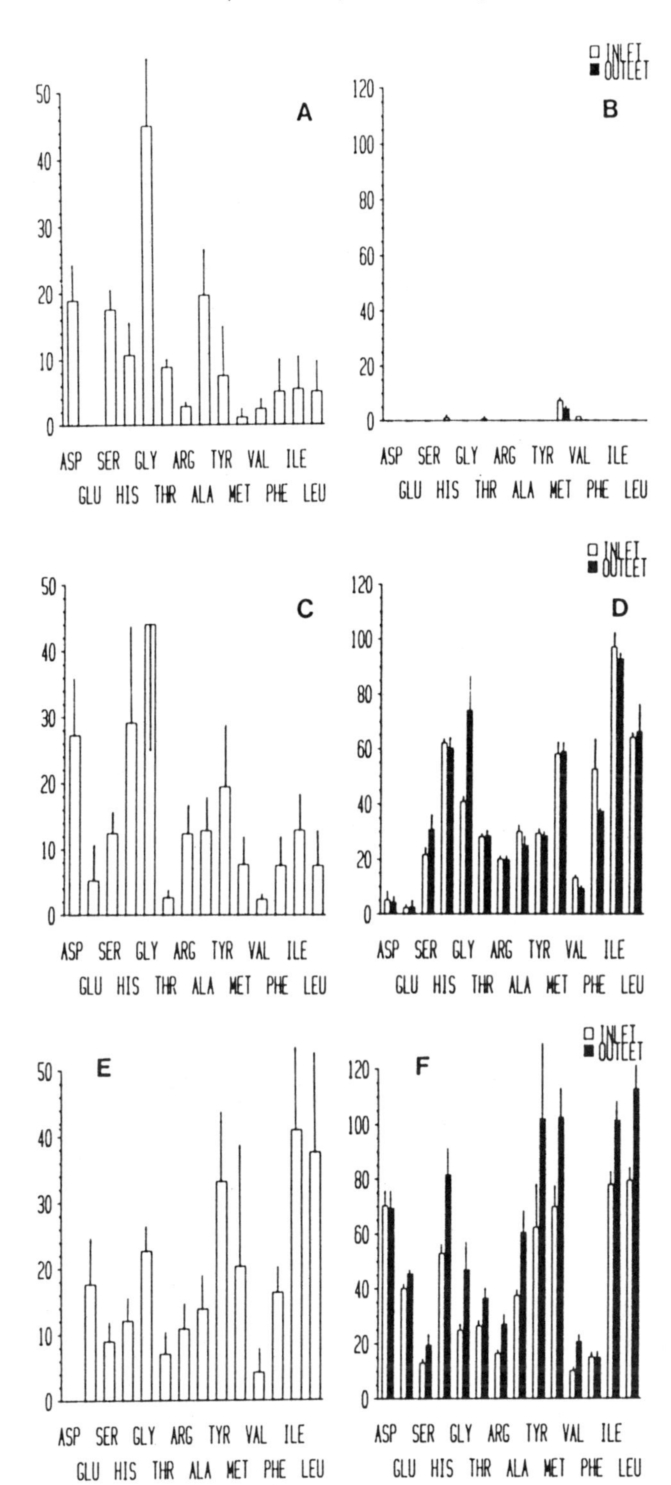
A
B
C
D
E
F
INLET
OUTLET
ASP SER GLY ARG TYR VAL ILE
GLU HIS THR ALA MET PHE LEU
0
10
20
30
40
50
60
80
100
120

Aminoacids (nmol liter^{-1})

High pressure liquid chromatography (HPLC) Samples for amino acid analyses were placed in organic-free glassware, filtered through 0.2-μm filters (Whatman), and held at 4°C until chromatographed (within 6 h). Amino acid analyses were carried out using an automated HPLC system (Fiebig, 1988), consisting of a pump (Phillips PU 4003) equipped with a solvent programmer for generating gradients and a Spark-Holland Promis autosampler with an automatic pre-column derivatization facility. Ortho-phthaldialdehyde (Sigma Chemicals Ltd.) was used as the derivatizing reagent. Amino acids were separated on a Perkin-Elmer Pecosphere 3 x 3 CR C18 column (50 mm length, 4.6 mm ID) in a sodium acetate-tetrahydrofuran/acetonitrile gradient and detected with a Perkin-Elmer LS-1 fluorometer.

Nonparametric ANOVA (Kruskal-Wallis) and two-sample tests (Mann-Whitney) were carried out using a Statgraphics Statistical package.

8.3 Results

Results of Experiment 1 are summarized in Figure 8.4 and Table 8.1. The amino acid spectrum of unenriched river water (sampling location A) was significantly different ($P < 0.5$) from that of the product loop (PL) (Figure 8.4A and B). In the product loop, nine amino acids were found that were originally absent from the perfusing medium, and the concentrations of four others present in the river water increased significantly ($P = 0.004$). In contrast, in the perfusate (sampling location B), the amino acids either disappeared (histidine, threonine) or were reduced in concentration (methionine, valine).

Perfusion with casein-enriched river water for 24 h and 48 h increased free amino acid concentrations in the product loop and perfusion medium inlet and outlets (Table 8.1). Significant differences in amino acid concentrations were found between product loop and perfusing medium ($P_{inlet} = 0.004$, $P_{outlet} = 0.002$ for 24 h perfusion with river waters; $P_{inlet} = 0.045$, $P_{outlet} = 0.029$ for 24 h per-

Table 8.1 Total amino acid concentrations in the product loop and perfusion medium inlet and outlets, corrected for abiotic generation of amino acids for Experiment 1

	Total amino acid concentration (nmole l^{-1})[1]		
	Product	Perfusion medium	
Biofilm perfused with	loop	Inlet	Outlet
River water for 24 h	149.6 (53.7)	10.7 (1.8)	4.8 (1.9)
River water + casein for 24 h	203.5 (49.1)	524.9 (21.7)	540.9 (19.3)
River water + casein for 48 h	245.4 (75.9)	540.9 (28.6)	841.8 (83.7)

[1]Mean values, n = 6 for river water, n = 5 for enriched river water; numbers in parenthesis are standard errors

fusion with river water + casein). No significant differences were found at the 5% level between product loop and perfusion medium at 48 h enriched perfusion. With extended protein enrichment (48 h), the proportion of hydrophobic amino acids to total amino acids doubled in the product loop from 25% to 54%.

Comparison of amino acids in the perfusion medium in Experiment 1 before and after passage across the biofilm showed no significant differences ($P < 0.05$) in concentrations after 24 h perfusion with river water + casein, with the exception of glycine ($P = 0.02$) and valine ($P = 0.04$). However, further enrichment produced significant increases ($P < 0.05$) between inlet and outlet in 10 out of the 14 amino acid concentrations (Figure 8.4F).

Results of Experiment 2 are shown in Table 8.2 and Table 8.3. Protein was present in the perfusion media (Table 8.2) but was below the level of detection in all four product loops. As in Experiment 1, elevated amino acid concentrations were found in perfusion inlets (Table 8.2) and product loops (Table 8.3) when river water was supplemented with casein, compared with the concentrations found in unamended river water. A 3.2-fold increase in amino acid

Table 8.2 Total amino acid and protein concentrations in perfusion cell inlet and outlets for Experiment 2

Biofilm perfused with	Inlet concentration		Outlet concentration	
	Amino acids[1]	Protein[2]	Amino acids[1]	Protein[2]
River water only	35.5	2.25	179.1	3.25
River water + casein	668.9	18.00	92.7	15.00

[1]Amino acid concentration is given in nmole l^{-1}. They were detected by HPLC; 11 amino acids analyzed.
[2]Protein concentration is given in mg l^{-1}. It was measured by Coomassie Blue Biorad assay.

Table 8.3 Increases in amino acid generation in the product loop and stimulation of peptidase activity as a consequence of elevated protein supply (Experiment 2; 11 amino acids were analyzed; n = 2; numbers in parenthesis are standard errors)

	Biofilm perfused with:	
	River water	Protein-enriched river water
Total amino acids in the product loop (nmole l^{-1})	127.7 (14.2)	393.1 (8.3)
Exopetidase activity (pmoles h^{-1} cm^{-2})	139.7 (3.6)	496.3 (33.2)
Endopeptidase activity (pmoles h^{-1} cm^{-2})	19.7 (1.9)	0

output to the product loop was associated with 3.9-fold increase in exopeptidase activity in the biofilm (Table 8.3). Unlike results from Experiment 1, total amino acid concentrations decreased in the perfusion outlet following enrichment with casein.

8.4 Discussion

Amino acids present in the casein amended perfusing medium (sampled at the inlet; Table 8.1) was probably due to microbial activity on the added casein, and not to amino acids introduced in the protein enrichment. The perfusion assembly was not maintained aseptically. Hence planktonic and sessile bacteria capable of degrading added protein were not excluded from the experiments, which resulted in elevated levels of amino acids in the inlet of amended river water perfusions. Evidence for a microbial role was indicated by lower amino acid levels in formaldehyde-treated perfusion media, compared to untreated perfusion media (208 and 200 nmole l^{-1}, compared to 733 and 795 nmole l^{-1} for 24- and 48-h perfusion inlets, respectively, Experiment 1). Results from Experiment 2 showed the same effect.

Natural river biofilms were shown to produce changes in amino acid composition and concentration (sample point B) in river water circulating over the surface of the biofilm. The disappearance of amino acids from river perfusate supports the work of Fiebig (1988) who demonstrated the ability of river biofilms to immobilize free amino acids. However, passage over the surface of the biofilms did not always reduce amino acid concentrations: casein enrichment in Experiment 1 (Table 8.1) and perfusion with river water only in Experiment 2 (Table 8.2) produced higher outlet than inlet concentrations. In Experiment 2, changes in amino acid concentrations were reflected by similar changes in protein concentrations. The conflicting results of biofilm effect on perfusing medium could be due to differences in perfusion times between these two experiments or the extent of surface ectoenzyme activity.

If amino acids appearing in the product loop and downstream perfusate are assumed to be derived, at least partially, from peptidase activity within the biofilm, elevated amino acid concentrations should be detected at these sampling locations. Results of both experiments generally confirmed this hypothesis. Higher amino acid concentrations in the product loop of river water only and protein-enriched perfusions were found in both experiments, compared with inlet concentrations. Elevated protein supply produced profound changes in biofilm functioning in both experiments. Significant increases in individual amino acid concentrations between inlet and outlet (Figure 8.4F) suggest possible stimulation of peptidase activity within the biofilm. Further evidence of enzyme participation was suggested in the second experiment by the observed stimulation of exopeptidase activities in biofilms perfused with protein-enriched river water (Table 8.3). However, perfusion with protein-enriched water inhibited endopeptidase activity (Table 8.3), perhaps due to feedback inhibition by products of stimulated exoproteolysis. Stimulation (as shown by increased amino acid concentrations) increased with enrichment contact time: in Experiment 1, 48-h perfusion with added casein showed that the total amino acid concentration was 20% higher in the product loop and 64% higher in the perfusate, compared with 24 h enrichment concentrations. An increased proportion of hydrophobic amino acids to total amino acids was noted in the product loop but not in the perfusate after 48 h enrichment. This phenomenon may have been due to lower solubilization of hydrophobic amino acids in the hydrophilic polysaccharide matrix, resulting in an apparent rise in that fraction of amino acids in the product loop.

The appearance of amino acids observed in the perfusate indicate (1) putative enzyme activity at the surface of the biofilm or (2) diffusion of amino acids (enzymatically derived?) out from the biofilm into the perfusion medium. Differences in amino acid spectra were also noted between the product loop and perfusate (Figure 8.4), indicating differences in composition of enzymes present at the surface and deeper within the biofilm or differential sorption/desorption processes.

The assumption has so far been made that the appearance of amino acids in the product loop and perfusate is a consequence of enzymatic degradation of proteins. Three other possible explanations for the amino acid increase are: cell lysis, diffusion of amino acids from the perfusing medium into the product loop, or amino acid exodus from cells (Payne, 1980). The first possibility was discounted since during the course of perfusion experiments, no increase of protein-type lysis indicators occurred in the product loop, as detected by Coomassie Blue stain for the proteins. If extensive lysis had occurred in the biofilm, protein, as well as amino acid, concentrations would have risen. The increase in amino acid concentrations in the product loop of river water + casein perfused cells was not considered to be due to ingress of protein from the perfusion medium since (1) protein was not detected in the product loop in Experiment 2, (2) correction for "abiotic" amino acid flow into the product loop was provided by subtraction of killed controls (assuming that amino acid passage across fixed biofilms is equivalent to living biofilms), and (3) the amino acid spectra of product loop and perfusing media were usually dissimilar. The possibility of amino acid exodus was tested indirectly in Experiment 2 by examining the response of the biofilm to elevated levels of protein while simultaneously measuring biofilm proteolytic activity. Parallel increases in amino acid concentrations and high peptidase activities strongly support the hypothesis that the appearance of amino acids in the product loop and perfusate was a consequence of enzymatic activity.

If the appearance of amino acids in the product loop and perfusate is indicative of biofilm enzyme activity, then it is likely that this represents a net estimate of the phenomenon. It seems likely that some portion of the presumptive enzymatically generated amino acids would be retained within the biofilm due to (1) cellular uptake and (2) immobilization within the biofilm matrix by adsorption or ion exchange. Future work should be directed to the apportionment of products between cells, biofilm and diffusional losses using radiolabelled substrates.

References

Fiebig, D.M. 1988. Riparian zone and stream water chemistries, and organic matter immobilization at the stream-bed interface. *Ph.D. Thesis, University of Wales.*

Hollibaugh, J.T. and F. Azam. 1983. Microbial degradation of dissolved proteins in seawater. *Limnology and Oceanography* 28: 1104–1116.

Hoppe, H.G. 1983. Significance of exoenzymatic activities in the ecology of brackish water: measurements by means of methylumbelliferyl-substrates. *Marine Ecology Progress Series* 11: 299–308.

Hoppe, H.G., Kim, S.J. and K. Gocke. 1988. Microbial decomposition in aquatic environments: combined process of extracellular enzyme activity and substrate uptake. *Applied and Environmental Microbiology* 54: 784–790.

Jones, J.G. 1979. Microbial activity in lake sediments with particular reference to electrode potential gradients. *Journal of General Microbiology* 115: 19–26.

Jones, S.E. and M.A. Lock. 1989. Hydrolytic extracellular enzyme activity in heterotrophic biofilms from 2 contrasting rivers. *Freshwater Biology* 22: 289–296.

Meyer-Reil, L.A. 1987. Seasonal and spatial distribution of extracellular enzymatic activities and microbial incorporation of dissolved organic substrates in marine sediments. *Applied and Environmental Microbiology* 53: 1748–1755.

Payne, J.W. 1980. Energetics of peptide transport in bacteria. pp. 359–377 in Payne, J.W. (editor), *Microorganisms and Nitrogen Sources*. John Wiley and Sons, Chichester.

Rosso, A.L. and F. Azam. 1987. Proteolytic activity in coastal oceanic waters: depth distribution and relationship to bacterial populations. *Marine Ecology Progress Series* 41: 231–240.

Sinsabaugh, R.L. and A.E. Linkins. 1988. Exoenzyme activity associated with lotic epilithon. *Freshwater Biology* 20: 249–261.

9

Aminopeptidase Activity in Lakes of Differing Eutrophication

Timothy R. Jacobsen and Hakumat Rai

9.1 Introduction

The important role of bacterioplankton in food webs suggested by Pomeroy (1974) has been partially quantified within the last decade. Recently, measurements of bacterial production have indicated that aquatic bacteria can convert a significant fraction of primary production into bacterial biomass (Fuhrman and Azam, 1980; 1982). The exact methodology needed to determine bacteria production in aquatic ecosystems may still be in question, but the overwhelming importance of bacterioplankton in microbial food webs cannot be denied.

We have made considerable progress in determining the dynamics of bacterial secondary production in aquatic food webs but we still have not delineated all pathways of organic matter to bacteria. Defining the role of bacteria in the transformations of organic matter in aquatic ecosystems is a difficult task for several reasons. First, bacteria use dissolved organic material, ranging from simple monomers to polymers and particles (Azam and Fuhrman, 1984). The utilization of a wide variety of carbon substrates makes understanding the total flux of organic material to bacteria a difficult task. Furthermore, the chemical forms of dissolved organic matter (DOM) in water are very diverse and only a small fraction of the total DOM is readily utilizable by aquatic bacteria (Williams, 1986).

Initially, aquatic ecologists studied bacterial utilization of small organic monomers such as amino acids. Free amino acids comprise less than 10% of the dissolved organic nitrogen pool available to aquatic bacteria. Their results along with others strongly indicate that the nitrogen demands of aquatic bacteria are not wholly met by dissolved free amino acids. Thus, bacterial utilization of polymeric organic nitrogen (proteins, polypeptides, and other forms) is important in supporting the growth of aquatic bacteria.

Defining the importance of polymer flux to bacteria will provide new insights in understanding the role of bacteria in microbial food webs. Hollibaugh and Azam (1983) demonstrated that marine bacteria rapidly use polymeric amino acids. They found that iodinated hemoglobin added to seawater was rapidly degraded and the amino acids produced were utilized by the bacterial size fraction. Some extracellular protease activity was found, but the bulk of the proteolytic activity occurred in the bacterial-size fraction. Recently, Somville and Billen (1983) presented a technique for the measurement of protease activity (aminopeptidases and some proteases) in seawater. Their method followed the formation of a fluorescent product from an aminopeptide analog. Hashimoto et al. (1985) described a method for the measurement of carboxypeptidases in seawater. Similar methods using different substrates have been used in marine ecosystems (Hoppe, 1983) and sediments (Meyer-Reil, 1981).

All the above methods require incubations of several hours to determine the rate of activity. We have developed a simple kinetic assay for the measurement of aminopeptidase in lake water that requires only 3 to 5 min (Jacobsen and Rai, 1988). We have used this method to determine aminopeptidase activity in lakes of differing eutrophication. The objective of this paper is to define some of the dynamics of the transformation of polymeric nitrogen to amino acids by microbial assemblages in lakes of differing eutrophication and to highlight the importance of microbial aminopeptidase (Amp) for the transformation of polymeric nitrogen compounds in microbial food webs.

9.2 Methods

Measurements of aminopeptidase activity Aminopeptidase (Amp) activity in size-fractionated water samples was measured by a kinetic assay method (Jacobsen and Rai, 1988). Sampling was conducted as part of a routine sampling program for northern German Baltic lakes conducted by H. Rai. Water samples were collected at 1-m intervals throughout the water column from six lakes. Samples were quickly returned to the laboratory, size fractionated (at 35, 10, 3, 1 and 0.2 μm), and held at in situ temperature until analysis.

Total Amp activity (35-μm filtrate) and dissolved aminopeptidase activity (0.2-μm filtrate) were determined at each depth. Substrate was added to a cuvette followed by 3 ml of lake water sample. Final L-leucyl-β-naphthylamide substrate concentration was 1 mM. The production of β-naphthylamine was followed with the Kontron spectrofluorometer (SFM-25). All assays were run at in situ temperature, and triplicate determinations were made on each sample. The incubation of samples was followed until a linear rate of β-naphthylamine production was obtained.

Amp activity was calculated in nmoles l^{-1} of β-naphthylamine per unit time. This is equivalent to nmoles of amino acid bonds cleaved by enzymatic activity per liter per hour. Aminopeptidase carbon and nitrogen flux was calculated from the hydrolysis rate normalized to the average mole percentage of carbon and nitrogen found in the 20 common amino acids.

Determination of other biological parameters Other biological parameters were determined in parallel with the aminopeptidase determinations. Chlorophyll$_a$ was measured by the method of Rai (1973). Bacterial cell numbers were determined by acridine orange direct counts (Hobbie et al., 1977). Primary production and bacterial uptake of glucose were determined by radiolabelled methods outlined by Rai (1984).

9.3 Results and Discussion

Evaluation of the aminopeptidase assay The successful use of fluorometric assays in lake water requires that close attention be paid to the details of analysis. For example, pH and ionic effects can drastically affect assay conditions (Guilbault, 1973). Fluorescent assays used to examine aminopeptidase activity in lake water have similar problems of pH optimum (Jacobsen and Rai, 1988). Additional errors arise from interferences from other organic compounds, such as humic and fulvic acids (Münster et al., 1989).

There are several biological and biochemical constraints that also need to be addressed when fluorometric assays are used to determine biological processes in lake water. One important question is whether the fluorescent substrate to be used is a good analog for the compound or compounds to be measured. A particularly important ecological question is whether the enzymatic assay employed is a measure of the ecological process (e.g., protein degradation) or only a measure of a particular class of enzymatic activity (e.g., aminopeptidases, carboxypeptidases).

Jacobsen and Rai (1988) addressed the question of whether the fluorescent assay procedure developed was a quantitative measure of protein degradation in lakewater. It is not surprising that the aminopeptidase assay underestimated the degradation of combined amino acids (e.g., proteins) in lake water, since the technique only measures aminopeptidase and some protease activity. Carboxypeptidase activity will not be detected by this technique. Another reason for the difference between the two techniques may be that L-leucyl-naphthylamide was not a suitable substrate analog for the combined amino acids found in lakewater. Thus, the technique we employed does not provide a completely quantitative picture of protein degradation in aquatic ecosystems, but it appears to provide a useful tool for examination of aminopeptidase activity in lake water. Critical evaluation of the shortcomings in current methodology will provide a solid basis for our further understanding of how to quantify the flux of organic matter in aquatic ecosystems.

Location of aminopeptidase activity The kinetic assay procedure that we used does allow the measurement of pmoles of product in the presence of mmoles of substrate and other components that can fluoresce at the excitation and emission wavelengths used. Figure 9.1 shows the production of β-naphthylamine with time for five different size fractions. The large difference in activity be-

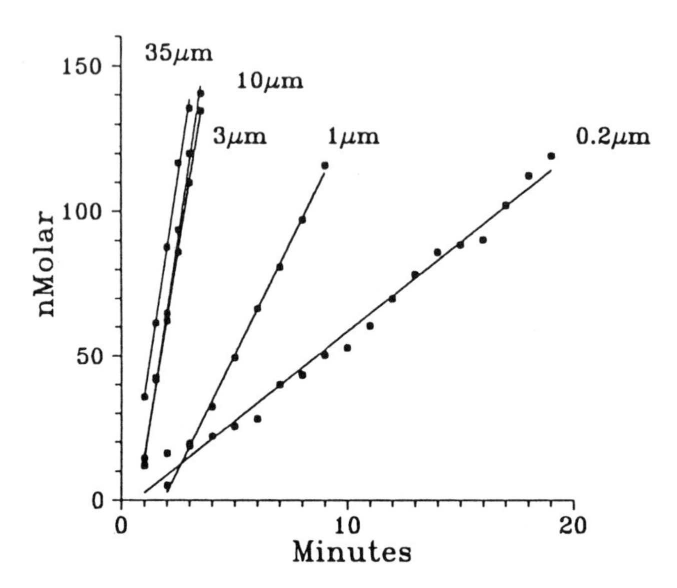

Figure 9.1 Production of β-naphthylamine with time in five size fractions of lake water samples from the Großer Binnensee. Samples were taken on 5 July 1989 at 1-m depth. Lines are the linear fit of the data by least squares regression analysis. Correlation coefficients (r^2) range from 0.999 to 0.987.

tween the 3-μm and the 1-μm filter results from a significant number of large free-living bacteria being retained on the 1-μm filter. Only a small fraction of the total bacteria were attached to particles in this sample. The Großer Binnensee is the most eutrophic lake we examined. Total aminopeptidase activity reached 5,000 nmole l^{-1} h^{-1}. Extracellular aminopeptidase activity in this sample was <10% of total activity. We found similar linear responses of β-naphthylamine production in less productive lakes (Jacobsen and Rai, 1988).

Our initial results indicated that the extracellular, i.e., dissolved in the water, aminopeptidase activity accounted for 10 to 30 % of the total activity measured (Jacobsen and Rai, 1988). We employed gentle gravity and pressure filtration with our samples to minimize cell breakage. The levels of dissolved aminopeptidase we found are comparable to the levels found by Somville and Billen (1983) for Belgian coastal waters and Rosso and Azam (1987) for the southern California Bight. Hashimoto et al. (1985) found similar levels of dissolved carboxypeptidase activity in the coastal waters of Japan.

Our recent studies of microbial proteolytic activity have yielded significantly higher dissolved aminopeptidase activity than we found in our initial studies. During a study of aminopeptidase activity after the fall overturn in the Schöhsee, we observed that the total aminopeptidase activity passed through a 0.2-μm filter (Table 9.1). The rate of aminopeptidase activity in this study was found to be correlated with temperature (r^2 = 0.72, df= 8). In addition, samples analyzed during 1989 also show high levels of dissolved aminopeptidase activ-

Table 9.1 Seasonal changes and fractionation of aminopeptidase activity in Lake Schöhsee at 2-m depth

		Aminopeptidase Activity		
Date (1975)	Temperature (°C)	Total (nmol l^{-1} h^{-1})	<1-μm Filtrate (%)[a]	<0.2-μm (%)[a]
27 September	13.1	168	90	92
9 October	13.4	106	120*	43
16 October	13.6	134	69	43
23 October	12.6	85	82	66
30 October	11.4	70	106*	74
6 November	9.1	60	100	82
13 November	8.0	70	70	70
27 November	5.2	49	71	71
4 December	4.1	25	84	180*
11 December	4.8	35	71	111*

[a]Calculated as % of the activity found in the 35 μm filtrate (total activity).
*Values in excess of 100% may represent alterations in community structure, cell breakage or liberation of enzymes into the system.

ity. The reason for the high dissolved extracellular aminopeptidase activity in lake water is not known. Perhaps stress or changes in the microbial assemblage responsible for the enzymatic activity combined with the prefiltration could account for the observed results. The reason for the high amounts of dissolved aminopeptidase activity requires further investigation.

Aminopeptidase activity was largely confined to the epilimnion in lakes like Plußsee (Figure 9.2). Maximum aminopeptidase activity was found at the surface. A second subsurface maximum of aminopeptidase activity that was associated with increased bacterial numbers was found at the metalimnion. The rates that we measured on 10 July 1989 were comparable with those found by Chróst et al. (1989) for the spring bloom period in Plußsee. Shallow well-mixed northern German lakes such as the Einfeldersee had rates of aminopeptidase activity of more than 1,000 nmole l^{-1} h^{-1} (Figure 9.3). Although the lake was not stratified, there was a 50% decrease in aminopeptidase activity with depth. This decrease in aminopeptidase activity corresponded to a 30% decrease in bacteria abundance.

We examined aminopeptidase activity in six lakes of differing eutrophic states. Table 9.2 presents a summary of biomass parameters (chlorophyll$_a$ and bacterial numbers) primary production, glucose uptake and aminopeptidase activity found during our sampling. The 29-fold variation in aminopeptidase activity in these lakes was roughly equal to the variation in chlorophyll$_a$ concentrations measured. However, aminopeptidase activity was not consistent with bacterial numbers. This would suggest, as pointed out by Rosso and Azam (1987), that the microbial population is capable of induction of aminopeptidase activity.

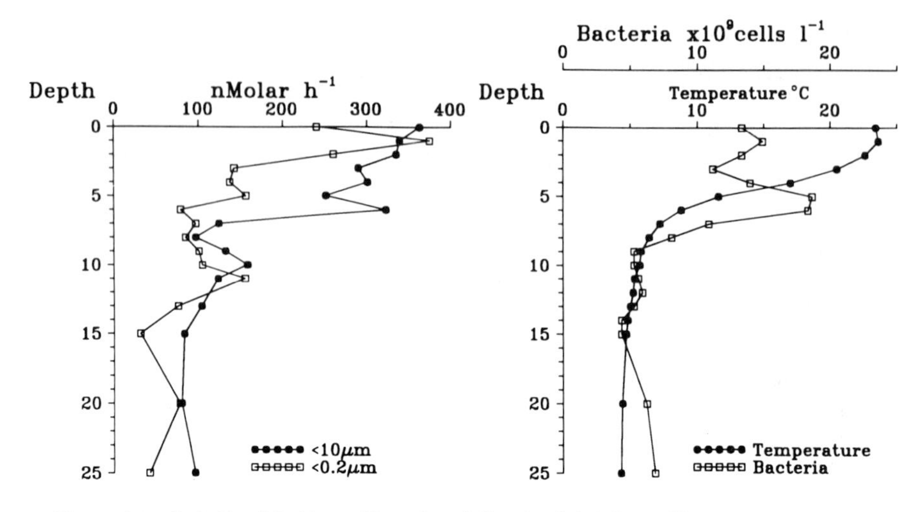

Figure 9.2 Left: Total (<10-μm filtrate) and dissolved (<0.2-μm filtrate) aminopeptidase activity with depth in lake Plußsee. Right: Temperature and bacterial abundance with depth.

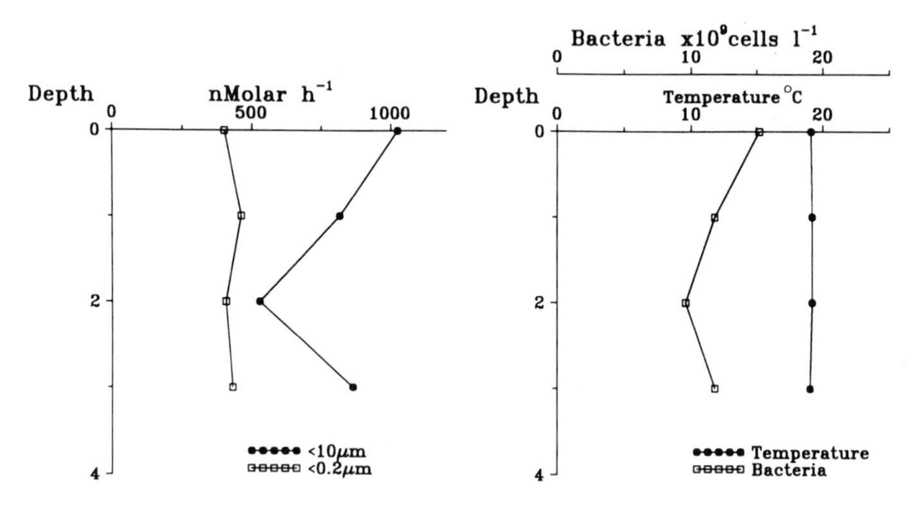

Figure 9.3 Left: Total and dissolved aminopeptidase activity with depth in Lake Einfeldersee. Right: Temperature and bacterial abundance with depth. Samples were taken on 13 July 1989.

Importance of aminopeptidase for hydrolysis of polymeric nitrogen compounds The importance of dissolved polymeric nitrogen compounds (e.g., peptides and proteins) in the carbon (energy) dynamics of a lake ecosystem can be suggested by calculation of the aminopeptidase activity on the basis of carbon flux (Table 9.3). Although the reader may question the assumptions made

Table 9.2 Characteristics of six lakes of differing eutrophic states[1]

Lake	Sampling date	Chlorophyll$_a$ (μg l^{-1})	Bacteria ($\times 10^6$ ml^{-1})	Primary production (μg C l^{-1} h^{-1})	Glucose uptake (nmol l^{-1} h^{-1})	Amp (nmol l^{-1} h^{-1})
Schöhsee	8.10.84	1.8 (0.5–8.0)	1.2 (0.3–5.1)	2.0 (0.4–5.8)	1.3 (0.06–3.8)	176
Plußsee[2]	10.7.89	18.0 (0.04–13.1)*	13.6 (2.5–17.0)*	— (0.1–183)*	— (0.2–1.2)*	315
Kellersee	25.9.85	— (0.5–30.0)	— (3.0–7.6)	— (2.9–14.7)	— (0.2–2.6)	352
Einfeldersee	13.7.89	35.8 —	12.1 —	— —	— —	1,070
Edebergsee	15.10.84	22.6 (26.6–98.2)	6.0 (4.5–11.2)	3.8 (6.2–64.3)	8.7 (1.3–31.2)	3,080
Großer Binnensee	5.7.89	60.4 —	52.8 —	— —	— —	5,041
	12.7.89	53.3 (1.8–230)	22.1 (4.6–52.8)	— (0.9–997)	— (0.2–7.4)	3,504

[1]The value in parentheses indicates the seasonal range.
[2]The ranges are from data of Chróst et al. (1989).

Table 9.3 Characteristics and carbon (C) and nitrogen (N) flux in two lakes

Depth (m)	Chlorophyll$_a$ (μg l^{-1})	Bacteria ($\times 10^6$ ml^{-1})	Primary production (μg C l^{-1} h^{-1})	Glucose uptake (μg C l^{-1} h^{-1})	Amp activity In μg C l^{-1} h^{-1}	Amp activity In μg N l^{-1} h^{-1}
Schöhsee, 8 August 1984						
0	1.2	2.3	3.1	1.0	12.0	3.9
1	1.5	1.3	2.7	1.4	15.0	4.9
2	2.2	2.0	2.2	1.6	12.0	3.9
3	2.1	1.1	1.8	1.2	12.3	4.0
5	1.5	0.5	1.3	1.5	9.3	3.1
11	2.0	0.2	0.6	0.8	9.7	3.2
16	1.4	0.1	0.3	0.3	0.7	0.2
Edebergsee, 15 August 1984						
0	13.6	9.5	6.2	11.3	333.6	109.2
1	18.4	7.5	4.0	10.4	193.5	63.3
2	20.6	7.1	4.0	8.2	173.5	56.8
3	27.6	8.6	2.9	13.3	193.5	63.3
5	26.8	5.9	2.1	8.1	133.4	43.7
7	25.2	1.5	0.2	7.4	20.0	6.6
8	26.7	1.6	0.2	6.5	6.7	2.2

in these calculations, the conclusion cannot be dismissed. Calculated carbon fluxes from aminopeptidase activity were significantly greater than primary production in both lakes (Schöhsee and Edebergsee). These calculations indicate that the carbon flux of amino-carbon, as measured by our assay, is in excess of primary production and is nearly equal to the standing crop of primary producers or bacteria. One reason for this discrepancy could be that the production of protein (e.g., primary production) and the degradation of protein were separated in time. Chróst et al. (1989) found that aminopeptidase activity increased dramatically during the decay of a spring bloom in Plußsee.

Another explanation for the discrepancy in the carbon flux might arise from changes in what the assay procedure being used is really measuring. Our initial experiments with the kinetic assay indicated that the aminopeptidase assay we were using actually underestimated the flux of polymeric nitrogen compounds to microbial components (Jacobsen and Rai, 1988). It is quite possible that, under some instances, the assay procedure may be measuring the maximum potential (V_{max}) of the microbial community and not the natural rate of protein degradation. Since no labelled natural proteins can be added to the system it is difficult to define what the natural rates of protein degradation are at all times. Molecular probing techniques could possibly be used to determine if we are indeed measuring natural rates of protein degradation in aquatic ecosystems.

The data presented here suggests that the combined amino acid-nitrogen pool is an important source of nitrogen for heterotrophic bacteria in lakes of differing levels of eutrophication. However, considerable work must be done before we can define the mechanisms and understand the role of polymeric nitrogen degradation by bacteria in aquatic ecosystems.

Conclusions Bacteria size particles and soluble extracellular enzymes (<0.2-μm filtrate) comprised the major size fractions of proteolytic activity. Dissolved proteolytic activity comprised roughly 10% of the total activity measured in most samples. However, in some determinations, dissolved activity was considerably higher. Aminopeptidase activity was elevated in the more eutrophic lakes studied. Measured rates of polymeric nitrogen degradation suggest that combined or polymer amino nitrogen is a major pathway for bacterial transformation of nitrogen in the lakes studied. Conversion of the measured aminopeptidase activity to units of carbon flux indicate that total proteolytic activity is greater than measured primary production. These results suggest that the total dissolved combined amino acid pool may be rapidly cycling in the lakes studied.

Acknowledgments We thank Dr. W. Lampert and the Max Planck Association for granting the fellowship to conduct this research at the Max-Planck Institute of Limnologie, Abteilung Ökophysiologie, Plön, Germany. We also thank the New Jersey Agricultural Experiment Station and state funds. This is Publication No. C–32403–1–89 of the New Jersey Agricultural Experiment Station.

References

Azam, F. and J.A. Fuhrman. 1984. Measurement of bacterioplankton growth in the sea and its regulation by environmental conditions. pp. 179–196 in Hobbie, J.E. and Williams, P.J.LeB. (editors), *Heterotrophic Activity in the Sea*. Plenum Press, New York.

Chróst, R.J., Münster, U., Rai, H., Albrecht, D., Witzel, P.K. and J. Overbeck. 1989. Photosynthetic production and exoenzymatic degradation of organic matter in the euphotic zone of a eutrophic lake. *Journal of Plankton Research* 11: 223–242.

Fuhrman, J.A. and F. Azam. 1980. Bacterioplankton secondary production estimates for coastal waters of British Columbia, Antarctica and California. *Applied and Environmental Microbiology* 39: 1085–1095.

Fuhrman, J.A., and F. Azam. 1982. Thymidine incorporation as a measure of heterotrophic bacterioplankton production in marine surface waters: Evaluation of field results. *Marine Biology* 66: 109–120.

Guilbault, G.G. 1973. *Practical Fluorescence: Theory, Methods and Techniques*. Marcel Dekker Inc., New York. 107 pp.

Hashimoto, S., Fujiwara, K., Fuwa, K. and T. Saino. 1985. Distribution and characteristics of carboxypeptidase activity in pond, river and seawaters in the vicinity of Tokyo. *Limnology and Oceanography* 30: 631–645.

Hobbie, J. E., Daley, R.J. and S. Jasper. 1977. Use of Nuclepore filters for counting bacteria by fluorescence microscopy. *Applied and Environmental Microbiology* 33: 1225–1228.

Hollibaugh, J.T. and F. Azam. 1983. Microbial degradation of dissolved protein in seawater. *Limnology and Oceanography* 28: 1104–1111.

Hoppe, H.G. 1983. Significance of exoenzyme activities in the ecology of brackish water: measurements by means of methylumbelliferyl-substrates. *Marine Ecology Progress Series* 11: 299–308.

Jacobsen, T.R. and H. Rai. 1988. Determination of aminopeptidase activity in lakewater by a short term kinetic assay and its application in two lakes of differing eutrophication. *Archiv für Hydrobiologie* 113: 359–370.

Meyer-Riel, L.A. 1981. Enzymatic decomposition of proteins and carbohydrates in marine sediments: methodology and field observations during spring. *Kieler Meeresforschung Sonderheft* 5: 311–317.

Münster, U., Eino, P. and J. Nurminen. 1989. Evaluation of the measurements of extracellular enzyme activities in a polyhumic lake by means of studies with 4-methylumbelliferyl-substrates. *Archiv für Hydrobiologie* 115: 321–337.

Pomeroy, L.R. 1974. The ocean's food web; a changing paradigm. *Bioscience* 24: 499–504.

Rai, H. 1973. Methods involving the determination of photosynthetic pigments using spectrophotometry. *Verhandlungen der Internationalen Vereinigung für Theoretische und Angewandte Limnologie* 18: 1864–1875.

Rai, H. 1984. Magnitude and heterotrophic metabolism of photosynthetically fixed dissolved organic carbon (PDOC) in Schöhsee, West Germany. *Archiv für Hydrobiologie* 102: 91–103.

Rosso, A.L. and F. Azam. 1987. Proteolytic activity in coastal oceanic waters: depth distribution and relationship to bacterial populations. *Marine Ecology Progress Series* 41: 231–240.

Somville, M. and G. Billen. 1983. A method for determining exoproteolytic activity in natural waters. *Limnology and Oceanography* 28: 190–193.

Williams, P.M. 1986. Chemistry of the dissolved and particulate phases in the water column. pp. 53–83 in Eppley, R.W. (editor), *Plankton Dynamics of the Southern California Bight*. Springer Verlag, New York.

10

Role of Ecto-Phosphohydrolases in Phosphorus Regeneration in Estuarine and Coastal Ecosystems

James W. Ammerman

10.1 Introduction

Primary production by phytoplankton is dependent on the availability of phosphorus (P), nitrogen (N), and other nutrients; and may be limited by the lack of these nutrients. P limitation of phytoplankton has been clearly demonstrated in freshwater environments, but the commonly accepted N limitation of phytoplankton in marine waters has been less clearly shown (Hecky and Kilham, 1988). Regardless of the presence or absence of nutrient limitation, nutrient regeneration is of fundamental importance in aquatic ecosystems. Over 99% of the total N and P consumed each year by marine primary producers is ultimately supplied by regeneration (Harrison, 1980), and 80% is provided by in situ regeneration. Phosphorus metabolism is particularly dependent on rapid regeneration of orthophosphate (P_i) from particulate and dissolved organic phosphate, since there is no biological P- input process analogous to N_2 fixation (Froelich, 1984).

Figure 10.1 shows a simplified water-column phosphorus cycle. Three of the major pathways are numbered as follows: (1) uptake of soluble reactive phosphate (SRP) by phytoplankton and bacteria; (2) phagotrophy of bacterial and algal biomass and P by protozoan zooplankton; and (3) regeneration of SRP from dissolved organic phosphate (DOP). Other pathways, some of which may also be important, are shown by arrows without numbers.

In this study, the term SRP will be used for concentration measurements

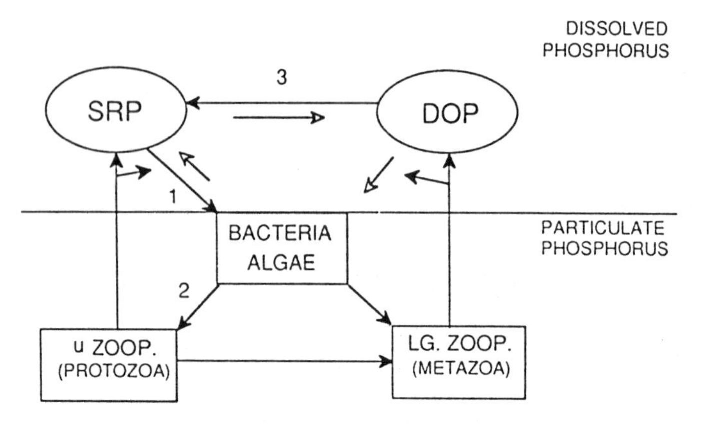

Figure 10.1 A simplified water-column phosphorus cycle (uZOOP-microzooplankton; LG.ZOOP-macrozooplankton).

in natural waters, but P_i will be used for P additions or when the chemical species is clearly orthophosphate. Phosphorus regeneration is defined as hydrolysis of P_i from organic or other complex P compounds, soluble or particulate, in which the hydrolyzed P_i is released outside the cell. The subsequent fate of this P_i could be immediate uptake by the same cell, complete mixing with the background dissolved P_i, or a combination of both. This definition is comparable to that in Harrison (1983) but differs from that in Anderson et al. (1986) and in Güde (1985), which both include release of dissolved organic phosphorus (DOP) as P regeneration. Phosphorus regeneration will be written as P_i regeneration throughout the rest of this paper to make this distinction clear. The emphasis in these studies was on the regeneration of P_i from DOP compounds by microbial ecto-phosphohydrolases (Figure 10.1, pathway no. 3).

There are at least four major agents of P_i regeneration in the water columns of aquatic ecosystems. These include bacteria, phytoplankton, and micro- and macrozooplankton. For the purposes of this discussion, the first three groups above are considered microbes. Input of P_i from the benthos can also be very important in shallow waters but will not be discussed here. The role of bacteria in carbon cycling has received much attention (Azam et al., 1983; Hobbie and Williams, 1984), but their role in P cycling has been little explored. Since the membranes of bacteria and other microorganisms have limited permeability to phosphorylated organic compounds, many bacteria have a suite of phosphohydrolytic ectoenzymes (usually periplasmic) which are induced by low P concentrations and regenerate P_i from specific classes of organic or complex phosphates. *Escherichia coli*, the best studied, has at least four such enzymes; acid phosphatase, alkaline phosphatase (APase), 2′,3′-cyclic phosphodiesterase, and 5′-nucleotidase (5′PN) (Lugtenberg, 1987). This would suggest that aquatic bacteria, if they have similar enzymes, could be important in P_i regeneration, but there has been little evidence for this. Recent studies, however, have demonstrated the activity of some of the above bacterial ecto-phosphohydrolases in the field. These include bacterial alkaline phosphatase (Chróst and Overbeck, 1987) and bacterial 5′PN (Ammerman and Azam, 1985). Other studies, however, suggest that bacteria compete with algae for uptake of P and then sequester it (Harrison et al., 1977; Krempin et al., 1981; Vadstein et al., 1988; Vadstein and Olsen, 1989). These two different roles of bacteria in the P cycle do not necessarily conflict, bacteria may take up P_i that they have just regenerated from DOP (Chróst, 1988; 1990). The relative rates of P_i hydrolysis and uptake, however, will determine the amount of P_i, if any, which is made available to phytoplankton by bacterial DOP hydrolysis.

Algal alkaline phosphatase is another potentially important mechanism of P_i regeneration, and in contrast to the bacterial enzymes, it has been extensively studied in aquatic ecosystems (Cembella et al., 1984). Until recently, in fact, most alkaline phosphatase activity measured in the field was assumed to be algal, which was clearly an oversimplification (Cembella et al., 1984; Chróst and Overbeck, 1987). Since alkaline phosphatases of algae and other organisms are often repressed by measurable SRP concentrations (Perry, 1972; Cembella et al., 1984), they are frequently used as indicators of apparent P limitation. Little attempt has been made, however, to consider their quantitative role in P_i regeneration. It is usually assumed that virtually all the P_i hydrolyzed by APase activity is immediately taken up by the same cell and thus the net regeneration (that available to other organisms) is zero. The fluorometric methods currently used, however, do not provide any information on this since the hydrolyzed P_i is not labeled. APase studies on P-depleted axenic cultures of the oceanic dinoflagellate *Pyrocystis noctiluca* (Rivkin and Swift, 1980) showed that the rate of enzymatic liberation of P_i from glucose-6-P and β-glycerol-P was 9 and 14 times greater, respectively, than the uptake rate of the released P_i. This contradicts the conventional wisdom above and suggests that some of the released P_i might be available to other organisms.

There has been considerable recent interest in microzooplankton (especially microflagellate) predation on bacteria and phytoplankton and the subsequent release of nutrients, including phosphorus. This protozoan phagotrophy releases the P sequestered by algae and bacteria and may be an important P_i regeneration mechanism. Laboratory studies by Güde (1985) and Andersen et al. (1986) suggest that little phosphorus regeneration occurs in cultures of phytoplankton and bacteria unless phagotrophic protozoans, which could graze on either, are present. Such phagotrophy may be an important mechanism of P_i regeneration in aquatic ecosystems, but corroborative field data may be difficult to obtain because of the fragile nature of most protozoa.

Macrozooplankton are clearly important potential sources of regenerated P_i, especially in lakes and the open ocean where particulate phosphorus is a large part of the total P pool (Harrison, 1980; Wetzel, 1983). Macrozooplankton appear to be less important as agents of P_i regeneration in productive coastal waters, however (Harrison, 1980). Even in open ocean and lake studies, the published P-excretion rates for field-collected, large zooplankton vary widely because of variations in experimental conditions (Cembella et al., 1984), and further work needs to be done.

10.2 Study Areas

This short review is an attempt to summarize and compare several recent studies by the author's laboratory in three different coastal and estuarine environments. These studies include: Ammerman and Azam (1991), which focused on the Southern California Bight (SCB), and as yet unpublished studies from the Chesapeake Bay (CB) and the Hudson River Estuary (HRE). These three environments are briefly described in the sections below.

The Southern California Bight The Southern California Bight is an open coastal embayment of the Pacific Ocean, bounded on the west by the California Current (Eppley, 1986). It is in an area with little freshwater runoff and hence little seasonal salinity variation. The bottom topography is very irregular, and the major flow pattern in the upper waters of the Bight is a large counterclockwise eddy (Eppley, 1986). There is considerable seasonal change in water temperature in the Bight, especially in the upper 100 m. The surface-mixed-layer depth generally varies from 15 to 30 m in the winter and from 5 to 10 m in the summer, with the smaller values inshore (Eppley, 1986).

There is a clear onshore-offshore gradient of nutrients and plankton biomass and activity in the Southern California Bight (Eppley, 1986). Inorganic nutrient isolines shoal in inshore waters, resulting in increased phytoplankton biomass (as determined by both $chlorophyll_a$ concentration and microscopic enumeration) and primary productivity in this zone (Eppley, 1986). The same pattern is also generally true for bacterial number, biomass, and production, the latter determined by thymidine incorporation (Fuhrman et al., 1980; Eppley, 1986). The coastal surface waters of the Southern California Bight are depleted in nitrate relative to SRP (Eppley, 1986) as are many marine environments. When just nitrate and SRP are considered, the N/P ratio is less than the Redfield ratio of 16. However, if dissolved organic nitrogen (DON) and DOP are included, then the N/P ratio in the Bight is close to 16 (Jackson and Williams, 1985). Surface concentrations of DOP and DON in the Bight can equal or exceed SRP and nitrate concentrations (Jackson and Williams, 1985).

The Chesapeake Bay The Chesapeake Bay is a large and very productive coastal plain estuary, with relatively low nutrient levels, which has been the focus of many studies of N and P cycling (Taft et al., 1977; Wheeler et al., 1982; Glibert et al., 1982). It is subject to significant nutrient loading from municipal and agricultural sources (Fisher and Doyle, 1987). There is a large freshwater input in the spring and a strong density stratification in the summer, which limits nutrient exchange and contributes to anoxia in the deep water of the midbay. There is disagreement as to the major causes for this anoxia and whether it has increased significantly in extent since 1950 (Officer et al., 1984; Seliger and Boggs, 1987). Both increased benthic respiration due to nutrient loading and increased summer water-column stratification have been implicated as possible causes (Officer et al., 1984). The P cycle in the Chesapeake Bay can be influenced by this anoxia, though that will not be addressed here. For example, summer anoxia in deep water can accelerate the release of P from the sediments, which can in turn increase primary production, creating a positive feedback loop (Officer et al., 1984).

Malone et al. (1986) showed that the biomass and production of bacteria in the water column averages from 20% to greater than 50% of the phytoplankton biomass and production. In addition, Fisher and Doyle (1987) have recently compiled data which suggest that 95% of the phosphorus and 78% of the nitrogen in the Chesapeake Bay are supplied by water-column recycling. The remainder of the P and N come from the sediments or outside inputs.

The Chesapeake Bay seems to display nutrient-limitation features intermediate between those of marine and freshwater systems. Results from both the Patuxent River, a tributary of the bay (D'Elia et al., 1986), and the main channel of the bay itself (Fisher and Doyle, 1987) show a seasonal variation in nutrient limitation of primary production. In general, phosphorus was more limiting in spring during high freshwater runoff. During low-runoff conditions in summer, the situation was reversed, and nitrogen was more limiting. Consistent with this, Taft et al. (1977) found that the highest APase activity, normalized to $chlorophyll_a$, occurred in the spring.

The Hudson River Estuary The Hudson River Estuary (the estuarine portion of the lower Hudson River) is a shallow temperate estuary with a mean depth of only 10 m and a marked seasonality of both temperature and fresh water input. It is also eutrophic, with high nutrient concentrations and moderate-to-high plankton biomass. Sewage input is the dominant source of P and other nutrients, though agricultural runoff is a significant input upstream (Deck, 1981). The major nutrient removal process is river flow, and the residence time of the water in the lower 25 km of the estuary is at most 10 days and may be less than half of that in the spring (Deck, 1981). Biological utilization has only minor effects on nutrient budgets because of the short residence time of the water (Malone, 1977) and the turbidity of the river (Simpson et al., 1975). These factors generally combine to prevent the formation of large blooms. The result is that P_i and some other nutrients are conservative tracers of salinity (Simpson et al., 1975).

10.3 Methods and Experimental Design

The goal of these studies was to determine whether microbial ecto-phosphohydrolases make a significant contribution to P_i regeneration in these ecosystems. I measured microbial alkaline phosphatase (APase) and 5′-nucleotidase (5′PN) activities in these three environments, and, along with several collaborators, also measured microbial biomass and P concentration and flux. Given the major assumptions made, such as the in situ substrate concentrations for the enzymes, the results must be treated with some caution. Nonetheless, they indicate that microbial ecto-phosphohydrolases can make important contributions to P_i regeneration in these ecosystems, though the largest percentage contributions were found in the non-P-limited ecosystems.

Sampling The Southern California Bight (SCB) data were collected in late August, 1984 and early May, 1985. Samples were collected up to 100 km offshore and to depths up to 200 m. The Chesapeake Bay (CB) data were collected on two cruises in mid-May and early August of 1987. Surface (1 m) and deep (10 to 32 m) samples were collected almost the entire length of the bay from the Susquehanna River to the mouth. Deep samples from August were not included in the data summary because the deep waters were anoxic and some of the data were unreliable. Hudson River Estuary (HRE) samples were collected from the estuarine portion of the lower Hudson River (roughly the lower 100 km) in late April and early August of 1988. Both surface (1 m) and deep (5 to 17 m) samples were collected.

Determination of 5′-nucleotidase activity The few available measurements of dissolved nucleotides in seawater (Azam and Hodson, 1977; Hodson et al., 1981a; 1981b; McGrath and Sullivan, 1981) suggest that AMP, ADP, and ATP each reach maximum concentrations of about 1 nM in coastal seawater. These data suggest that a conservative value for total dissolved nucleotide-P concentrations in seawater is 5 nM. In these studies we measured 5′-nucleotidase (5′PN) activity by following the hydrolysis of 50 pM to 2 nM of added [γ-^{32}P]ATP (Ammerman and Azam, 1985; 1991). Therefore, these were true tracer assays, especially those in the CB and HRE where the lower concentrations were used. Tracer assays yield true in situ hydrolysis rates, and when multiplied by the in situ substrate concentration should yield correct P_i regeneration values. Since the hydrolyzed P_i in this assay method is labeled, the fate of this P_i can be followed.

Determination of alkaline phosphatase activity Alkaline phosphatase (APase) was measured by the use of fluorescent analogs. The method of Perry (1972) was used for SCB samples, and the method of Hoppe (1983) for CB and HRE samples. The major difference between these two methods is the fluorogenic substrate used, 3-*O*-methylfluorescein phosphate in the first technique and 4-methylumbelliferyl phosphate in the second. The same concentration of substrate, 100 nM, was used in both methods. This concentration is probably higher than the in situ APase substrate concentration in all areas studied, and it is about half the average total DOP concentration found in the SCB. Therefore, unlike the 5′PN assay, the APase assay is not a true tracer assay, and the measured hydrolysis rates are probably underestimates of the true rates. Furthermore, the hydrolyzed P_i is not labeled, so its fate cannot be followed.

Physicochemical analyses of samples Temperature was measured by a laboratory thermometer, reversing thermometer, expendable bathythermograph, conductivity-temperature-depth sensor (CTD), or a combination of the above. Salinity was measured by a CTD or salinometer. Soluble reactive phosphorus (SRP, determined by absorbance of the phospho-molybdate complex), and chlorophyll$_a$ and phaeopigments (determined by fluorescence), were all measured according to the methods of Strickland and Parsons (1972). Bacterial numbers were determined by epifluorescence microscopy of acridine orange-stained samples (Hobbie et al., 1977), or 4,6-diamidino-2-phenylindole-stained (DAPI) samples (Porter and Feig, 1980). Uptake of SRP was determined by incubating water samples with $^{32}P_i$, followed by filtration and rinsing of the filter with particle-free sample water. To calculate the actual SRP flux, the uptake rate determined above was multiplied by the measured SRP concentration.

All the readily available data for each measurement from each location and season was collected, and the median, range, and total number of values was determined. These data are shown in many of the following figures. The median was used rather than the mean because it is much less subject to the effects of extreme values.

10.4 Physicochemistry of the Three Ecosystems

Spring and summer data were collected from all three environments in order to make seasonal comparisons as well as comparisons among the three different environments. Surface temperatures in the upper 40 m of the SCB were 12 to 16°C in May and 16 to 24°C in August. In the CB, temperatures were about 15°C in May and 27 to 28°C in August. In the HRE, water temperatures were about 9 to 10°C in April and 20 to 30°C in August. Salinity was not measured in the SCB, but typical values in the upper 200 m are 33 to 34 ppt (Eppley, 1986). Measured salinities in the CB and HRE spanned most of the range from 0 to 30 ppt.

The highest median chlorophyll$_a$ concentrations were found in the spring in the CB and in the summer in the HRE (Figure 10.2A). Both had wide ranges, with that in the HRE the widest, in part due to a large localized bloom. The next highest chlorophyll concentrations were in the CB in summer and the HRE in spring, and the ranges were also much narrower than those in the earlier groups. The lowest chlorophyll$_a$ concentrations, by far, were found in the SCB, though the range in the spring was much greater than the summer range because of an apparent red tide concentrated at one station.

Bacterial numbers (Figure 10.2B) showed roughly the same pattern as chlorophyll$_a$, with the highest counts found in the CB. The median was higher in the summer even though the range was narrower. The bacteria in the HRE in

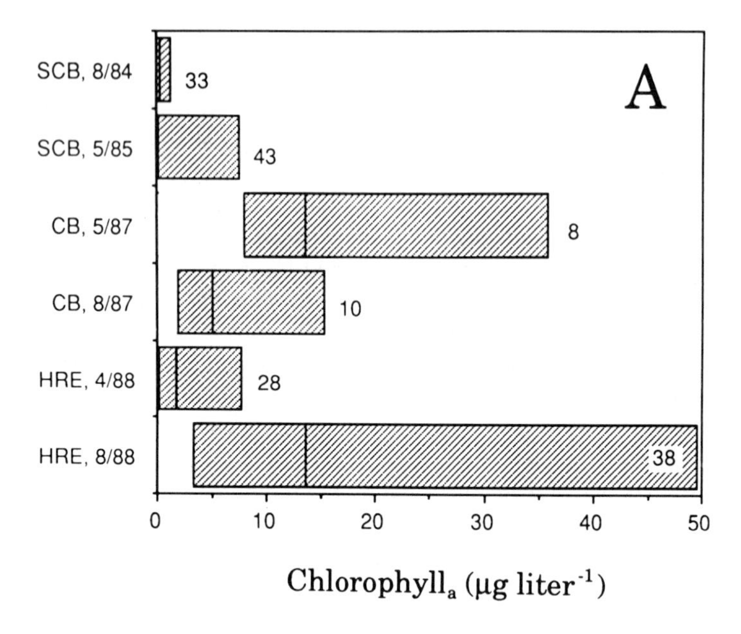

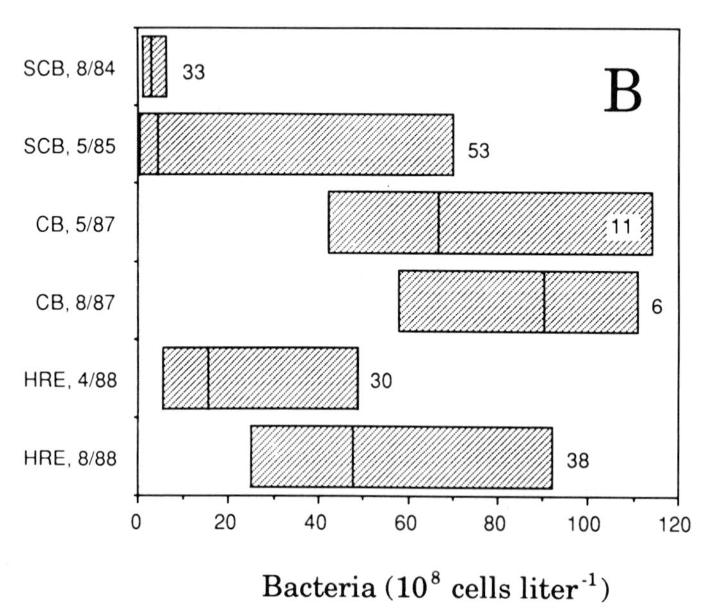

Figure 10.2 Median, range, and number of samples of (A) the chlorophyll$_a$ concentration and (B) the bacterial number in spring and summer samples from the Southern California Bight (SCB), Chesapeake Bay (CB), and Hudson River Estuary (HRE). The summer SCB chlorophyll data include phaeopigments; the spring CB chlorophyll data are from vertical profiles, not longitudinal sections like the other CB data.

summer were somewhat less abundant than in the CB, though still much greater the than in the spring in the HRE. The lowest bacterial number was in the SCB, though the range in the spring was large because of high bacterial populations associated with the red tide.

SRP concentrations are shown in Figure 10.3A. The highest median concentrations were found in the HRE, with little change between the two seasons. The range was greater in the summer due to the decrease in river flow (which increased the maximum sewage-derived SRP concentration) and to a localized algal bloom which lead to an unusual SRP depletion. The median SRP concentration in summer in the SCB was much less than in the HRE (no SRP determinations were made in the spring in the SCB). The range was much greater at the high end than at the low end because measurements were made down to 200 m. The lowest SRP concentrations were found in the CB in May, and they were significantly higher in August. The CB also had the narrowest concentration range of all the locations, with all concentrations at about 0.5 μM or less. These SRP concentrations generally reflect the concentrations of other nutrients in these environments, with the possible exception of the CB (see Section 10.1).

The median SRP uptake, or turnover rate, expressed as % h^{-1}, was far higher in the CB in spring than at any other time or place (Figure 10.3B), and the range was also broad. Summer CB SRP turnover was a distant second and had a much smaller range. All other median rates were less than 1% h^{-1} with summer in the SCB the lowest (spring in the SCB was not measured).

10.5 Phosphohydrolytic Activity and Phosphorus Regeneration

APase activity (Figure 10.4A) was distributed similarly to SRP turnover, though the differences were not quite as extreme. By far the highest activity and the greatest spread was in the CB in spring, and this was distantly followed by summer samples in the CB and HRE. The highest values for the latter were due to a localized SRP depletion in a bloom. Summer APase activities in the SCB and spring activities in the HRE were very low (<2% h^{-1}), and spring SCB values were not determined.

5′PN activity (Figure 10.4B), in contrast, was high in both seasons in the CB and in the summer in the HRE. The highest median value was spring in the CB, followed by summer in the CB and the HRE. All three periods had large ranges of activity. Median spring 5′PN activity in the HRE was much less and had a much narrower range than in the summer. The lowest activity was in the SCB, particularly in the summer.

The percent of the P_i regenerated by 5′PN which was immediately taken up by a P_i transport system (coupled uptake; Ammerman and Azam, 1991) is shown in Figure 10.5. The highest median uptake was in the CB in spring, and the range was wide (including complete uptake). This was followed by the

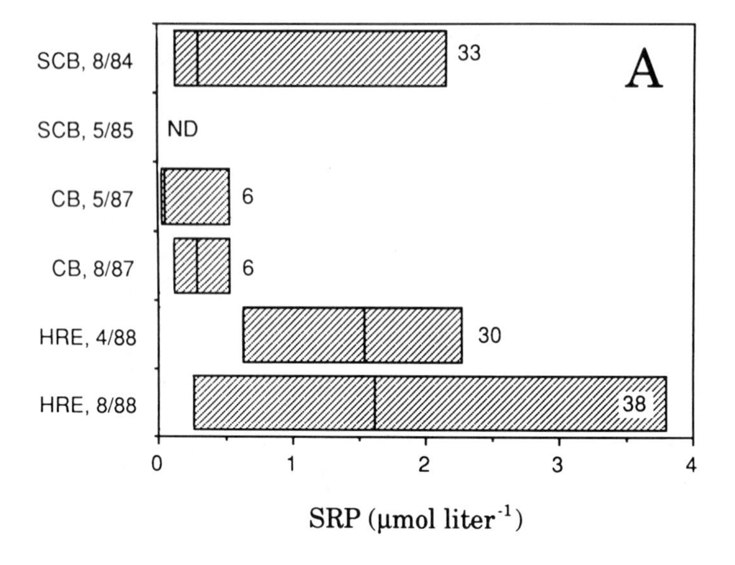

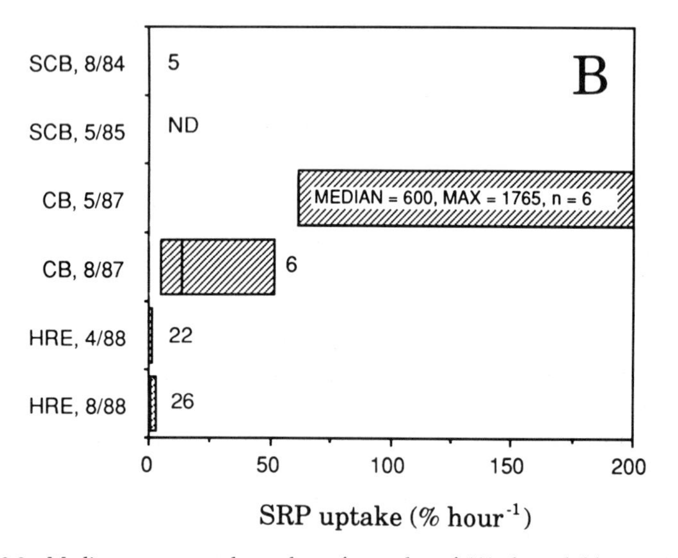

Figure 10.3 Median, range, and number of samples of (A) the soluble reactive phosphate (SRP) concentration and (B) the SRP uptake or turnover rate in spring and summer samples from the Southern California Bight (SCB), Chesapeake Bay (CB), and Hudson River Estuary (HRE).

summer CB median value which had a considerably narrower range. Summer HRE samples had the next highest median, trailed by the two SCB samples which each had wide ranges. The lowest mean coupled uptake was in the spring HRE samples, though the high end of this range was several times the median value.

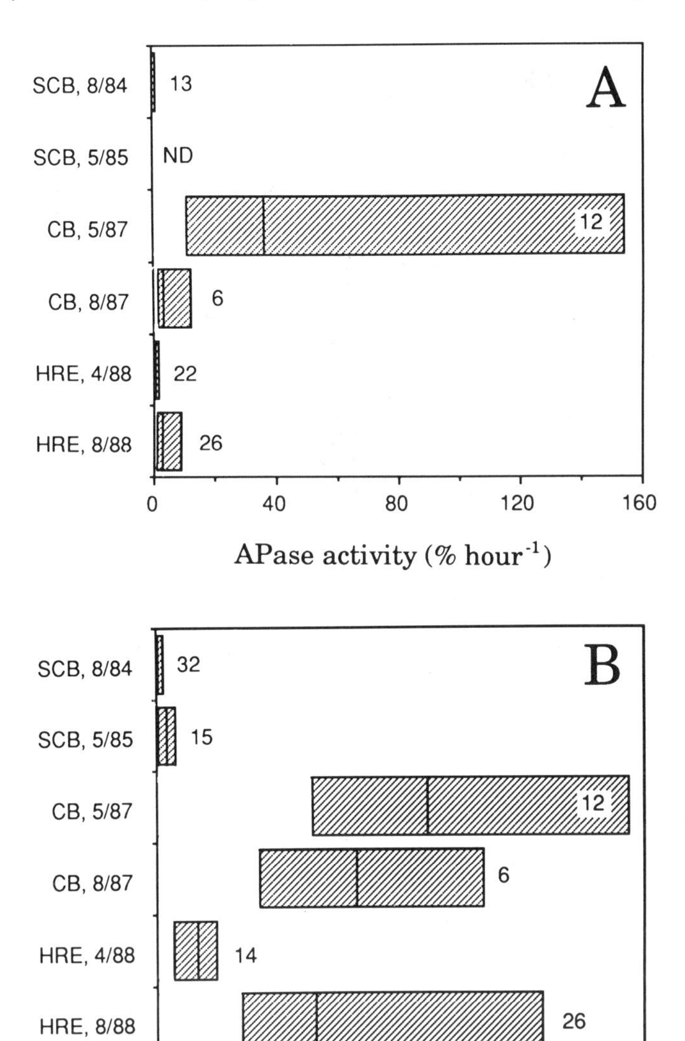

Figure 10.4 Median, range, and number of samples of (A) the alkaline phosphatase activity and (B) the 5′-nucleotidase in spring and summer samples from the Southern California Bight (SCB), Chesapeake Bay (CB), and Hudson River Estuary (HRE).

Figure 10.6A shows a compilation of all the measured APase activities plotted against the in situ SRP concentrations. APase activity was low (about 5% h^{-1} or less) at all SRP concentrations above 0.5 μM. Some of this may be due to an APase activity which is often present regardless of the SRP concentration.

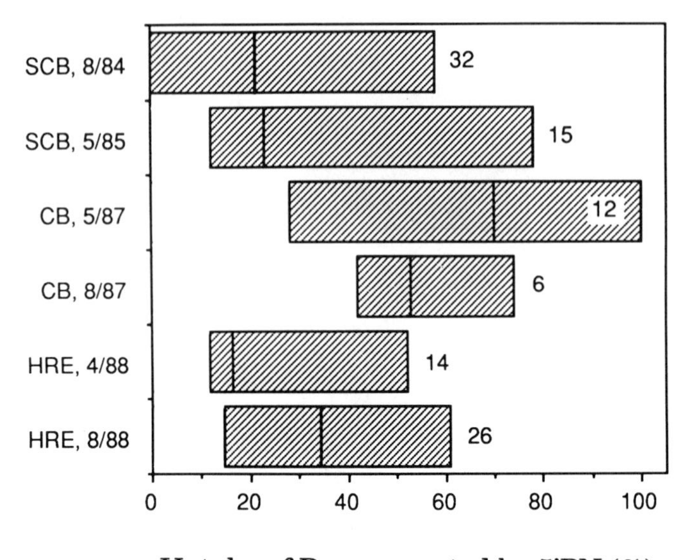

Figure 10.5 Median, range, and number of samples of the uptake of P_i regenerated by 5′PN or coupled uptake in spring and summer samples from the Southern California Bight (SCB), Chesapeake Bay (CB), and Hudson River Estuary (HRE).

Somewhat higher activities were found at lower SRP concentrations, but the highest APase activities (>25% h^{-1}) all occurred in the spring in the CB where SRP concentrations were close to zero.

Figure 10.6B shows a similar figure for 5′PN activity vs. SRP concentration. It is obvious that this relationship was not as clear as the one for APase vs. SRP. However, with the exception of the HRE samples, most of the data points with significant 5′PN activity were found at SRP concentrations of 1 μM or less. In fact, these data almost form a right triangle with the hypotenuse running from the maximum 5′PN activity to an SRP concentration of about 1 μM. The 5′PN activity in the HRE, in contrast, appeared to be largely independent of the SRP concentration and was much greater than one would expect if it was regulated by SRP.

Coupled uptake of the P_i regenerated by 5′PN is plotted against SRP concentration in Figure 10.7. This figure is similar to Figure 10.6B. The major difference between this figure and Figure 10.6B is that far fewer data points are on or near the abscissa, since in most samples at least 10% of the P_i was taken up by coupled uptake. The coupled P_i uptake in the HRE, as with 5′PN activity, was high and largely independent of the SRP concentration.

Since 5′PN activity was not as clearly tied to the SRP concentration as APase activity was, 5′PN activity was plotted against $chlorophyll_a$ and bacterial number to look for other controls on 5′PN activity. Both $chlorophyll_a$ and bacterial number were significantly and positively correlated with 5′PN, but the more

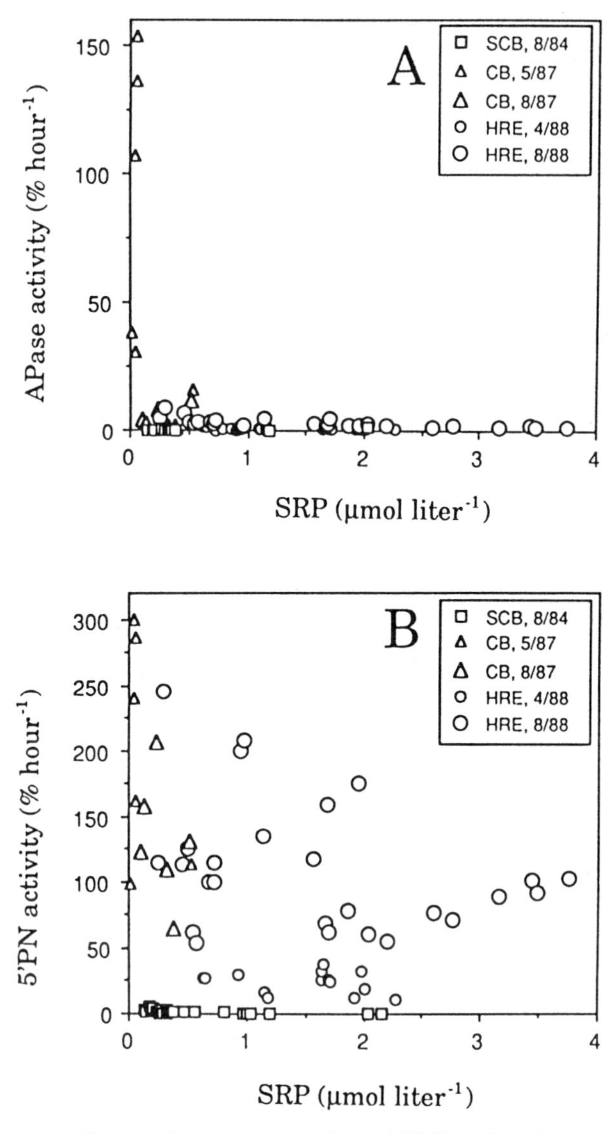

Figure 10.6 (A) Alkaline phosphatase activity and (B) 5′-nucleotidase vs. SRP concentration in spring and summer samples from the Southern California Bight (SCB), Chesapeake Bay (CB), and Hudson River Estuary (HRE).

significant correlation ($p < 0.001$) was with chlorophyll$_a$, and this is shown in Figure 10.8.

The SRP flux (SRP uptake multiplied by SRP concentration) is shown in Figure 10.9A. The greatest SRP flux was in the spring in the CB; the median value was many times the median of the next highest samples, those from the

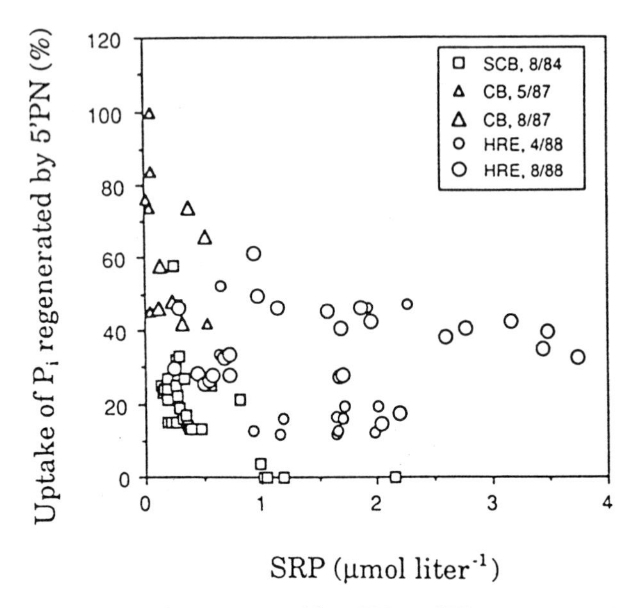

Figure 10.7 The uptake of P_i regenerated by 5′PN vs. SRP concentration in spring and summer samples from the Southern California Bight (SCB), Chesapeake Bay (CB), and Hudson River Estuary (HRE).

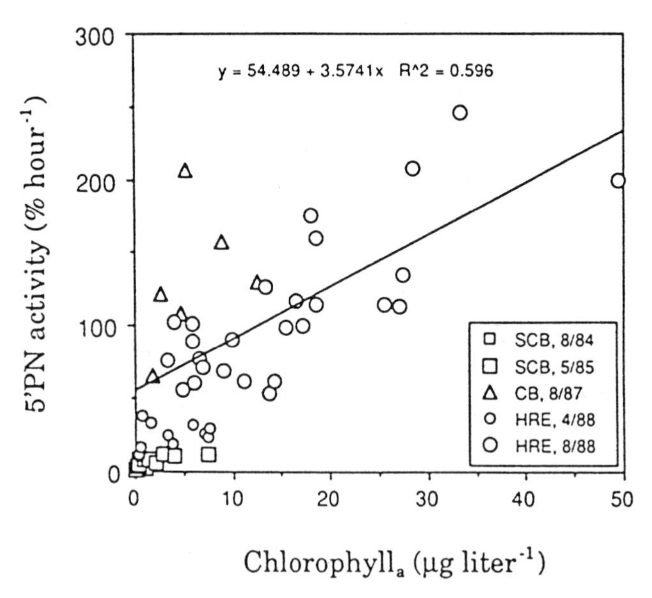

Figure 10.8 5′-Nucleotidase (5′PN) activity vs. chlorophyll$_a$ concentration in spring and summer samples from the Southern California Bight (SCB), Chesapeake Bay (CB), and Hudson River Estuary (HRE). This regression was highly significant ($P < 0.001$).

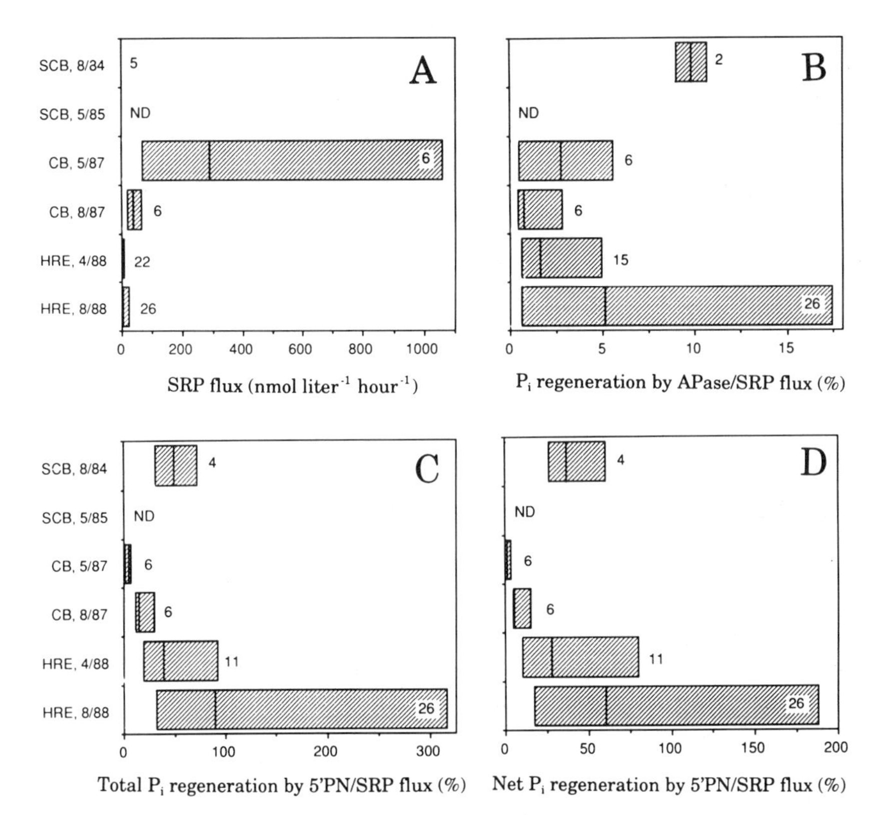

Figure 10.9 (A) Median, range, and number of samples of the SRP flux; (B) the P_i regeneration by alkaline phosphatase (APase) activity divided by the SRP flux; (C) the total and (D) the net P_i regeneration by 5′PN activity divided by the SRP flux in spring and summer samples from the Southern California Bight (SCB), Chesapeake Bay (CB), and Hudson River Estuary (HRE).

CB in summer. The spring CB samples also covered a much wider range of values than any of the others. Summer and spring HRE samples, respectively, had the next highest SRP flux, and the lowest were found in the summer SCB samples (the values were too low to appear distinct from the ordinate). SRP flux was not measured in the spring SCB samples.

Figure 10.9B shows the ratio in percent of P_i regeneration by APase to the total SRP flux. In other words, this shows what percent of the SRP demand could be met by P_i regeneration by APase. The P_i regeneration by APase was calculated from the APase hydrolysis rate multiplied by an assumed APase substrate-P concentration of 10 nM. This assumed concentration is a conservative value extrapolated from the nucleotide measurements discussed in methods (see Section 10.3). Apparently no seawater concentrations of APase substrates other than nucleotides have been measured. Surprisingly, the percentages found in Chesapeake Bay were among the lowest, even though APase activity was high, SRP flux was very large, and therefore the percentage contribution by APase was small. If much of this regenerated P_i was taken up immediately by a coupled uptake similar to that for 5′PN, then this percentage would be even smaller. Since APase is normally induced by a P deficiency, coupled uptake could be even more important than with 5′PN, but the current APase assays don't follow the fate of the hydrolyzed P_i, so little data is available. Information discussed in Section 10.1 minimizes the importance of coupled P_i uptake for APase, but further work is needed to clarify the issue. The importance of P_i regeneration by APase to the total SRP flux was generally larger in the SCB and HRE, which had much lower SRP fluxes than the CB, but the median regeneration by APase was nowhere greater than 10% of the total flux.

Figure 10.9C is similar to Figure 10.9B but instead of P_i regeneration by APase it shows what percentage of the SRP demand could be met by total P_i regeneration by 5′PN. The P_i regeneration by 5′PN was calculated from the 5′PN hydrolysis rate multiplied by an assumed nucleotide-P concentration of 5 nM. This assumed concentration is a conservative value based on the nucleotide measurements discussed in Section 10.3. Though the relative pattern of locations and seasons was similar to that for APase, regeneration of P_i by 5′PN accounted for a far greater percent of total SRP flux than did APase (compare the scales on the abscissa in Figures 10.9B and 10.9C). Though the highest median 5′PN activity was in the CB (Figure 10.4B), it supplied the smallest percentage of the total SRP demand of all the locations studied, because the CB also had the largest SRP flux (Figure 10.9A). The percentage was higher in the SCB and especially the HRE, where in the summer the median 5′PN activity could supply virtually all the SRP flux.

The data in Figure 10.9C is for total P_i regeneration by 5′PN and was not corrected for coupled P_i uptake. Figure 10.9D shows what happens when coupled uptake was subtracted from the total P_i regeneration, in other words the net P_i regeneration. This correction had little effect on the overall pattern for location and season, but just lowered the mean values by roughly 40%. 5′PN activity still made its smallest percent contribution to the SRP flux in the CB, but supplied over 60% of the total SRP demand (median value) in the summer in the HRE.

10.6 Significance of Microbial Ecto-phosphohydrolases in Estuarine and Coastal Waters

This short review compares and contrasts the role of microbial ecto-phosphohydrolases in three very different coastal and estuarine environments, the Southern California Bight, the Chesapeake Bay, and the Hudson River Estuary. It also compares the spring and summer seasons in all three environments. This review demonstrates that these enzymes make important contributions to P_i regeneration in these environments, but the percent of total P_i regeneration attributable to them varied with time and location.

The three environments discussed present gradients of both SRP concentrations and algal and bacterial biomass. SRP concentrations were clearly lowest in the CB, followed by the SCB and the HRE. Only the CB showed an important seasonal change in median SRP concentration. The gradient of algal and bacterial biomass was different than the SRP gradient, the lowest biomass was found in the SCB, followed by the HRE and the CB. (This assumes constant and equal carbon/chlorophyll$_a$ ratios and bacterial cell sizes in the three environments and at different seasons.) All except the SCB showed large seasonal changes in median values. It is notable that the CB had the lowest SRP concentrations and yet the highest algal and bacterial biomass. The CB was clearly the most productive environment and though apparently N or P limited at certain times (Fisher and Doyle, 1987), it cycled the available nutrients rapidly. The extremely rapid cycling of SRP in the CB compared to the other environments was evident in the SRP uptake data. The SCB and the HRE were less productive environments, even though they had higher SRP concentrations. Their productivity and biomass was limited by low surface N concentrations, residence time of the water, turbidity, and other factors (see Section 10.1).

The activities of the two microbial ecto-phosphohydrolases, APase and 5′PN, were also greatest in the CB, though summer 5′PN activity in the HRE was also high. APase activity in all three environments was clearly controlled by the SRP concentration, or some other manifestation of it, such as cellular P pools. 5′PN activity showed some evidence of control by SRP concentration, if the HRE samples were excluded, but the coupling was not as tight. This makes sense because the gene for 5′PN (at least in *Escherichia coli*) is part of the *pho* regulon, which responds to P deficiency (Lugtenberg, 1987). In coastal and estuarine environments, however, 5′PN activity appeared to be largely constitutive (in contrast to the inducible APase) and was correlated with biomass, both bacterial numbers and chlorophyll$_a$ concentrations. Previous studies (Ammerman and Azam, 1985; 1991) have shown that 5′PN is a bacterial enzyme, this work does not contradict this, it just suggests that more 5′PN activity is found in eutrophic areas which have greater chlorophyll$_a$ concentrations.

Due to the rapid SRP turnover, the total SRP flux into the microbes in the CB in spring far exceeded that in any other place or season, including summer in the CB. This was true despite the very low SRP concentrations. In attempting to compare this total SRP demand with the P$_i$ regenerated by APase and 5′PN, two factors are important. These are the total enzyme activity (the APase activity measurements are probably underestimates, see Section 10.3) and the degree of coupled uptake. The APase assay does not permit the determination of coupled uptake (see Section 10.3), so the discussion below concerns only 5′PN activity. In non-P-limited environments like the SCB and the HRE, coupled uptake was relatively low, much of the P$_i$ hydrolyzed by bacterial 5′PN was released into the environment. The relationship between the percentage of coupled uptake and the SRP concentration was similar to that between 5′PN activity and SRP. The coupled uptake in the SCB seemed to be largely determined by isotope dilution by background SRP (Figure 10.7, this study; Ammerman and Azam, 1991). The HRE was again anomalous, though lower than in the CB, the percent of coupled uptake in the HRE was greater than would be predicted from the SRP concentration. The released P_i in these environments was presumably available to any organism with the need and the transport system for it.

In more P-limited systems like the Baltic Sea (Ammerman and Azam, 1985) and the CB, the coupled uptake may reach 100%, meaning that none of the hydrolyzed P_i was released. While much of this P_i was probably taken up by the bacteria that hydrolyzed it, there is a suggestion in some size-fractionation studies that phytoplankton were responsible for much of the uptake of P_i released by bacterial 5′PN activity (Ammerman and Azam, 1985). This must remain an open question for now, but it is clear that even if the bacteria in P-limited environments took up 100% of the hydrolyzed P_i by the coupled uptake system, they still make a contribution to the total P_i regeneration. In fact, this regenerated P_i, if never released to the environment, would not show up in conventional isotope-dilution measurements of P_i regeneration such as those made by Harrison (1983), but could be a significant part of the total P budget, especially if the bacterial-P demand is as high as some estimates suggest (Vadstein et al., 1988; Vadstein and Olsen, 1989). The terms total and net have been applied to 1) P_i regeneration by 5′PN and 2) that fraction of the regenerated P_i remaining after coupled P_i uptake, respectively.

Though APase and 5′PN activity were highest in the CB, these two ecto-phosphohydrolases supplied the smallest percentage of the total SRP demand in the CB compared to the SCB and HRE, because the SRP flux was so high in the CB. This was true whether the flux of P_i due to 5′PN activity was corrected for coupled P_i uptake or not (coupled uptake could not be measured for APase activity, see Section 10.3). While it is difficult to believe that the low levels of APase activity in the SCB and HRE were really significant in the P cycle, they accounted for up to 10% of the total SRP demand in these environments. The P_i contribution from 5′PN activity in the SCB and HRE was much greater, however, nearly 100% of the SRP demand in some cases, and correction for coupled uptake only decreased this to 60%. Based on these data and the assumed nucleotide concentrations above, coupled uptake of P_i hydrolyzed by 5′PN alone equaled nearly 40% of the total flux of SRP in the summer in the HRE. Summer HRE samples, of course, had the highest SRP concentrations, up to 4 μM. The explanation for this rapid utilization of organic P, in an environment with very high SRP concentrations, may be that much of the SRP is biologically unavailable due to complex formation with iron or other chemical processes (Froelich, 1988; Fox, 1989).

The major surprise of this study was the relatively small contribution (20% or less) of the ecto-phosphohydrolases to P_i regeneration in the CB compared to the other environments. Spring in the CB was the one sampling period in all these studies that the plankton biomass was clearly P limited (Fisher and Peele, unpublished observations). Yet the SRP flux was so rapid that additional regeneration mechanisms were required to supply the necessary SRP. However, the picture would change drastically if we substituted "biologically available phosphate" (BAP) for SRP. BAP was measured by an isotope dilution bioassay similar to that of Levine and Schindler (1980) and was typically about 10% of the SRP concentration in the spring in the CB (Fisher and Peele, unpublished). If BAP was the only fraction of the SRP which was really available to the biota, then the total P demand would be 10 times less. The relative contribution of microbial ecto-phosphohydrolases to the required P supply (which is unaffected by the change from SRP to BAP) would then be 10 times greater and could in some cases supply all of it. Thus the biological availability of the SRP in coastal and marine waters appears to be a critical unresolved factor.

Another potentially important source of substrates for APase and 5′PN activity, which is not considered above, is dissolved DNA and RNA. Data on the concentrations of these macromolecules in aquatic environments have recently been compiled by Karl and Bailiff (1989). In estuarine and coastal water columns, the concentrations of dissolved DNA ranged from 0.05 to 21 μg per liter (with one very high value of 81) and the concentrations of dissolved RNA ranged from 7 to 51 μg per liter. Since roughly 10% of the weight of nucleic acids is in the form of phosphate, the total of the higher values of the dissolved DNA and RNA could provide over 230 nM P_i.

In fact, the major function of 5′PN activity, as distinct from the general phosphate-ester hydrolysis by APase, may be to hydrolyze the ribo- and deoxyribo-nucleotide monophosphates produced by rapid RNA and DNA breakdown. If nucleic acid breakdown occurs mostly on cell surfaces and is closely coupled to 5′PN activity, than many of the nucleotides produced may never show up in measurements of bulk nucleotide concentrations. Thus the few available measurements of dissolved nucleotides (see Section 10.3) may significantly underestimate their importance.

Acknowledgments I wish to thank Drs. T. Fisher and E. Peele for the use of their unpublished data from the Chesapeake Bay, Dr. D. Angel for collaboration on the Hudson River Estuary study, and Dr. F. Azam for helpful discussions. This work was supported by National Science Foundation Grants OCE83-00360 to Dr. F. Azam and OCE85-18401 to the author. Further support was provided by Hudson River Foundation Grant 001/87A/002 to the author.

References

Ammerman, J.W. and F. Azam. 1985. Bacterial 5′-nucleotidase in aquatic ecosystems: a novel mechanism of phosphorus regeneration. *Science* 227: 1338–1340.

Ammerman, J.W. and F. Azam. 1991. Bacterial 5′-nucleotidase activity in estuarine and coastal marine waters: 1. Characterization of enzyme activity. 2. Role in phosphorus regeneration. *Limnology and Oceanography* (in revision).

Andersen, O.K., Goldman, J.C., Caron, D.A. and M.R. Dennett. 1986. Nutrient cycling in a microflagellate food chain: III. Phosphorus dynamics. *Marine Ecology Progress Series* 31: 47–55.

Azam, F., Fenchel, T., Field, J.G., Gray, J.S., Meyer-Reil, L.A. and F Thingstad. 1983. The ecological role of water column microbes in the sea. *Marine Ecology Progress Series* 10: 257–263.

Azam, F. and R.E. Hodson. 1977. Dissolved ATP in the sea and its utilization by marine bacteria. *Nature* 267: 696–698.

Cembella, A.D., Antia, N.J. and P.J. Harrison. 1984. The utilization of inorganic and organic phosphorus compounds as nutrients by eukaryotic microalgae: a multidisciplinary perspective. *CRC Critical Reviews in Microbiology,* Part 1, 10: 317–391; Part 2, 11: 13–81.

Chróst, R.J. 1988. Phosphorus and microplankton development in an eutrophic lake. *Acta Microbiologica Polonica* 37: 205–225.

Chróst, R.J. 1990. Microbial ectoenzymes in aquatic environments. pp. 47–78 in Overbeck, J. and Chróst, R.J. (editors), *Aquatic Microbial Ecology: Biochemical and Molecular Approaches.* Springer Verlag, New York.

Chróst, R.J. and J. Overbeck. 1987. Kinetics of alkaline phosphatase activity and phosphorus availability for phytoplankton and bacterioplankton in Lake Plußsee (North German eutrophic lake). *Microbial Ecology* 13: 229–248.

Deck, B.L. 1981. Nutrient-element distributions in the Hudson Estuary. *Ph.D. dissertation, Columbia University,* 396 pp.

D'Elia, C.F., Sanders, J.G. and W.R. Boynton. 1986. Nutrient enrichment studies in a coastal plain estuary: Phytoplankton growth in large-scale, continuous cultures. *Canadian Journal of Fisheries and Aquatic Sciences* 43: 397–406.

Eppley, R.W. 1986. *Plankton Dynamics of The Southern California Bight.* Springer Verlag, New York, 373 pp.

Fisher, T.R. and R.D. Doyle. 1987. Nutrient cycling in Chesapeake Bay. pp. 49–53 in Mackiernan, G.B. (editor), *Dissolved Oxygen in the Chesapeake Bay, Processes and Effects.* Maryland Sea Grant Publication, College Park.

Fox, L.E. 1989. A model for inorganic control of phosphate concentrations in river waters. *Geochimica et Cosmochimica Acta* 53: 417–428.

Froelich, P.N. 1984. Interactions of the marine phosphorus and carbon cycles. pp. 141–176 in Moore, B. and Dastoor, M.N. (editors), *The Interaction of Global Biochemical Cycles.* NASA, Jet Propulsion Laboratory Publication, Pasadena, California.

Froelich, P.N. 1988 Kinetic control of dissolved phosphate in natural rivers and estuaries: A primer on the phosphate buffer mechanism. *Limnology and Oceanography* 33: 649–668.

Fuhrman, J.A., Ammerman, J.W. and F. Azam. 1980. Bacterioplankton in the coastal euphotic zone: Distribution, activity and possible relationships with phytoplankton. *Marine Biology* 60: 201–207.

Glibert, P.A., Lipschultz, F., McCarthy, J.J. and M.A. Altabet. 1982. Isotope dilution models of uptake and remineralization of ammonium by marine plankton. *Limnology and Oceanography* 27: 639–650.

Güde, H. 1985. Influence of phagotrophic processes on the regeneration of nutrients in two-stage continuous culture systems. *Microbial Ecology* 11: 193–204.

Harrison, W.G. 1980. Nutrient regeneration and primary production in the sea. pp. 433–460 in Falkowski, P.G. (editor), *Primary Productivity in the Sea.* Plenum Press, New York.

Harrison, W.G., Azam, F., Renger, E.H. and R.W. Eppley. 1977. Some experiments on phosphate assimilation by coastal marine phytoplankton. *Marine Biology* 40: 9–18.

Harrison, W.G. 1983. Uptake and recycling of soluble reactive phosphorus by marine microplankton. *Marine Ecology Progress Series* 10: 127–135.

Hecky, R.E. and P. Kilham. 1988. Nutrient limitation of phytoplankton in freshwater and marine environments: A review of recent evidence on the effects of enrichment. *Limnology and Oceanography* 33: 796–822.

Hobbie, J.E., Daley, R.J. and S. Jasper. 1977. Use of Nuclepore filters for counting bacteria by fluorescence microscopy. *Applied and Environmental Microbiology* 33: 1225–1228.
Hobbie, J.E. and P.J.LeB. Williams. 1984. *Heterotrophic Activity in the Sea*. Plenum, New York. 569 pp.
Hodson, R.E., Azam, F., Carlucci, A.F., Fuhrman, J.A., Karl, D.M. and 0. Holm-Hansen. 1981a. Microbial uptake of dissolved organic matter in McMurdo Sound, Antarctica. *Marine Biology* 61: 89–94.
Hodson, R.E., Maccubbin, A.E. and L.R. Pomeroy. 1981b. Dissolved adenosine triphosphate utilization by free-living and attached bacterioplankton. *Marine Biology* 64: 43–51.
Hoppe, H.G. 1983. Significance of exoenzymatic activities in the ecology of brackish water: measurements by means of methylumbelliferyl-substrates. *Marine Ecology Progress Series* 11: 299–308.
Jackson, G.A. and P.M. Williams. 1985. Importance of dissolved organic nitrogen and phosphorus to biological nutrient cycling. *Deep-Sea Research* 32: 223–235.
Karl, D.M. and M.D. Bailiff. 1989. The measurement and distribution of dissolved nucleic acids in aquatic environments. *Limnology and Oceanography* 34: 543–558.
Krempin, D.W., McGrath, S.M., SooHoo, J.B. and C.W. Sullivan. 1981. Orthophosphate uptake by phytoplankton and bacterioplankton from the Los Angeles Harbor and Southern California coastal waters. *Marine Biology* 64: 23–33.
Levine, S.N. and D.W. Schindler. 1980. Radiochemical analysis of orthophosphate concentrations and seasonal changes in the flux of orthophosphate to seston in two Canadian Shield lakes. *Canadian Journal of Fisheries and Aquatic Sciences* 37: 479–487.
Lugtenberg, B. 1987. The *pho* regulon in *Escherichia coli*. pp. 1–2 in Torriani-Gorini, A., Rothman, F.G., Silver, S., Wright, A. and E. Yagil. (editors), *Phosphate Metabolism and Cellular Regulation in Microorganisms*. American Society for Microbiology, Washington, D.C.
Malone, T.C. 1977. Environmental regulation of phytoplankton productivity in the Lower Hudson Estuary. *Estuarine and Coastal Marine Sciences* 5: 157–171.
Malone, T.C., Kemp, W.M., Ducklow, H.H., Boynton, W.R., Tuttle, J.H. and R.B. Jonas. 1986. Lateral variation in the production and fate of phytoplankton in a partially stratified estuary. *Marine Ecology Progress Series* 32: 149–160.
McGrath, S.M., and C.W. Sullivan. 1981. Community metabolism of adenylates by microheterotrophs from the Los Angeles Harbor and Southern California coastal waters. *Marine Biology* 62: 217–226.
Officer, C.B., Biggs, R.B., Taft, J.L., Cronin, L.E., Tyler, M.A. and W.R. Boynton. 1984. Chesapeake Bay anoxia: origin, development, and significance. *Science* 223: 22–27.
Perry, M.J. 1972. Alkaline phosphatase activity in subtropical Central North Pacific waters using a sensitive fluorometric method. *Marine Biology* 15: 113–119.
Porter, K.G. and Y.S. Feig. 1980. The use of DAPI for identifying and counting aquatic microflora. *Limnology and Oceanography* 25: 943–948.
Rivkin, R.B. and E. Swift. 1980. Characterization of alkaline phosphatase and organic phosphorus utilization in the oceanic dinoflagellate *Pyrocystis noctiluca*. *Marine Biology* 61: 1–8.
Seliger, H.H. and J.A. Boggs. 1987. Anoxia in the Chesapeake Bay. *EOS* 68: 1693.
Simpson, H.J., Hammond, D.E., Deck, B.L. and S.C. Williams. 1975. Nutrient budgets in the Hudson River Estuary. pp. 618–635 in Church, T.M. (editor), *Marine Chemistry in the Coastal Environment*. American Chemical Society, Washington, D.C.
Strickland, J.D.H. and T.R. Parsons. 1972. *A Practical Handbook of Seawater Analysis*. Bulletin 167, 2nd edition. Fisheries Research Board Canada, Ottawa, 310 pp.
Taft, J.L., Loftus, M.E. and W.R. Taylor. 1977. Phosphate uptake from phosphomonoesters by phytoplankton in the Chesapeake Bay. *Limnology and Oceanography* 22: 1012–1021.
Vadstein, O., Jensen, A., Olsen, Y. and H. Reinertsen. 1988. Growth and phosphorus

status of limnetic phytoplankton and bacteria. *Limnology and Oceanography* 33: 489–503.

Vadstein, O. and Y. Olsen. 1989. Chemical composition and phosphate uptake kinetics of limnetic bacterial communities cultured in chemostats under phosphorus limitation. *Limnology and Oceanography* 34: 939–946.

Wetzel, R.G. 1983. *Limnology*, 2nd edition. Saunders, Philadelphia, 767 pp.

Wheeler, P.A., Glibert, P.M. and J.J. McCarthy. 1982. Ammonium uptake and incorporation by Chesapeake Bay phytoplankton: Short term uptake kinetics. *Limnology and Oceanography* 27: 1113–1128.

11

Bacterial Phosphatases from Different Habitats in a Small, Hardwater Lake

James B. Cotner, Jr. and Robert G. Wetzel

11.1 Introduction

The significance of alkaline phosphatase (APase) in aquatic environments has been examined extensively in both freshwater and marine systems. Interest in this enzyme historically has been in its capacity as a predictor of the degree of phosphorus (P) limitation in the phytoplankton (Kuenzler and Perras, 1965; Fitzgerald and Nelson, 1966; Healey and Hendzel, 1980; Wetzel, 1981). This predictive capacity, however, has been challenged by several studies because not all APase production in the pelagic zone is by phytoplankton (Stevens and Parr, 1977; Cembella et al., 1984; Jansson et al., 1988). APase hydrolyzes dissolved organic phosphorus compounds (DOP) to an organic moiety and inorganic phosphate (see Figure 13.1, Chapter 13). The hydrolyzed phosphorus is then available for uptake either by the organisms that hydrolyze it or by other organisms (Ammerman and Azam, 1985; see Chapter 2). Because biological production in many lakes has been shown to increase with increased phosphorus inputs, regeneration of phosphorus through APase could increase productivity at times when allochthonous inputs of P are minimal.

Although it has been suggested that bacteria are generally limited for growth by the availability of organic carbon (Currie and Kalff, 1984; Atlas and Bartha, 1987), most aquatic studies have been performed in pelagic freshwater or marine environments where organic carbon loading is dampened relative to the littoral zone and the sediments. Potentially, bacteria can be limited by phosphorus or other nutrients in environments where organic carbon is plentiful (Haines and Hanson, 1979; Newell et al., 1983; Benner et al., 1988). In this chapter, we suggest that the ability of bacteria to utilize a wide variety of DOP compounds in different habitats is reflected in the amount of enzyme produced and the kinetic characteristics of APase. Specifically, we postulate that organisms in phosphorus-rich habitats have less capacity to hydrolyze DOP through APase than organisms that have evolved in habitats poor in phosphorus. A total of three strains of bacteria from pelagic, epiphytic, and epipelic environments were isolated and their APase characterized by their kinetic constants, V_{max}, K_m, and K_i, using an artificial APase substrate, methylumbelliferyl-phosphate (MUP). We assumed that these parameters are important descriptors of the ecological significance of this enzyme to phosphorus acquisition by the different bacterial strains.

11.2 Methods and Experimental Design

Isolation of bacterial strains and species identification Lake water was collected from the surface or the 3-m stratum in Third Sister Lake (Washtenaw Co., Michigan) in spring or fall 1988 in sterile 1-liter bottles. Approximately 10 μl of water was pipetted onto 2% agar plates made with Third Sister Lake water with or without 2.8 μM glucose. Plates were incubated at 20°C for 2 to 5 days and individual colonies were re-inoculated twice onto sterile agar plates to insure that pure cultures were obtained.

Once isolated, pure isolates were screened for APase production using the low phosphorus medium of Torriani (1968). APase activity was qualitatively observed by dropping 250 μM *p*-nitrophenylphosphate in 0.1 M Tris-HCl buffer (pH 8.0) onto colonies and noting the development of the yellow product, *p*-nitrophenol (Echols et al., 1961). Pure cultures were maintained on nutrient agar slants at 4°C.

Species identifications were made with the API-20E Rapid Identification System (API Laboratory Products, Basingstoke; see Schneider and Rheinheimer, 1988 for a brief description) that utilizes 21 biochemical tests. Pure strains were grown in Guillard's WC medium (Guillard, 1975) with either 5 g l^{-1} glucose alone or 2.5 g l^{-1} glucose and 2.5 g l^{-1} peptone as a carbon source. KH_2PO_4 was added to 25 μM in the glucose medium but no phosphorus was added to the glucose/peptone medium other than that which contaminated the peptone.

APase production relative to soluble reactive phosphorus (SRP) concentrations was determined in glucose medium at an initial KH_2PO_4 concentration of 5 μM with *Acinetobacter* sp. Soluble reactive phosphorus was measured by the molybdate-ascorbic acid colorimetric method (APHA, 1985) with a Beckman DU-8 spectrophotometer using cuvettes with a 5-cm path-length.

Occurrence of APase in situ On 3 July 1989, APase activity in surface samples was determined along a transect in Third Sister Lake, which is a moderately eutrophic lake in southeastern Michigan. Samples were taken at evenly spaced intervals from the littoral region to the deepest portion of the lake. Lake water was filtered through 1.0-μm Nuclepore filters to separate algae and bacteria. Usually, less than 15% of the chlorophyll passed the 1.0-μm filter and greater than 90% of the bacteria passed the filter (Cotner, 1990). APase activity was measured in whole lake water and filtered lake water samples using the fluorescent substrate methylumbelliferyl-phosphate (MUP) (Pettersson and Jansson, 1978; Chróst et al., 1989). Because enzyme substrates can be inhibitory at high concentrations (Suelter, 1985), an initial estimate of K_m was obtained by the methods of Suelter (1985), and total activity was measured at 10 times K_m which produces a rate that is greater than 95% of V_{max}.

Enzyme purification A relatively rapid but effective methodology for purification of bacterial APase was developed. Much of the procedure is similar to methods developed for the purification of *Escherichia coli* APase by Malamy and Horecker (1964). Cells were inoculated into media and grown at 20°C to stationary phase in a 48-h period. Cells were centrifuged at 3,000 x g for 25 minutes, and the pellet was washed three times with 0.01 M Tris-HCl, 0.001 M $MgSO_4$ (Tris-Mg) buffer and was suspended in 50–100 ml of TES buffer (0.03 M Tris-HCl pH 8.0, 0.5 mM EDTA and 0.5 M sucrose). Lysozyme (from chicken egg white, Sigma grade VI) was added to a final concentration of 0.10 mg ml^{-1}. Deoxyribonuclease (4 μg ml^{-1}), EDTA (0.6 mM), and $MgSO_4$ (0.6 mM) were added after 2 minutes. Completeness of spheroplast formation was monitored by changes in absorbance at 590 nm in a 1-cm cell (Malamy and Horecker, 1964). When spheroplast formation was complete, the preparation was centrifuged for 20 min at 12,000 x g to remove particulate matter. The supernatant was decanted and phenylmethyl sulfonyl fluoride (1 mM) was added to inhibit proteases.

Prior to column chromatography, the APase preparation was washed with Tris-Mg buffer using ultrafiltration with a PM30 filter (Amicon, molecular weight cutoff 30,000 daltons). This filter retained greater than 95% of APase activity in all strains. For isolates C-2 and S-1, 0.1 mM $ZnSO_4 \cdot 7H_2O$ was added to the Tris buffer (Tris-Zn) instead of $MgSO_4$ because it was required to maintain optimal enzyme activity. Once the effluent pH was stable at 8.0, the enzyme preparation was loaded onto a 20 x 1.5 cm column of Sepharose Q Fast Flow (Pharmacia) anion exchange resin that had been equilibrated with either Tris-Mg or Tris-Zn buffer. A step elution program was used where NaCl concentrations in the buffer were increased in four equal steps from 0.0 M to 0.2 M in the column buffer. The APase eluted with either the 0.15 M or 0.2 M NaCl fraction. Fractions with the greatest APase activity were pooled, washed and concentrated with ultrafiltration. Strain 14-1 APase was purified in the same manner except it was eluted with gel permeation chromatography (Biorad P-450 gel) and Tris-Mg buffer at pH 8.0. A 1.5 x 20-cm column was used to elute this protein. Protein concentrations were measured by the bicinchoninic acid modification (Smith et al., 1985) of the Lowry et al. (1951) method.

Gel electrophoresis Polyacrylamide gels were prepared for sodium dodecyl sulfate (SDS) and native gel electrophoresis (Suelter, 1985). The enzyme preparation was concentrated by precipitation with a 90% saturated solution of ammonium sulfate (0.603 g ml^{-1}), and centrifuged at 40,000 x g for 30 minutes.

SDS gels were 13.5% with respect to acrylamide/bis acrylamide and were run with Tris-glycine-SDS buffer at pH 8.3 (Suelter, 1985). Twenty μl of sample were loaded onto the gel along with Biorad SDS-low-molecular-weight standards and Sigma *E. coli* alkaline phosphatase. Samples were chromatographed overnight, and protein bands were developed with Biorad's silver stain kit (Merril et al., 1981). Native gels were run with 10% acrylamide/bis acrylamide and no stacking gel at pH 9.0. These gels were stained with Biorad alkaline phosphatase conjugate substrate kit.

Kinetics of bacterial APase Kinetic experiments for estimating K_m, V_{max}, and K_i were designed to follow the protocol of Allison and Purich (1979). Enzyme assays were performed on a Turner 111 fluorometer with MUP as substrate (Pettersson and Jansson, 1978). To insure that each enzyme preparation conformed to the Michaelis-Menten model, the assumption of minimal enzyme concentration relative to the substrate was tested prior to all assays by insuring that doubling the enzyme concentration also doubled the velocity of the saturated reaction. An initial estimate of K_m (Suelter, 1985) was obtained and concentrations of MUP from 0.2 to 5 times this estimate were used to describe the shape of the Michaelis-Menten curve (Allison and Purich, 1979).

The SAS (SAS Institute, Inc., Cary, NC) nonlinear regression program was used to solve for the kinetic constants V_{max} and K_m in the Michaelis-Menten equation:

$$v = (V_{max} \times [S])/(K_m + [S])$$

where S is substrate (MUP) concentration, v is the velocity of the reaction, V_{max} is the rate of the reaction at substrate saturation, and K_m is the half-saturation constant.

The phosphate inhibition constant, K_i, was estimated by regressing the slope of a Lineweaver-Burk plot against phosphate concentration to determine the x-intercept (Dixon and Webb, 1979).

11.3 Results

Several species of bacteria were isolated from different habitats in Third Sister Lake. Both Gram-negative and -positive organisms were found in pelagic, epiphytic, and epipelic (on sediment) habitats. All three habitats also supported bacteria that could produce APase when grown on low phosphorus agar (Table 11.1). Three of the four strains isolated from the pelagic region tested positive for APase production with nitrophenylphosphate (Table 11.1). Both strains isolated from the sediments and two of the six strains isolated from macrophytes in Third Sister Lake also tested positive for APase activity. Among the macrophytic strains, only the strains isolated from *Ceratophyllum demersum* (C-1 and C-2) exhibited APase activity.

The spatial distribution of bacterial APase was further examined by mea-

Table 11.1 Relative APase activity in strains of bacteria isolated from different habitats in Third Sister Lake.

Habitat	Strain[1]	Relative APase activity[2]
Pelagic	14–1	+++
	9	+
	3	−
	5	++
Sediments	S1	++
	S2	++
Macrophyte	C1	++
	C2	+
	P1	−
	P2	−
	N1	−
	N2	−

[1]Letters in front of the numbers of the macrophyte strains indicate the macrophyte genus from which the strains were isolated. C = *Ceratophyllum*, P = *Potamogeton*, and N = *Nuphar*.
[2]Activity was determined with *p*-nitrophenylphosphate.

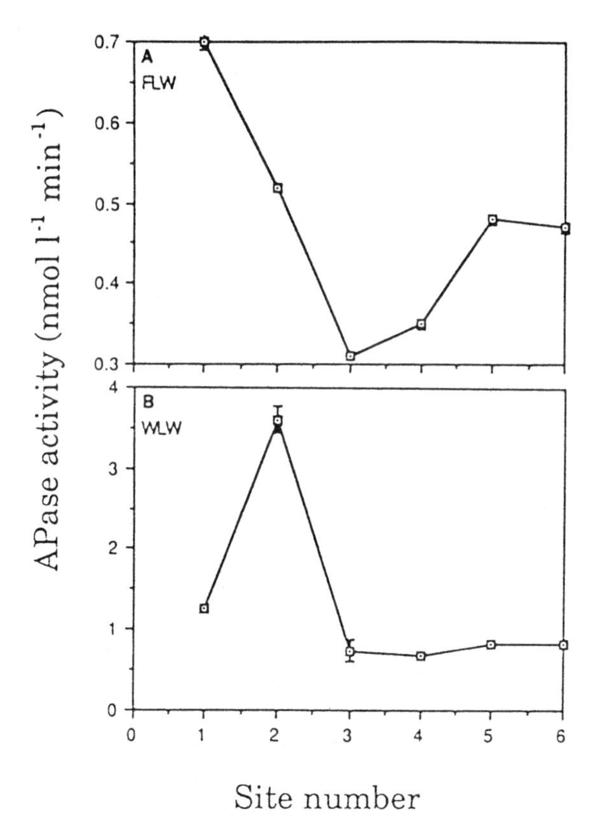

Figure 11.1 Alkaline phosphatase activity on a transect from the littoral zone (site 1) to the pelagic zone (site 6) using:(A) 1.0-μm filtered lake water and (B) whole lake water.

suring the APase activity in the lake water which passed a Nuclepore 1.0-μm filter on a transect from the littoral zone to the pelagic region in early July, 1989 (Figure 11.1A). Although some of the APase in this fraction may be dissolved (Stewart and Wetzel, 1982; see Chapter 2), most of the activity is associated with bacterial-sized particles. The transect from the littoral to the pelagic region of the lake demonstrated higher relative bacterial APase activity in the littoral zone, i.e., sites 1 and 2, where *Nuphar variegatum*, *Ceratophyllum demersum*, and *Potamogeton nodosus* were dominant (Figure 11.1A). Activity in this bacterial fraction increased moderately at the stations furthest from apparent littoral influence. There was no significant correlation ($r^2 = 0.11$) between APase activity in the <1.0-μm fraction and whole lake water (Figure 11.1A and 11.1B), which suggests that algal and bacterial APase production were not related at this time.

APase production Production of APase by an *Acinetobacter* sp. from the pelagic zone was characterized in WC medium with 5 g l^{-1} glucose. APase activity was minimal until SRP decreased below approximately 0.1 μM. (Figure 11.2). There was a lag in APase production after SRP decreased to these levels, perhaps because internal cell quotas were not yet exhausted. APase activity per absorbance unit increased one-thousand-fold after phosphate levels were depleted (Figure 11.3).

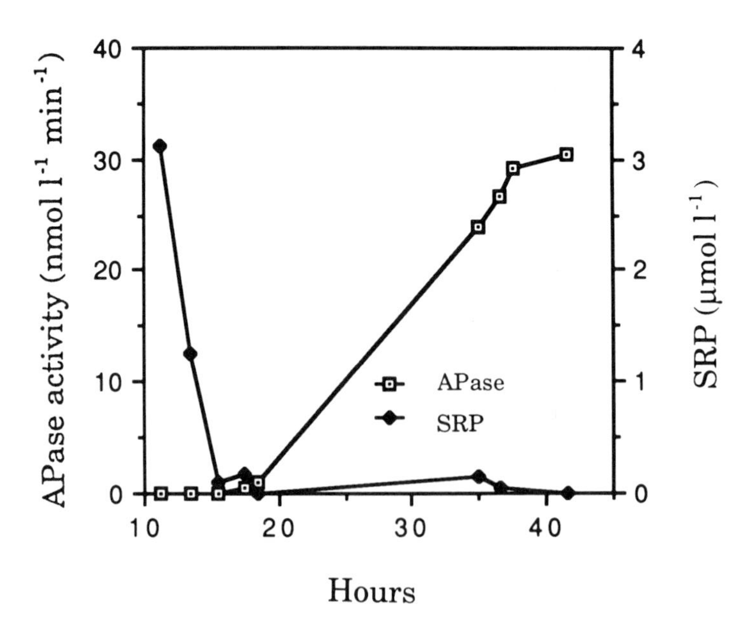

Figure 11.2 Alkaline phosphatase activity and soluble reactive phosphorus concentration as a function of time in a culture of *Acinetobacter* (strain 14-1) grown in 5 g l^{-1} glucose WC medium.

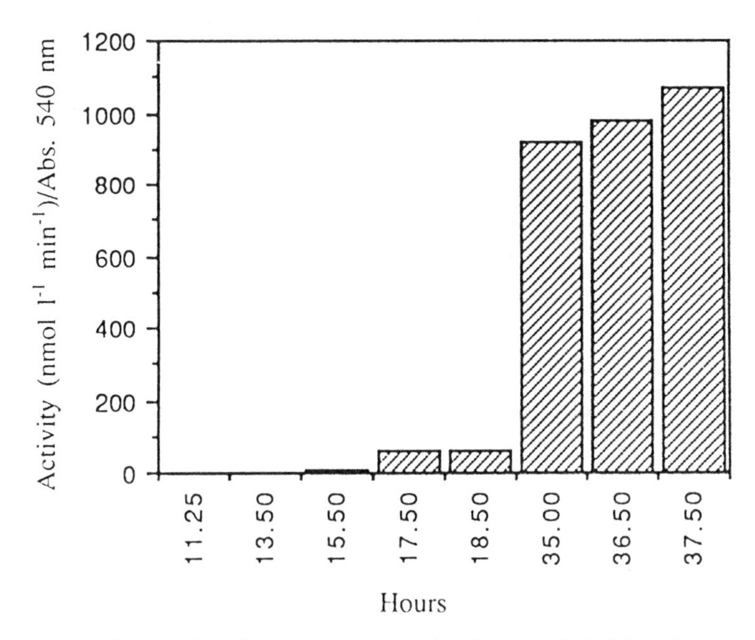

Figure 11.3 Alkaline phosphatase activity per absorbance unit (λ 563 nm) in a culture of *Acinetobacter* (strain 14-1) grown in 5 g l^{-1} glucose WC medium.

Table 11.2 Purification table for alkaline phosphatases from three different strains of bacteria isolated from Third Sister Lake. Data for the three main purification steps are given.

Bacterial strain	Purification step	Specific activity (nmol min^{-1} mg^{-1})	Factor[1]
Acinetobacter 14-1	Crude	16,124	
	PM30	49,348	3.06
	Q Sephadex	542,529	33.65
Pseudomonas C-2	crude	5,870	
	PM30	7,135	1.22
	Q Sephadex	135,228	23.04
Enterobacter S-1	crude	1,713	
	PM30	5,663	3.31
	Q Sephadex	46,964	27.42

[1]This represents the fold increase of crude-extract enzyme activity after purification.

Enzyme purification APase of strains from pelagic (*Acinetobacter* sp., isolate 14-1), epiphytic (*Pseudomonas* sp., isolate C-2) and epipelic (*Enterobacter* sp., isolate S-1) habitats were purified prior to kinetic measurements. There was considerable variation in the specific activities of the APase from these three bacterial strains (Table 11.2). Total purification varied from 23-fold in the *Pseudomonas* strain to 34-fold in the *Acinetobacter* strain (Table 11.2). Minimal amounts (0.01 mg ml^{-1}) of lysozyme were required to release the majority of the APase activity into solution, which indicates that this enzyme was, indeed, periplasmic. Zinc sulfate had to be added at 0.1 mM for *Pseudomonas* (strain C-2) and *Enterobacter* (strain S-1) to maintain activity but it inhibited activity in the *Acinetobacter* enzyme. In both the *Pseudomonas* and the *Enterobacter* strains, most APase activity eluted in one major fraction on Sepharose Q anion exchange chromatography (Figure 11.4). A similar peak in enzyme activity was found when *Acinetobacter* was eluted with size-exclusion chromatography.

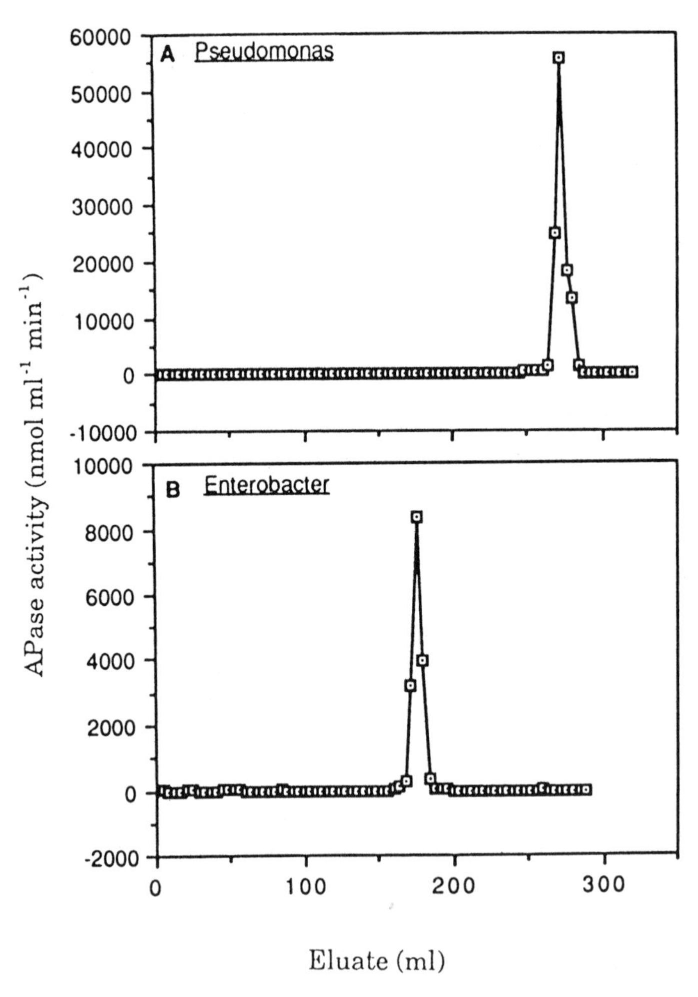

Figure 11.4 Sepharose Q Fast Flow anion exchange chromatography of APase activity in the effluent of (A) *Pseudomonas* (strain C-2) and (B) *Enterobacter* (strain S-1). Elution programs were not identical so the relative positions of the peaks cannot be directly compared.

Denaturing polyacrylamide gel electrophoresis on two of the three strains indicated that there were several proteins contaminating the purified enzyme (Figure 11.5). Both gels had a band migrating below the ovalbumin standard at a molecular weight of about 40,000 daltons, which was near the dominant band in the *E. coli* APase standard. This band appeared to be a doublet in *Pseudomonas*. Native gel electrophoresis on the purified *Acinetobacter* APase indicated only one band of APase activity, which suggests that, although several proteins were found in this preparation, only one phosphatase enzyme was present (Figure 11.6). Attempts to perform native gel electrophoresis on *Pseudomonas* APase resulted in no discernable bands of APase in the gel. If the pK of this enzyme was above the pH of the buffer used in this technique (pH 9.0), this enzyme would have migrated out of the gel toward the anode rather than the cathode since the preparation was loaded at the anode end of the gel.

All three purified APase enzymes showed Michaelis-Menten kinetics with MUP as substrate (Figure 11.7). Each enzyme was competitively inhibited by phosphate, which suggests that inhibition occurred at the active site of the enzyme. However, there were notable differences in the kinetic constants of the APase enzymes (Table 11.3). There were differences in both the K_m and V_{max} for all three strains, which resulted in differences in the turnover rates (V_{max}/K_m) of these three enzymes of over four orders of magnitude. The three strains exhibited notable differences in K_i, also. The K_i for the Acinetobacter strain was

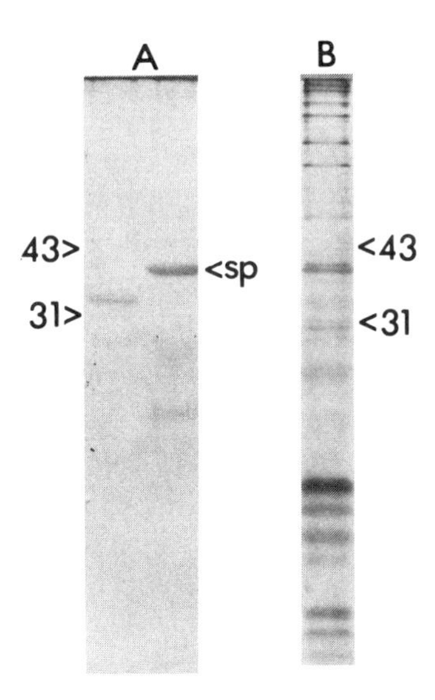

Figure 11.5 Polyacrylamide SDS gel electrophoresis of partially purified APase from (A) *Acinetobacter* and (B) *Pseudomonas* run in a 13.5% acrylamide gel. Sigma SDS-PAGE low-molecular-weight standards were run with each gel and the 43,000 and 31,000 daltons positions are indicated with arrows. sp represents the elution position of *E. coli* alkaline phosphatase (from Sigma).

Figure 11.6 Native gel electrophoresis of *Acinetobacter* alkaline phosphatase in 10% acrylamide gel without SDS at pH 8.5. Sigma *E. coli* alkaline phosphatase (sp) was run as a standard.

lower than the K_m, which indicates that this enzyme had a higher affinity for phosphate than the DOP substrate, MUP. However, this was not the case for *Pseudomonas* or *Enterobacter* (Table 11.3). In *Pseudomonas* the K_i was almost six times greater than the K_m, and the K_i for *Enterobacter* was slightly greater than K_m (Table 11.3).

Phosphate did not inhibit *Enterobacter* APase by a simple competitive mechanism. There was a proportional effect of phosphate concentration on the slope of the Lineweaver-Burk plot at concentrations below 4 μM but not above this concentration (Figure 11.7).

11.4 Discussion

APase production Alkaline phosphatase activity has been measured intra- and extracellularly in aquatic organisms ranging from bacteria (Jones, 1972; McComb et al., 1979) to zooplankton (Rigler, 1961; Boavida and Heath, 1984) and protozoa (McComb et al., 1979). The occurrence of this enzyme in lakes has been negatively correlated to phosphate concentrations (Heath and Cooke, 1975; Stevens and Parr, 1977; Wetzel, 1981). However, Chróst and Overbeck (1987) suggested that aquatic bacterial phosphatases may, in fact, be constitutive, and Reid and Wilson (1971) concluded that most bacterial acid phosphatases are constitutive. Our data on the bacterial strain isolated from the pelagic region of a lake do not support the hypothesis that bacterial APase is constitutive. *Acinetobacter* APase activity increased by one-thousand-fold when nor-

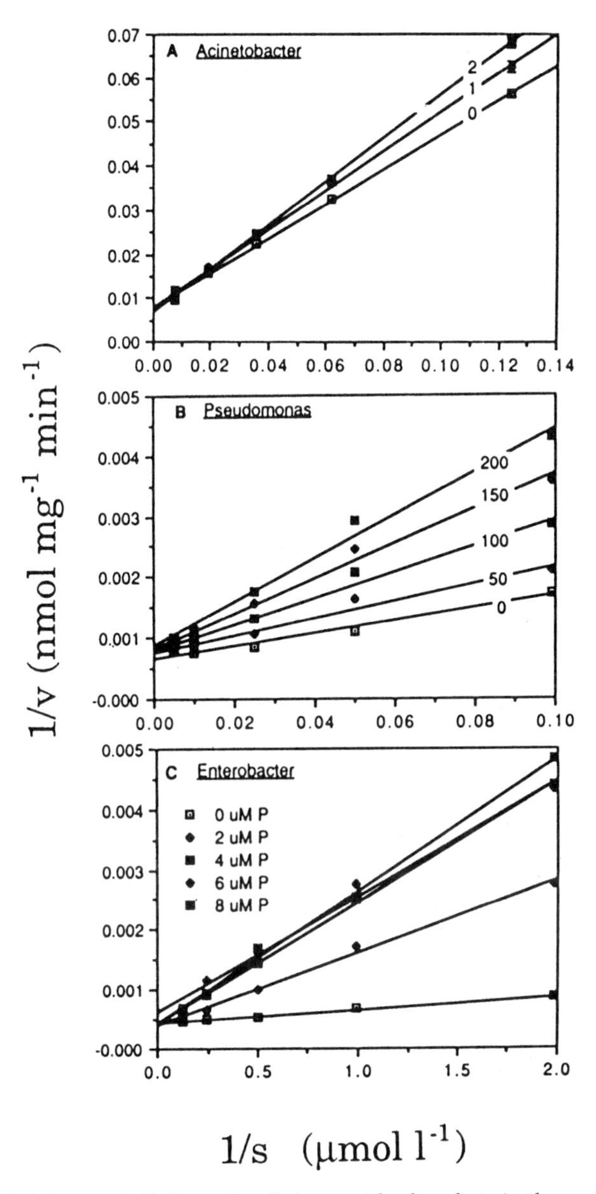

Figure 11.7 Inhibition of alkaline phosphatases with phosphate in three strains of bacteria. (A) *Acinetobacter* was isolated from the pelagic region, (B) *Pseudomonas* was isolated from the macrophyte *Ceratophyllum*, and (C) *Enterobacter* was isolated from sediment. Numbers on lines in (A) and (B) represent the concentration (μM) of KH_2PO_4 in the reaction flask.

Table 11.3 Kinetic parameters determined for three strains of bacteria isolated from different habitats in Third Sister Lake[1]

Strain	Habitat	K_m (μM)	V_{max} (nmol l^{-1} mg^{-1})	V_{max}/K_m	K_i (μM)
14-1	Pelagic	74.69 (±7.1)	161.52 (±7.8)	2.1	18.58
TC-2	Macrophyte	11.07 (±2.0)	1,471.65 (±80.8)	132.9	41.95
TS-1	Sediment	0.51 (±0.03)	22,069.50 (±340)	43,443.9	0.51[2]

[1]Numbers in parentheses are the standard errors.
[2]For strain TS–1, K_i was determined graphically only for 0, 2, and 4 μM KH_2PO_4, which was the linear range of inhibition.

malized for biomass. Other strains did not produce measurable APase until growth approached stationary phase. van Gemerden and Kuenen (1984) speculated that the enzymes that are required for transport of essential nutrients in low and discontinuous supply must be constitutive so that the machinery to use the nutrient is always available when the nutrient becomes available. Because APase is a catabolic enzyme and usually does not play a role in phosphorus transport, it would be predicted to be an inducible (or derepressible) enzyme in oligotrophs (van Gemerden and Kuenen, 1984). If a bacterium can obtain sufficient phosphorus for growth through its phosphate permease(s), synthesizing APase would not increase the fitness of that organism unless it is produced to increase the availability of organic carbon (see below).

In Third Sister Lake, phosphatase activity was detected in the bacterial fraction in all samples, both littoral and pelagic, on a seasonal basis. The ubiquity of bacterial APase activity indicates that this enzyme may be important in phosphorus regeneration processes in the lake and also suggests that there is potential for some of the bacteria to be P-limited. It is not the absolute concentration of phosphorus that determines whether organic carbon or phosphorus will limit growth but rather the ratio of these nutrients to each other relative to their cellular needs (Sommer, 1989). Benner et al. (1988) demonstrated that bacteria, when supplied excess lignocellulose, were either limited by nitrogen or phosphorus. Epiphytic bacteria may be carbon sufficient because of the release of DOC from the plant, and therefore may compete strongly for other nutrients such as phosphorus and nitrogen.

Therefore, it was initially surprising that bacteria associated with *Nuphar* and *Potamogeton* did not produce APase because most macrophytes have been demonstrated to excrete measurable amounts of DOC (see also Chapter 3). However, it is probable that proportional amounts of phosphorus are also excreted from rooted macrophytes. McRoy et al. (1972) estimated P excretion by *Zostera marina* as 58% of P absorbed from the roots. Moeller et al. (1988) demonstrated significant movement of ^{32}P-phosphate from the sediments, up through

Najas, and out to adnate epiphytes. Track autoradiography showed that adnate algae obtained as much as 60% of their P from the macrophyte *Najas*. Variation among epiphytic algal species in P acquisition from *Najas* was attributed to spatial differences, with adnate species obtaining a larger proportion of P from the macrophyte than loosely attached species. No similar studies have been performed on floating macrophytes, but if these rooted macrophytes excrete organic carbon and phosphorus in a ratio significantly less than that of the floating macrophyte, *Ceratophyllum*, i.e., proportionately more phosphorus, the epiphytes on the rooted macrophytes would be less severely P-limited. Differences in phosphorus excretion by these two types of macrophytes would be consistent with what is already known about phosphorus availability to macrophytes through organic sediments. The lowered redox potential found in organic sediments mobilizes phosphate (Mortimer, 1941; 1942; Carlton and Wetzel, 1988), which then can be taken up by the rooted macrophytes and presumably excreted to the epiphytes. However, floating macrophytes must obtain all of their P from the open water. Because concentrations of phosphate are usually below the limit of detection in open water, these floating macrophytes must retain internal P more effectively than rooted macrophytes.

In addition, the growth form of the macrophyte may contribute to differences in the phosphorus dynamics of the associated epiphytes. *Ceratophyllum* grows in dense beds at the surface where water flow is severely restricted by plant material (Losee and Wetzel, 1988; Spencer and Bowes, 1989). Diffusion is the dominant mechanism for movement of nutrients to the leaf under these low-flow conditions (Losee and Wetzel, 1988). Epiphytic algae and bacteria may have adapted to this stagnant, nutrient-depleted environment by increasing the effective nutrient concentration using hydrolytic enzymes such as alkaline phosphatase. Because the *Enterobacter* strain was isolated from littoral sediments where we anticipated significant phosphate release (Mortimer, 1941; 1942; Carlton and Wetzel, 1988; Gächter et al., 1988), we did not expect these strains to be P-limited. However, the V_{max}/K_m of both *Acinetobacter* and *Pseudomonas* were quite modest compared to this ratio in *Enterobacter*, and both strains isolated from the sediments did produce APase when grown on low P media.

There are several possible explanations for this apparent high capacity to utilize DOP in epipelic bacteria. The first explanation suggests that these bacteria are extremely leaky for phosphorus compounds and, therefore, may employ APase to recover lost P. Gächter et al. (1988) proposed that bacteria release phosphorus through the hydrolysis of internally stored polyphosphate under anaerobic conditions in organically rich sediments. Dephosphorylation enables electron transport to continue in the absence of oxygen and nitrate. Phosphate accumulates intracellularly and is eventually transported out of the cell. In shallow sediments where there is sufficient light penetration to support epipelic photosynthesis, the sediments can alternate between aerobic and anaerobic conditions (Carlton and Wetzel, 1988). Under these conditions, high-affinity phosphate transport and alkaline phosphatase enzymes may facilitate growth by recovering some of the phosphorus released during anaerobiosis. The comparable K_m and K_i for this enzyme is consistent with this hypothesis because it suggests that the enzyme does not discriminate between organic and inorganic phosphorus substrates. The second explanation suggests that the habitat from which these epipelic bacteria were isolated was not very P-rich. Because the *Enterobacter* strain was isolated from very shallow sediment in the littoral zone, sorption of phosphorus by calcium carbonate, formed as a result of macrophyte photosynthesis (Otsuki and Wetzel, 1972), may maintain very low phosphate concentrations on these sediments. A third possibility, which will be discussed below, is that these bacteria produce APase in response to organic carbon limitation rather than P-limitation, and therefore the regulation of APase is controlled by the internal organic carbon status of the cell rather than internal P status (see also Chapter 3). All of these processes may have played a role in the evolution of the *Enterobacter* APase.

APase in different habitats The kinetic differences measured in the three purified APase enzymes suggest that the habitat in which these bacteria grow has affected their ability to acquire P. The large degree of variation in the kinetic parameters of the enzymes was not surprising considering the diversity of the environments which were sampled. The V_{max} of the *Pseudomonas* enzyme was much higher than that of *Acinetobacter*. The ratio V_{max}/K_m reflects the capacity of the enzyme to function at high and low substrate concentrations. At high ambient substrate concentrations, it would be adaptive to have a high V_{max}, if growth was P-limited, so that excess substrate could be hydrolyzed, and the product, phosphate, could be assimilated and stored. Bacteria are capable of storing phosphorus as polyphosphate when phosphate concentrations are in excess of cellular needs (Kulaev, 1979; Deinema et al., 1980). However, at low substrate concentrations, it is usually advantageous to have a high affinity for the decreased substrate concentrations. Because the V_{max}/K_m was 50 times greater in the macrophytic strain than the pelagic strain, the strain isolated from *Ceratophyllum* may be better adapted to use DOP as a phosphorus source than the strain isolated from the pelagic region. This argument assumes the capacity for uptake of the hydrolyzed phosphate in the *Ceratophyllum* bacterial strain is equal to or greater than that of the pelagic strain. Data from this study (Cotner, 1990) and data from continuous culture of a planktonic bacterium and alga (Currie and Kalff, 1984) indicate that planktonic bacteria can be saturated for phosphate uptake at ambient phosphate concentrations. This suggests that these bacteria have little capacity for phosphate uptake above ambient concentrations. Unfortunately no comparable data are available for phosphate uptake capacity by epiphytic bacterial communities, but if this community is more P-limited than pelagic bacterial communities, as suggested by the APase kinetic data and gradient measurements of bacterial APase, a greater phosphate-uptake capacity would be expected in addition to the apparently greater capacity for DOP hydrolysis.

The littoral-to-pelagic-gradient data (Figure 11.1) showed that bacterial APase activity was greatest in the littoral and lower in the pelagic region. This gradient may indicate either that the bacteria in the littoral zone may be more P-limited or that there was greater bacterial biomass per unit volume in the littoral region compared to the pelagic region. Because no biomass data were obtained at this time, this issue cannot be completely resolved, but acridine-orange direct counts of the bacterial fraction in the littoral and pelagic regions of Third Sister Lake in August 1989 showed no significant differences in bacterial numbers (Cotner, 1990). Also, our study has demonstrated that some strains of bacteria can increase APase production per unit biomass by a factor of a thousand when ambient phosphate concentrations are low. Since numerical differences were small, but APase activity in the bacterial fraction differed by more than a factor of two in the littoral and pelagic zones, it is probable that the differences in APase activity were the result of greater APase production per unit biomass in the bacterial fraction of the littoral zone.

Phosphate inhibition The pattern of phosphate inhibition demonstrated in APase from the *Enterobacter* strain is consistent with the hypothesis that the organism may occasionally encounter relatively high phosphate concentrations as hypothesized to occur under anaerobic conditions. Unlike the other two enzymes studied, the *Enterobacter* APase was inhibited in a manner that is typical of partial competitive inhibition (Dixon and Webb, 1979). Beyond 4 μM, increasing the phosphate concentration in the assay had no effect on enzyme activity (Figure 11.7). The ecological significance of this type of inhibition is that some basal rate of DOP hydrolysis will occur even at relatively high external concentrations of phosphate, such as under anaerobic conditions. An additional consequence of this type of inhibition is that these bacteria may be able to utilize a wider variety of organic compounds for growth. Chróst and Overbeck (1987) speculated that APase may be as important to carbon uptake as phosphorus uptake in aquatic bacteria (see Chapter 3). Because some organic compounds must be dephosphorylated prior to assimilation of the organic moiety (Bengis-Garber and Kushner, 1982), APase hydrolysis may enable epipelic bacteria to utilize a wider variety of organic compounds, such as DOP, even at high phosphate concentrations. This type of response would be predicted to occur if the bacteria are organic carbon limited and encounter high phosphate concentrations relative to DOP concentrations.

The differences we measured in patterns of phosphate inhibition among bacterial phosphatases in this study and the constitutive bacterial phosphatases observed by Chróst and Overbeck (1987) suggest that different phosphatases may catalyze phosphomonoester hydrolysis. High concentrations of phosphate (greater than 140 $\mu g\ l^{-1}$) occur in Plußsee in spring (Chróst and Overbeck, 1987, Chróst et al., 1989). If pelagic bacteria in this lake produced a competitively inhibited APase in response to organic carbon limitation rather than P limitation (Chróst et al., 1989), high concentrations of phosphate would inhibit enzyme activity and decrease DOC availability, assuming phosphate hydrolysis was a prerequisite for DOC uptake. Phosphatases other than APase could be responsible for hydrolysis of DOP, however. Touati et al. (1987) described a periplasmic acid phosphatase produced by *E. coli* in alkaline media. Ammerman and Azam (1985) described a 5′-nucleotidase that hydrolyzed phosphate from nucleotides. Enzyme activity was not inhibited by phosphate at 100 μM concentrations. The substrate used by Chróst and Overbeck (1987) should not be a substrate for 5′-nucleotidase, but at the high concentrations of MUP used (10–200 μM), some residual activity could lead to measurable rates in the bacterial fraction. If a significant proportion of the bacterial hydrolytic activity in this lake is from 5′-nucleotidase, then the lack of inhibition of this fraction by phosphate is not surprising. The implication is that different types of bacterial phosphatases may predominate in different habitats or ecosystems. These different enzymes that catalyze similar reactions, but with different mechanisms of regulation, may be produced in response to the varying nutrient status of bacterial communities in different systems. In habitats with high DOC loading and a P-limited bacterial community, the bacteria would produce a DOP-hydrolyzing enzyme which would be regulated by P availability. However, in habitats with low DOC loading and a low ratio of DOP to phosphate, the bacteria might produce a DOP-hydrolyzing enzyme which is relatively insensitive to P concentrations because of organic carbon limitation. By coupling our knowledge of the physiology of these bacteria to data on external nutrient concentrations and ratios, a much greater understanding of the control of bacterial growth in situ will be gained.

Conclusion APase of three strains of bacteria isolated from pelagic, epiphytic, and epipelic habitats in a small moderately eutrophic lake had varied kinetic constants, K_m, V_{max}, and K_i with the artificial phosphatase substrate methylumbelliferyl-phosphate. The epipelic strain had the lowest K_m and the highest V_{max} which indicated that it had the greatest affinity for substrate at low substrate concentrations and the greatest capacity to hydrolyze substrate at elevated concentrations. The APase of the pelagic strain had a relatively high K_m and V_{max} and therefore had the least potential of the strains studied to hydrolyze phosphatase substrate. All three phosphatases were competitively inhibited by phosphate at low concentrations. The epipelic strain demonstrated partial competitive inhibition at 4 to 8 μM concentrations of phosphate. In addition, in situ measurements of APase activity in bacteria from littoral and pelagic environments suggested that enzyme activity is greater in the littoral environment. It was suggested that differential phosphorus loading in the habitats in which these organisms grow affects their ability to utilize DOP. The loading ratio of organic carbon to phosphorus in the habitat was hypothesized to be important to the ability of a bacterium to compete for the phosphorus moiety of DOP.

References

Allison, R.D. and D.L. Purich. 1979. Practical considerations in the design of initial velocity enzyme rate assays. pp. 3–22 in Purich, D.L. (editor), *Methods in Enzymology*, vol. 63. Academic Press, New York.

Ammerman, J.W. and F. Azam. 1985. Bacterial 5′-nucleotidase in aquatic ecosystems: a novel mechanism of phosphorus regeneration. *Science* 227: 1338–1340.
APHA. 1985. *Standard Methods For The Examination of Water and Wastewater*. American Public Health Association. 1268 pp.
Atlas, R.M. and R. Bartha. 1987. *Microbial Ecology: Fundamentals and Applications*. Benjamin/Cummings, Menlo Park, California. 533 pp.
Bengis-Garber, C. and D.J. Kushner. 1982. Role of membrane-bound 5′- nucleotidase in nucleotide uptake by the moderate halophile *Vibrio costicola*. *Journal of Bacteriology* 149: 808–815.
Benner, R., Lay, J., K'Nees, K. and R.E. Hodson. 1988. Carbon conversion efficiency for bacterial growth on lignocellulose: implications for detritus-based food webs. *Limnology and Oceanography* 33: 1514–1526.
Boavida, M.J. and R.T. Heath. 1984. Are the phosphatases released by *Daphnia magna* components of its food? Limnology and Oceanography 29: 641–645.
Carlton, R.G. and R.G. Wetzel. 1988. Phosphorus flux from lake sediments: effect of epipelic algal oxygen production. *Limnology and Oceanography* 33: 562–570.
Cembella, A.D., Antia, N.J. and P.J. Harrison. 1984. The utilization of inorganic and organic phosphorus compounds as nutrients by eukaryotic microalgae: a multidisciplinary perspective: Part 1. *CRC Critical Reviews in Microbiology* 10: 317–391.
Chróst, R.J., Münster, U., Rai, H., Albrecht, D., Witzel, P.K. and J. Overbeck. 1989. Photosynthetic production and exoenzymatic degradation of organic matter in the euphotic zone of a eutrophic lake. *Journal of Plankton Research* 11: 223–242.
Chróst, R.J. and J. Overbeck. 1987. Kinetics of alkaline phosphatase activity and phosphorus availability for phytoplankton and bacterioplankton in Lake Plußsee (north German eutrophic lake). *Microbial Ecology* 13: 229–248.
Cotner, J.B. 1990. Utilization of dissolved phosphorus compounds by bacteria and algae in lakes. *Ph.D. dissertation, University of Michigan*. 137 pp.
Currie, D.J. and J. Kalff. 1984. A comparison of the abilities of freshwater algae and bacteria to acquire and retain phosphorus. *Limnology and Oceanography* 29: 298–310.
Deinema, M.H., Habets, L.H.A., Scholten, J., Turkstra, E. and H.A.A.M. Webers. 1980. The accumulation of polyphosphate in *Acinetobacter* spp. *Federation of European Microbiology Societies, Microbiology Letters* 9: 275–279.
Dixon, M. and E.C. Webb. 1979. *Enzymes*. Academic Press, New York. 1116 pp.
Echols, H., Garen, A. and A. Torriani. 1961. Genetic control of repression of alkaline phosphatase in *E. coli*. *Journal of Molecular Biology* 3: 425–438.
Fitzgerald, G.P. and T.C. Nelson. 1966. Extractive and enzymatic analyses for limiting or surplus phosphorus in algae. *Journal of Phycology* 2: 305–309.
Gächter, R., Meyer, J.S. and A. Mares. 1988. Contribution of bacteria to release and fixation of phosphorus in lake sediments. *Limnology and Oceanography* 33: 1542–1558.
Guillard, R.R.L. 1975. Culture of phytoplankton for feeding marine invertebrates. pp. 29–60 in Smith, W.L. and Chanley, M.H. (editors), *Culture of Marine Invertebrate Animals*. Plenum Press, New York.
Haines, E.B. and R.B. Hanson. 1979. Experimental degradation of detritus made from the salt marsh plants *Spartina alterniflora* Loisel., *Salicornia virginica* L., and *Juncus roemerianus* Scheele. *Journal of Experimental Marine Biology and Ecology* 40: 27–40.
Healey, F.P. and L.L. Hendzel. 1980. Physiological indicators of nutrient deficiency in lake phytoplankton. *Canadian Journal of Fisheries and Aquatic Sciences* 37: 442–453.
Heath, R.T. and G.D. Cooke. 1975. The significance of alkaline phosphatase in a eutrophic lake. *Verhandlungen der Internationalen Vereinigung für Theoretische und Angewandte Limnologie* 19: 959–965.
Jansson, M., Olsson, H. and K. Pettersson. 1988. Phosphatases; origin, characteristics and function in lakes. *Hydrobiologia* 170: 157–175.
Jones, J.G. 1972. Studies on freshwater bacteria: association with algae and alkaline phosphatase activity. *Journal of Ecology* 60: 59–75.

Kuenzler, E.J. and J.P. Perras. 1965. Phosphatases of marine algae. *Biological Bulletin* 128: 271–284.
Kulaev, I.S. 1979. *The Biochemistry Of Inorganic Polyphosphates*. Wiley and Sons, New York. 225 pp.
Losee, R.F. and R.G. Wetzel. 1988. Water movement within submersed littoral vegetation. *Verhandlungen der Internationalen Vereinigung für Theoretische und Angewandte Limnologie* 23: 62–66.
Lowry, O.H., Rosenbrough, N.J., Farr, A.L. and R.J. Randall. 1951. Protein measurement with the folin phenol reagent. *Journal of Biological Chemistry* 193: 265–275.
Malamy, M.H. and B.L. Horecker. 1964. Release of alkaline phosphatase from cells of *Escherichia coli* upon lysozyme spheroplast formation. *Biochemistry* 3: 1889–1893.
McComb, R.B., Bowers, G.N. and S. Posen. 1979. *Alkaline Phosphatases*. Plenum Press, New York. 986 pp.
McRoy, C.P., Barsdate, R.J. and M. Nebert. 1972. Phosphorus cycling in an eelgrass (*Zostera marina* L.) ecosystem. *Limnology and Oceanography* 17: 58–67.
Merril, C.R., Goldman, D., Sedman, S.A. and M.H. Ebert. 1981. Ultrasensitive stain for proteins in polyacrylamide gels shows regional variation in cerebrospinal fluid proteins. *Science* 211: 1437.
Moeller, R.E., Burkholder, J.M. and R.G. Wetzel. 1988. Significance of sedimentary phosphorus to a rooted submersed macrophyte (*Najas flexilis* (Willd.) Rostk. and Schmidt) and its algal epiphytes. *Aquatic Botany* 32: 261–281.
Mortimer, C.H. 1941. The exchange of dissolved substances between mud and water in lakes, Parts 1 and 2. *Journal of Ecology* 29: 280–329.
Mortimer, C.H. 1942. The exchange of dissolved substances between mud and water in lakes, Parts 3 and 4. *Journal of Ecology* 30: 147–201.
Newell, R.C., Linley, E.A.S. and M.I. Lucas. 1983. Bacterial production and carbon conversion based on saltmarsh plant debris. *Estuarine and Coastal Shelf Sciences* 17: 405–419.
Otsuki, A. and R.G. Wetzel. 1972. Coprecipitation of phosphate with carbonates in a marl lake. *Limnology and Oceanography* 17: 763–767.
Pettersson, K. and M. Jansson. 1978. Determination of phosphatase activity in lake water—a study of methods. *Verhandlungen der Internationalen Vereinigung für Theoretische und Angewandte Limnologie* 20: 1226–1230.
Reid, T.W. and I.B. Wilson. 1971. *E. coli* alkaline phosphatase. pp. 373–415 in Boyer, P. (editor), *The Enzymes*. Academic Press, New York.
Rigler, F.H. 1961. The uptake and release of inorganic phosphorus by *Daphnia magna* Straus. *Limnology and Oceanography* 24: 107–116.
Schneider, J. and G. Rheinheimer. 1988. Isolation methods. pp. 73–94 in Austin, B. (editor), *Methods in Aquatic Bacteriology*. Wiley and Sons, New York.
Smith, P.K., Krohn, R.I., Hermanso, G.T., Mallia, A.K., Gartner, F.H., Provenza, M.D., Goeke, N.N., Olson, B.J. and D.C. Klenk. 1985. Measurement of protein using bicinchoninic acid. *Analytical Biochemistry* 150: 76–85.
Sommer, U. 1989. Nutrient status and nutrient competition of phytoplankton in a shallow, hypertrophic lake. *Limnology and Oceanography* 34: 1162–1173.
Spencer, W. and G. Bowes. 1989. Ecophysiology of the world's most troublesome aquatic weeds. pp. 124–138 in Peiterse, A.H. and Murphy, K.J. (editors), *Aquatic Weeds*. Oxford University Press, Oxford.
Stevens, R.J. and M.P. Parr. 1977. The significance of alkaline phosphatase in Lough Neagh. *Freshwater Biology* 7: 351–355.
Stewart, A.J. and R.G. Wetzel. 1982. Phytoplankton contribution to alkaline phosphatase activity. *Archiv für Hydrobiologie* 93: 265–271.
Suelter, C.H. 1985. *A Practical Guide to Enzymology*. Wiley and Sons, New York. 288 pp.
Torriani, A. 1968. Alkaline phosphatase of *Escherichia coli*. pp. 212–218 in Grossman, L. and Moldave, K. (editors), *Methods in Enzymology*, vol. 12. Academic Press, New York.

Touati, E.D., Dassa, J. and P.L. Boquet. 1987. Acid phosphatase (pH 2.5) of *Escherichia coli*: Regulatory characteristics. pp.31–40 in Torriani-Gorini, A., Rothman, F.G., Silver, S., Wright, A. and Yagil, E. (editors), *Phosphate Metabolism and Cellular Regulation In Microorganisms*. American Society for Microbiology, Washington, D.C.
van Gemerden, H. and J.G. Kuenen. 1984. Strategies for growth and evolution of microorganisms in oligotrophic habitats. pp. 25–54 in Hobbie, J.E. and Williams, P.J.LeB. (editors), *Heterotrophic Activity In The Sea*. Plenum Press, New York.
Wetzel, R.G. 1981. Longterm dissolved and particulate alkaline phosphatase activity in a hardwater lake in relation to lake stability and phosphorus enrichments. *Verhandlungen der Internationalen Vereinigung für Theoretische und Angewandte Limnologie* 21: 369–381.

12

Phosphatase Activity in an Acid, Limed Swedish Lake

Håkan Olsson

12.1 Introduction

In Lake Gårdsjön (pH 4.6), situated in southwestern Sweden, the level of acid phosphatase activity was very high before the lake was limed (Olsson, 1983). The characteristics of the acid phosphatase activity in Lake Gårdsjön were similar to those of alkaline phosphatases, with respect to the K_m value and inhibition by orthophosphate. The results presented by Jansson et al. (1981) and Olsson (1983) indicate that small plankton, mainly Ochromonadaceae spp. (small *Ochromonas* and *Chromulina* spp.), produced most of the acid phosphatase activity in this lake. Furthermore, Jansson (1981) found that aluminum can block phosphatase substrates in acid water, and that the acid phosphatase activity was elevated during incubation of aluminum-enriched water from Lake Gårdsjön. Jansson (1981) suggested that high aluminum concentrations in Lake Gårdsjön decreased the availability of phosphate esters, and thereby increased the planktonic phosphatase production.

The phosphatase activity in Lake Gårdsjön was monitored for one and a half years after the liming of the lake. The liming, which took place in April 1982, increased the lake pH from approximately 4.7 to 7.6 (Broberg, 1987), and it was expected that this whole-lake manipulation of pH would increase the availability of phosphorus in the lake. The interactions between aluminum and phosphatase substrates in an acid environment, as shown by Jansson (1981), were competitive in relation to phosphatases. These interactions were therefore expected to be less effective if either the aluminum concentration or the fraction of free aluminum ions decreased after liming. Aluminum hydroxides should be more frequent, and free aluminum ions less frequent, at pH 7.6 compared with pH 4.7 (Stumm and Morgan, 1981). Therefore, the total concentration of aluminum in the lake water was expected to decrease after liming due to the increased sedimentation of aluminum hydroxides.

After the liming of Lake Gårdsjön, the potential phosphatase activity was measured at pH 4.5, and the result was compared with the corresponding results obtained before liming. The phosphatase activity was also measured at the pH obtained in the lake water after liming in order to test if the ambient lake pH affected the pH maximum of the phosphatase activity within the time of the investigation. On one occasion, the phosphatase activity was measured in homogenates of several size fractions of seston in order to get information about the origin of the phosphatases.

12.2 Methods and Environmental Analyses

Sampling and basic analytical program Water samples were collected from Lake Gårdsjön, an oligotrophic, clear-water lake, acidified by acid rain and limed. The lake is situated in southwestern Sweden (latitude 58°3′34″N, longitude 12°1′36″E; altitude 113 m). The surface area of the lake is 0.312 km^2, and constitutes 15% of the drainage area. The mean and maximum depths of Lake Gårdsjön are 4.9 and 18.5 m, respectively, and the water retention time is about one year (Broberg, 1987). Further characteristics of the lake and drainage area are given by Olsson et al. (1985).

Water samples for phosphatase assay were taken on 20 occasions from May 1980 to May 1981 (pre-liming), and on 33 occasions from May 1982 to November 1983. During winter, when the lake was ice-covered, composite samples were collected from two strata, the epi- and the hypolimnion. During summer stratification, composite samples were taken biweekly from three strata: epi-, meta-, and hypolimnion. When the lake was isothermal, one composite sample was taken from the whole water column. Samples were frozen at minus 20°C and stored until analysis, which took place within 150 days after sampling. On three occasions after the liming of Lake Gårdsjön (8 July 1982, 23 September 1982, and 14 June 1983), unfrozen samples were brought to the laboratory in Uppsala for various special investigations.

Monitoring of total and particulate phosphorus fractions, total aluminum, and phytoplankton biomass and community composition were made with the above sampling strategy and sampling dates. Sampling, as well as analytical procedures, were identical before and after liming (Nilsson, 1985; Broberg, 1987; Larsson, 1988).

Phosphatase activity assay Phosphomonoesterase activity (phosphatase activity) was analyzed at 20°C using the substrate 4-methylumbelliferyl phosphate in Tris-malate-acetate buffer (1 mM with respect to each component) as described by Jansson et al. (1981). The assay was performed at pH 4.6 on samples taken before liming, and at pH 4.5 and 7.5 on samples taken after liming. The product of substrate hydrolysis (methylumbelliferone) was detected with a Turner 111 fluorometer described by Olsson (1983) and Jansson et al. (1981). The phosphatase activity is given in nmol 4-methylumbelliferyl phosphate hydrolyzed per liter of sample and per minute.

Freezing correction The effect of freezing on the phosphatase activity was tested on samples taken in Lake Gårdsjön for special investigations. Subsamples (20 ml) of lake water were frozen at minus 20°C immediately after analysis of phosphatase activity in the original sample. Remaining phosphatase activity in the subsamples was analyzed at pH 4.5 and 7.5 in time series of up to 215 days. The tests made in 1982 to 1983 showed no significant difference between values obtained 2 days after freezing and values obtained in samples stored frozen for a longer time (SAS GLM Least squares means, $P < 0.05$). Therefore, it was assumed that freezing and thawing caused the main losses of activity, and that the effect of storage longer than 2 days was of minor importance. The average remaining activity for each sampling event varied between 56 and 68% (SD ±10–14). Calculated as an average for all measurements made on samples taken in July and September 1982 and June 1983, the phosphatase activity in deep-frozen samples was 61% of the activity in unfrozen water (CV = 22, n = 158). There were no significant differences between the effects of freezing on phosphatase activity measured at pH 4.5 and pH 7.5 or between samples taken from different strata in the lake (SAS GLM Least squares means, $P < 0.05$). A 39% correction for freezing losses was therefore applied to freezer-stored water sampled after liming.

Olsson (1983) measured a rapidly declining phosphatase activity in frozen samples taken in Lake Gårdsjön before liming (1980–1981), and corrected frozen samples according to a curve obtained with nonlinear least squares summary statistics. The model predicted rapid initial losses, with the remaining activity reaching an asymptote of approximately 40% after two weeks. The average activity was 44% of the activity in unfrozen water, calculated as above, from all observations made on water stored frozen for 2 days or more (CV = 23, n = 22). Consequently, the freezing correction made for samples taken before liming was approximately 15% higher then the correction made for samples taken after liming.

Size fractionation of seston and phosphatases In order to relate the phosphatase activity to different size fractions of seston, this experiment was performed using a composite sample representing the whole water column. Samples taken on 14 June 1983 from the epi-, meta-, and hypolimnion in Lake Gårdsjön were pooled together in proportion to the volume of each strata in the lake. The composite sample was sequentially filtered on nylon meshes: 15 liters on 100-μm mesh, 7 liters on 30-μm mesh, 3 liters on 5-μm mesh and 2 liters on a Whatman GF/F Glassfiber filter (approx. 0.5-μm pore size). The filters were homogenized in Tris-malate-acetate buffer, and after centrifugation, the phosphatase activity in the supernatant was analyzed at pH 4.5 and 7.5. The phosphatase activity was also measured in the water which had passed the meshes and the filters. Quantitative phytoplankton analysis was made on the composite sample taken 14 June 1983. The density of the cells was assumed to be 1 g cm^{-3}.

In order to characterize the phosphatase activity of different sestonic size fractions, a technique often used for separation of biomolecules with respect to molecular weight ratio was applied. Gel filtrations (using Sephadex G-200) were made with a homogenate of particles collected on a Whatman GF/F filter, water passing a 0.2-μm membrane filter (concentrated by ultrafiltration, Diaflo filter PM 10), and with homogenates from the sequential size fractionation. Gel filtrations were performed on columns with a total volume of 118–120 ml and a void volume of 50 ml (determined with blue dextran). Elution positions were characterized by K_{av} values (Jansson et al., 1981). Tris-malate-acetate buffers of pH 4.5 and 7.5 were used as eluants, and 5-ml fractions were collected and analyzed for phosphatase activity at pH 4.5 and 7.5, respectively. Acetate buffer (0.1 M, pH 4.5) was used to test the effect of the choice of eluant. This buffer was used by Jansson et al. (1981) during the corresponding experiment performed before liming of Lake Gårdsjön.

The recovery of phosphatase activity after gel filtrations of the dissolved fraction varied between 90% and 100%, while the recovery was about 200% after gel filtration of phosphatases from the particulate fraction. A high recovery can be a result of a separation of phosphatases and an inhibitor, e.g., orthophosphate, during passage of the gel. After ultrafiltration of dissolved phosphatases, about 50% of the expected phosphatase activity was recovered in the concentrate. This may be due to aggregation and/or inactivation of enzymes, or to an effect of a nonlinear relationship between the enzyme concentration and the enzyme activity at the substrate concentration used.

12.3 Results

Phosphorus and aluminum concentrations Total and particulate phosphorus concentrations in Lake Gårdsjön were low before and after liming (Figure 12.1). Phosphorus was the growth-limiting nutrient for plankton before liming (Jansson et al., 1986), and it is also considered to be the most important growth-limiting nutrient after liming, since no significant changes in phosphorus and nitrogen concentrations were detected before and after liming (Broberg, 1987).

Aluminum concentrations were lower in Lake Gårdsjön after liming. This is especially noticeable when the data from 1980 is compared with 1983. During the stratified summer period in 1983, the mean and median concentrations in all strata were about 50% (maximum 75%, minimum near 0%) of the concentrations obtained during the same period in 1980. During the stratified period in 1982, the aluminum concentrations were about 70% (maximum 110%, minimum 40%) of the concentrations in 1980 (I. Nilsson, unpublished observations).

Phytoplankton biomass and composition Before liming, the total phytoplankton biomass in the epilimnion (whole water column during circulation) was below 0.5 mg l^{-1} (Figure 12.2). The dominating groups were Dinophyceae and Chrysophyceae (Lydén and Grahn, 1985; Larsson, 1988). After liming, in

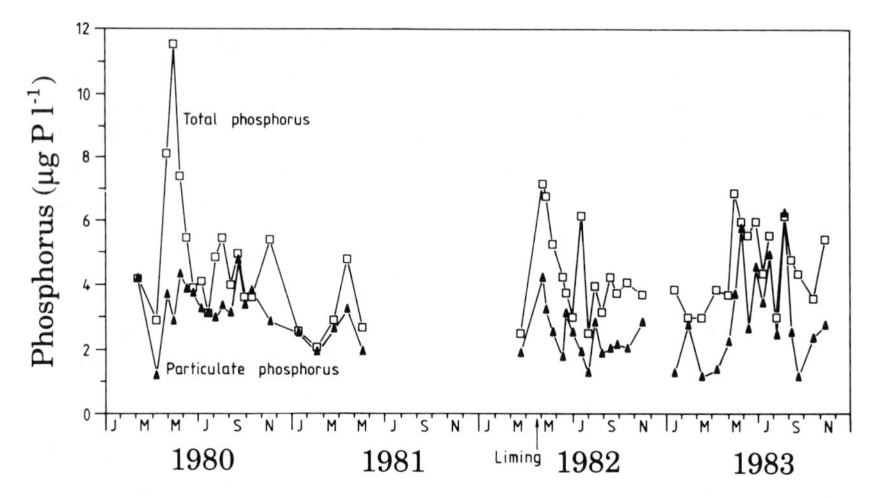

Figure 12.1 Total phosphorus and particulate phosphorus concentrations (data from Broberg, 1987) in the epilimnion (and during homothermy) of Lake Gårdsjön.

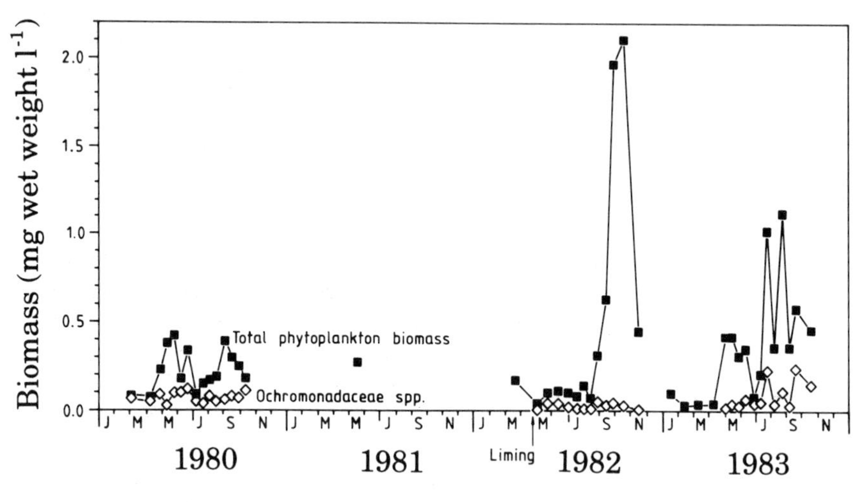

Figure 12.2 Total phytoplankton biomass and biomass of Ochromonadaceae in the epilimnion (and during homothermy) of Lake Gårdsjön (data from Larsson, 1988).

spring 1982, the phytoplankton biomass declined but Chrysophyceae and Dinophyceae still dominated. During summer 1982, the biomass was low and dominated by *Chrysidiastrum catenatum*. In autumn 1982, the biomass reached high values due to the presence of high levels of *Synedra nana* and *Dinobryon bavaricum* (Larsson, 1988).

In 1983, the phytoplankton biomass was often near 0.5 mg l^{-1}, and the composition was more diverse as compared with previous years. In spring,

Dinobryon sociale var. *americanum* and *Gymnodinium uberrimum* were most frequent, while *Ochromonas* sp. and *Monoraphidium minutum* dominated in late June and early July when the biomass was relatively low. Later, during summer and autumn, the phytoplankton biomass was above or near 0.5 mg l^{-1}. Cryptophyceae were dominating in late July and August; Ochromonadaceae and *Dinobryon* spp. in September; and Ochromonadaceae, Chlorophyceae, *Dinobryon* spp. and *Cosmocladium perissum* in October (Larsson, 1988). On average, Ochromonadaceae contributed 42% (SD ±21) to the total biomass (volume-weighted average for the whole water column) in 1980 (before liming), and 18% (SD ±15) in 1982 and 1983 (after liming).

Phosphatase activity measured at pH 4.5 and 7.5 The potential phosphatase activity measured at low pH was highest in samples taken during early summer 1980 (Figure 12.3). During other periods of the year, the differences in the phosphatase activity measured at low pH before and after liming were small in relation to the possible error in the freezing correction (see Section 12.2). The phosphatase activity at pH 4.5 was relatively high in the month of May 1980, 1981, and 1982. After liming, the phosphatase activity measured at pH 4.5 was usually higher than the activity measured at pH 7.5.

After liming, in spring 1982, the phosphatase activity measured at pH 4.5 declined and the activity at pH 7.5 was low (Figure 12.3). The phosphatase activity increased in autumn 1982, as did the phytoplankton biomass (Figure 12.2). Low phosphatase activities during winter 1983 were followed by higher activities in May and June, when the phytoplankton biomass had increased. The phosphatase activity and the phytoplankton biomass were low in late June to early July, and both variables were higher later in summer and in autumn.

The degree of covariation between variables is presented as a correlation matrix in Table 12.1. Before liming, the covariation between Ochromonadaceae spp. and the acid phosphatase activity was better than the covariation between total phytoplankton biomass and acid phosphatase activity. After liming, the covariation between the biomass of Ochromonadaceae and the acid phosphatase activity was weaker. The covariation between the phosphatase activity measured at pH 7.5 after liming and phytoplankton variables was stronger.

During summer stratification, the specific phosphatase activity measured at low pH (calculated per total phytoplankton biomass) often was lower after liming (Figure 12.4). Means and medians were lowest in 1983, the second summer after liming, and highest in 1980, prior to liming (Table 12.2).

Size separation of seston-bound and dissolved phosphatases Phosphatase activities in particles trapped on 100-μm and 30-μm meshes were very low, near the detection limit, and there were no detectable decreases in the activity in water passing the 100-μm and 30-μm meshes. The sum of the phosphatase activity in all fractions was 80% and 60% of the activity in the original sample at pH 4.5 and pH 7.5, respectively. It is not likely that this discrepancy was due to losses of phosphatases from particles during filtration since the activity

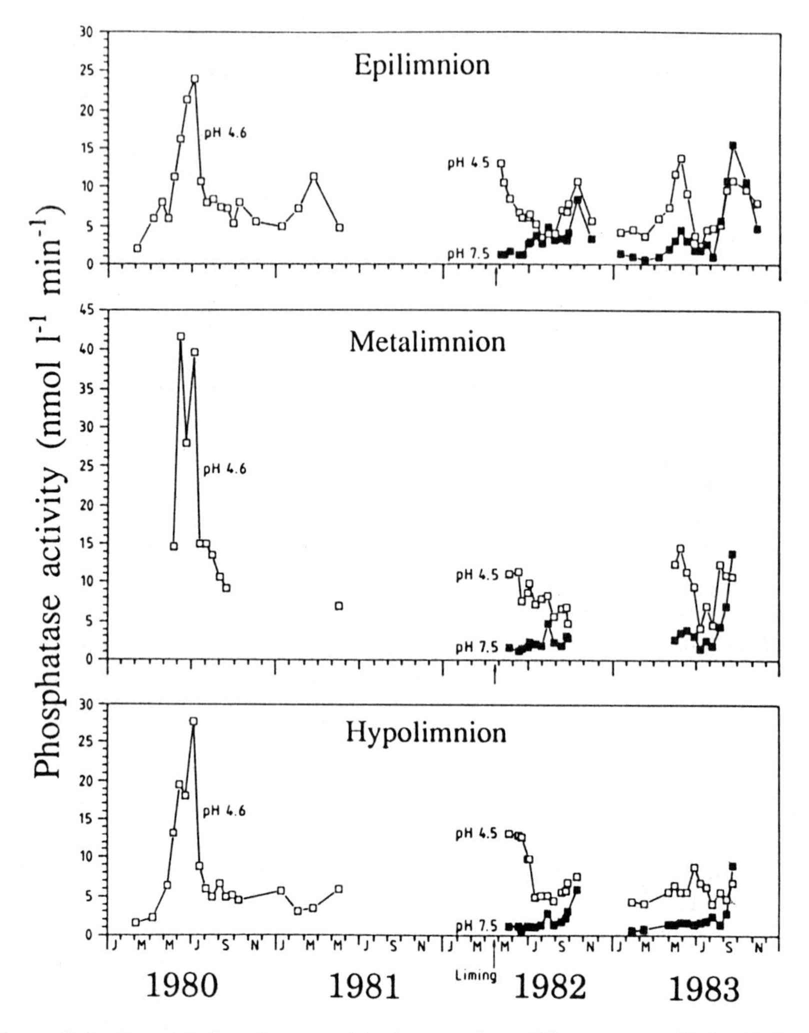

Figure 12.3 Potential phosphatase activity in water from different strata of Lake Gårdsjön. "Epilimnion" data includes data from periods when the lake is homothermal. Phosphatase activity measured at pH 4.5 (open squares) and at pH 7.5 (solid squares).

in the <0.5-μm fraction (Whatman GF/F) after subsequent filtration was equal to the activity in the filtrate obtained after filtration directly on the Whatman GF/F filter. It was therefore assumed that phosphatases bound to particles were partly inactivated during the preparation of the particulate fractions.

Based on this assumption, the distribution among the size fractions of the total phosphatase activity in the original sample was calculated. The observed loss of phosphatase activity during fractionations was added to the activity in particulate fractions in proportion to the measured activity in the fractions. According to this calculation, over 50% of the phosphatase activity was found in

Table 12.1 Correlation matrix for observations in Lake Gårdsjön before and after liming (volume-weighted means for the whole watercolumn)[1]

	PA 4.5	PA 7.5	Phyto	Ochro	TP	PP
				Before liming		
PA 4.5	—	—	0.08	0.48	−0.04	0.14
PA 7.5	0.34	—	—	—	—	—
Phyto	0.07	0.35	—	0.55	0.78	0.36
Ochro	0.16	0.57	0.32	—	0.28	0.42
TP	0.14	−0.09	−0.09	0.11	—	0.30
PP	0.01	−0.07	0.04	0.16	0.50	—
	After liming					

[1]Number of observations: before liming = 15; after liming = 28. Abbreviations: PA 4.5 and PA 7.5 = phosphatase activity measured at pH 4.5 and 7.5, respectively; Phyto = total phytoplankton biomass; Ochro = biomass of Ochromonadaceae species; TP = total phosphorus; PP = particulate phosphorus. No measurements were done at pH 7.5 before liming.

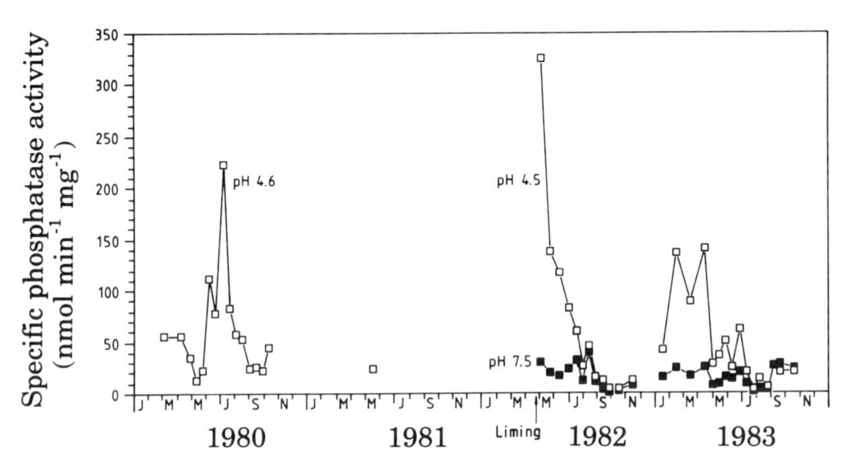

Figure 12.4 Specific phosphatase activity (phosphatase activity per phytoplankton biomass) in Lake Gårdsjön during ice-free periods. Volume-weighted mean values for the whole water column. Phosphatase activity measured at pH 4.5 and 4.6 (open squares) and at pH 7.5 (solid squares).

water passing the Whatman GF/F filter and the activity in this fraction was approximately equal to the activity measured in water passing a 0.2-μm membrane filter (Table 12.3). The results for the fractions of >5 μm are approximations because the recovery on filters was below 100%, and there were small differences in activity in water passing different meshes.

Quantitative analysis of the phytoplankton composition in the sample taken on 14 June 1983 showed that the total biomass (0.467 mg wet weight l^{-1}) was dominated by *Gymnodinium* spp. (39%) and *Dinobryon sociale* var. *americanum*

Table 12.2 Descriptive statistics of the phosphatase activity measured at pH 4.5 to 4.6 during 1980, 1982, and 1983[1]

	Specific phosphatase activity (nmol min^{-1} mg^{-1})		
	1980	1982	1983
Mean	70	56	27
Median	55	46	24
Maximum	222	138	62
Minimum	22	4	4

[1]The statistics were calculated per total phytoplankton biomass based on 10 observations (volume-weighted means for the whole water column of Lake ĞUrdsjön) per year within the period mid-May to September.

Table 12.3 Size distribution of phosphatase activity measured at pH 4.5 and 7.5[1]

	Phosphatase activity			
Size fraction	At pH 4.5		At pH 7.5	
(μm)	In nmol l^{-1} min^{-1}	In %	In nmol l^{-1} min^{-1}	In %
>30	0.4	5	0.05	2
5–30	0.7	8	0.25	9
0.5–5.0	1.8	20	0.95	35
<0.5	6.0	67	1.50	54
<0.2	6.0	67	1.30	47
Total	8.9	100	2.75	100

[1]Volume-weighted composite samples were collected from the whole water column of Lake Gårdsjön on 14 June 1983.

(30%). Individuals and colonies of both species were large enough to be trapped on a 30-μm mesh. Approximately 20% of the phytoplankton biomass were Ochromonadaceae species of <10 μm in size, one-fourth of which could pass a 5-μm mesh. Calculated in relation to the number of individuals, Ochromonadaceae of <5 μm contributed 29% to the total number of phytoplankton while species of >30 μm contributed only 12%.

Gel filtration Elution diagrams from gel filtration performed at pH 4.5 and 7.5 showed differences between free dissolved phosphatases (<0.2 μm) and seston-bound phosphatases (>0.5 μm). At pH 4.5, dissolved phosphatases gave a single peak at K_{av} 0.2–0.3, irrespective of whether tris-malate-acetate or acetate buffer was used as eluant (acetate buffer was used in gel filtration done before liming of Lake Gårdsjön; Jansson et al., 1981), while phosphatases in the particulate fraction eluted at the void volume (K_{av} 0) and around K_{av} 0.5–0.7 (Figure 12.5). At pH 4.5, sestonic subfractions showed detectable activities exclusively in the void fraction. Elution of the particulate fraction at pH 7.5

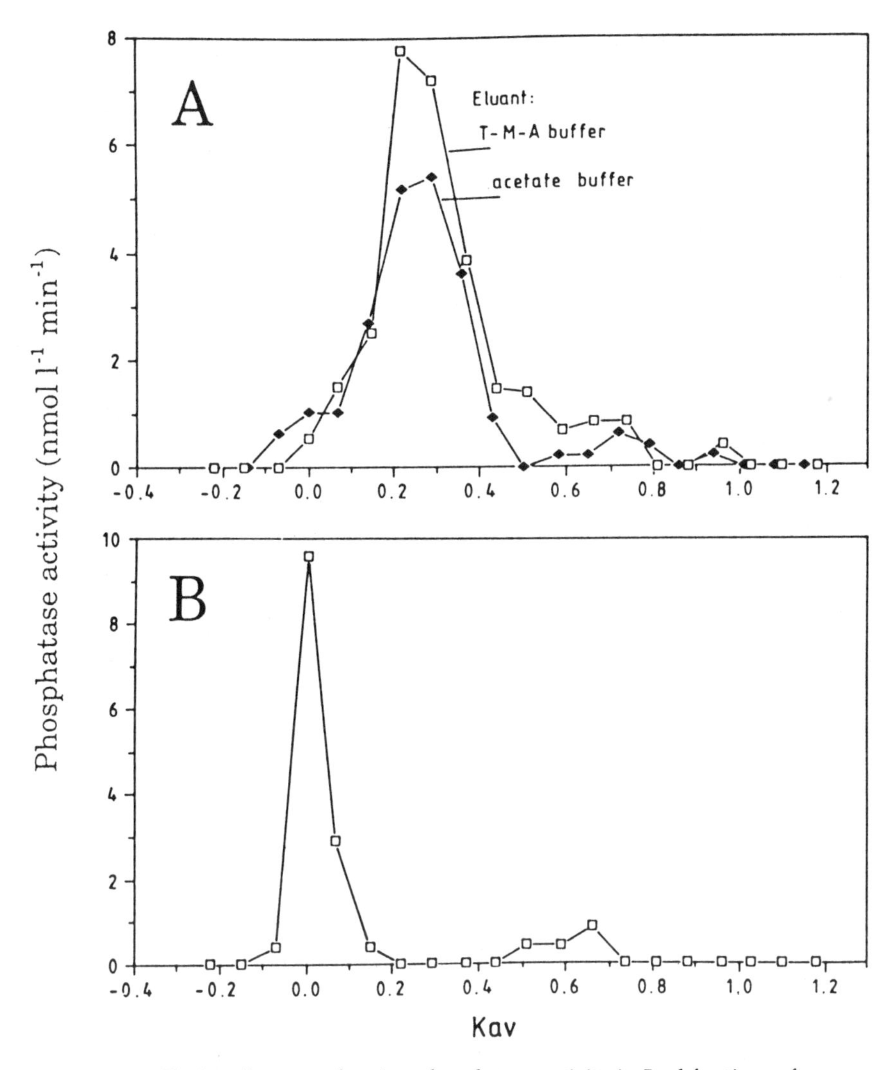

Figure 12.5 Elution diagrams showing phosphatase activity in 5-ml fractions after passage through a Sephadex G-200 gel. Sample water was taken in Lake Gårdsjön on 14 June 1983. Elution and analysis were performed at pH 4.5. (A) Separation of concentrated (ultrafiltration) membrane-filtered water (0.2-μm pore size). T-M-A buffer = Tris-malate-acetate buffer; (B) Separation of a homogenate of the particulate fraction trapped on a Whatman GF/F filter (approximate pore size 0.5 μm).

gave the lowest phosphatase activities at K_{av} 0.3–0.45, a position at which the activity was high after gel-filtering the dissolved fraction. At pH 7.5, sestonic subfractions showed detectable activities in the void and between K_{av} 0.5 and 0.8. A void peak also appeared after gel filtration of dissolved phosphatases at pH 7.5.

12.4 Discussion

Plankton organisms of various types can produce phosphatase activity in lake water (Jansson et al., 1988). In the investigation of phosphatases in Lake Gårdsjön before liming, small species of Ochromonadaceae appeared to be the main producers of phosphatases (Olsson, 1983). After liming, Ochromonadaceae became a minor or less dominating part of the phytoplankton biomass compared with the pre-liming period, and the covariation between Ochromonadaceae spp. and phosphatase activity was not much better than the covariation between total phytoplankton biomass and phosphatase activity (Table 12.1). Phytoplankton species other than Ochromonadaceae spp. and bacteria may therefore be the more important phosphatase producers after liming. Bacterial biomass was not analyzed in Lake Gårdsjön after liming, and therefore the covariation with phosphatase activity could not be tested.

The result of the size fractionation of phosphatase carriers was similar to the result obtained in the corresponding experiment performed before liming (Jansson et al., 1981). A major part of the phosphatase activity was found in the filtrate and in the fraction of seston passing a 5-μm mesh (Table 12.3). According to these results, small species of Ochromonadaceae and bacteria could be important contributors to the phosphatase activity measured in Lake Gårdsjön after liming. The major part of the phosphatase activity was, however, found in the dissolved fraction, and the origin of these phosphatases is unknown.

Gel filtration characteristics of dissolved phosphatases and phosphatases in sestonic size fractions did not give any clue revealing the origin of the dissolved phosphatases. Differences in the elution diagrams obtained from gel filtration of filtrate and seston preparations indicate that seston-bound phosphatases and dissolved phosphatases are of different types. It is possible that the dissolved phosphatases are of a type actively excreted by plankton, for the purpose of making phosphate from organic compounds available for algae and the organic compounds more available for bacteria (Chróst and Overbeck, 1987; Chróst et al., 1989).

The average specific phosphatase activity measured at pH 4.5–4.6 and calculated per total phytoplankton biomass was highest during the summer period in the unlimed Lake Gårdsjön. If the total phytoplankton biomass is the most relevant biomass estimator of the phosphatase producing organisms, the results may indicate a more severe phosphorus deficiency before liming (Berman, 1970; Healey and Hendzel, 1979; Pettersson, 1980; 1985; Gage and Gorham, 1985). The observed differences in the specific phosphatase activity could, however, be a result of a change in the plankton composition to species or groups of organisms with the ability to produce phosphatases with optimum activity at higher pH (Olsson, 1990).

Within two and a half years after the liming of Lake Gårdsjön, the phosphatase activity measured at pH 4.5 was usually higher than the activity measured at pH 7.5 (Figure 12.3). The differences between the measurements were greatest in spring and early summer, a period during which species of Dinophyceae and Chrysophyceae were dominant. Chrysophyceae species were also frequently found during periods when there were small differences between phosphatase activities measured at pH 4.5 and 7.5. The dinophyceans in Lake Gårdsjön were dominated by *Peridinium* spp. in spring 1982 and by *Gymnodinium uberrimum* in spring 1983 (Larsson, 1988).

Species of Dinophyceae are potential producers of acid phosphatases in Lake Gårdsjön, but other producers of acid phosphatases must also exist, since high phosphatase activity measured at pH 4.5 was also detected in Lake Gårdsjön during periods when Dinophyceae were not found in the water. According to the results from Lake Gårdsjön and from other Swedish lakes (Olsson, 1990), the acid phosphatase activity is higher than the alkaline when species of Ochromonadaceae and Dinophyceae are dominating the phytoplankton biomass. The phytoplankton biomass in Lake Gårdsjön was more dominated by species of Ochromonadaceae and Dinophyceae before liming. This change in the phytoplankton composition may be the explanation for the observed difference in the specific phosphatase activity calculated per total phytoplankton biomass before and after liming. Therefore, the results from this study cannot justify the theory that phytoplankton in this lake were more phosphorus deficient before liming than after.

The pH optimum of the phosphatase activity can also change due to an "adaptation" of the planktonic production of phosphatases to the new lake pH (Boavida and Heath, 1986). Since the acid phosphatase activity still was highest two years after liming of the lake, the phosphatase production was not adapted to the new lakewater pH. Also, in other Swedish forest lakes of neutral pH the optimum activity of phosphatases was found at low pH (Olsson, 1990). This fact could be due to frequent abundance of algal species or bacteria with the ability to produce acid phosphatases.

Acid phosphatase activity may preferably be produced by species of Chrysophyceae (Reichardt and Overbeck, 1969; Patni et al., 1974; Patni and Aaronson, 1974; Healey and Hendzel, 1979) and by heterotrophic bacteria (Siuda, 1984). Species of Chrysophyceae were often abundant in Lake Gårdsjön even after liming, as bacteria probably were. Before liming, the numbers of bacteria in the lake were within the range found in other oligotrophic lakes, but the mean cell volume was large (Andersson, 1983). After liming of Lake Gårdsjön, the transparency of the lake water decreased (Larsson, 1988), and during 1982 and 1983 the phytoplankton biomass in the hypolimnion was approximately 50% of the biomass in the epilimnion. The relatively high acid phosphatase activity in the hypolimnion detected after liming could therefore mainly be of bacterial origin.

Acknowledgments Financial support was received from the Swedish Environmental Protection Agency and the study was carried out within the integrated Lake Gårdsjön project. I thank the members of the Lake Gårdsjön project, especially for their help in supplying data used in this investigation. Ola Broberg, Stefan Larsson, and Ingvar Nilsson supplied the data on phosphorus concentrations, phytoplankton composition, and aluminum concentrations, respectively. This work was done during my stay at the Institute of Limnology, University of Uppsala, Sweden.

References

Andersson, I. 1983. Bacterioplankton in the acidified Lake Gårdsjön. *Hydrobiologia* 101: 59–64.

Berman, T. 1970. Alkaline phosphatases and phosphorus availability in Lake Kinneret. *Limnology and Oceanography* 15: 663–674.

Boavida, M.J. and R.T. Heath. 1986. Phosphatase activity of *Chlamydomonas acidofila* Negoro (Volovcales, Chlorophyceae). *Phycologia* 25: 400–404.

Broberg, O. 1987. Nutrient responses to the liming of Lake Gårdsjön. *Hydrobiologia* 150: 11–24.

Chróst, R.J., Münster, U., Rai, H., Albrecht, D., Witzel, P.K. and J. Overbeck. 1989. Photosynthetic production and exoenzymatic degradation of organic matter in the euphotic zone of a eutrophic lake. *Journal of Plankton Research* 11: 223–242.

Chróst, R.J. and J. Overbeck. 1987. Kinetics of alkaline phosphatase activity and phosphorus availability for phytoplankton and bacterioplankton in Lake Plußsee (North German eutrophic lake). *Microbial Ecology* 13: 229–248.

Gage, M.A. and E. Gorham. 1985. Alkaline phosphatase activity and cellular phosphorus as an index of phosphorus status of phytoplankton in Minnesota lakes. *Freshwater Biology* 15: 227–233.

Healey, F.P. and L.L. Hendzel. 1979. Fluorometric measurement of alkaline phosphatase activity in algae. *Freshwater Biology* 9: 429–439.

Jansson, M. 1981. Induction of high phosphatase activity by aluminum in acid lakes. *Archiv für Hydrobiologie* 93: 32–44.

Jansson, M., Olsson, H. and O. Broberg. 1981. Characterization of acid phosphatases in the acidified Lake Gårdsjön, Sweden. *Archiv für Hydrobiologie* 92: 377–395.

Jansson, M., Olsson, H. and K. Pettersson. 1988. Phosphatases; origin, characteristics and function in lakes. *Hydrobiologia* 170: 157–175.

Jansson, M., Persson, G. and O. Broberg. 1986. Phosphorus in acidified lakes: The example of Lake Gårdsjön, Sweden. *Hydrobiologia* 139: 81–96.

Larsson, S. 1988. Effects of lime treatment on species composition and biomass of phytoplankton in Lake Gårdsjön. pp. 245–280 in Dickson, W. (editor), *Liming of Lake Gårdsjön. An Acidified Lake in SW Sweden*, Swedish Environment Protection Board. Report No 3426.

Lydén, A. and O. Grahn. 1985. Phytoplankton species composition, biomass and production in Lake Gårdsjön—an acidified clearwater lake in SW Sweden. pp. 195–202 in Andersson, F. and Olsson, B. (editors), *Lake Gårdsjön. An Acid Forest Lake and its Catchment*. vol. 37, Ecological Bulletins, Stockholm.

Nilsson, I. 1985. Budgets of aluminum species, iron and manganese in the Lake Gårdsjön catchment in SW Sweden. pp. 120–132 in Andersson, F. and Olsson, B. (editors), *Lake Gårdsjön. An Acid Forest Lake and its Catchment*, vol. 37, Ecological Bulletins, Stockholm.

Olsson, H. 1983. Origin and production of phosphatases in the acid Lake Gårdsjön. *Hydrobiologia* 101: 49–58.

Olsson, H. 1990. Phosphatase activity in relation to phytoplankton composition and pH in Swedish lakes. *Freshwater Biology* 23: 353–362.

Olsson, B., Hallbäcken, L., Johansson, S., Melkerud, P.A., Nilsson, S.I. and T. Nilsson. 1985. The Lake Gårdsjön area—physiographical and biological features. pp. 10–28 in

Andersson, F. and Olsson, B. (editors), *Lake Gårdsjön. An Acid Forest Lake and Its Catchment*. vol. 37, Ecological Bulletins, Stockholm.

Patni, N.J. and S. Aaronson. 1974. Partial characterization of the intra- and extracellular acid phosphatase of an alga, *Ochromonas danica*. *Journal of General Microbiology* 83: 9–20.

Patni, N.J., Aaronson, S., Holik, K.J. and R.H. Davis. 1974. Existence of acid and alkaline phosphohydrolase activity in the phytoflagellate *Ochromonas danica*. *Archiv für Mikrobiologie* 97: 63–67.

Pettersson, K. 1980. Alkaline phosphatase activity and algal surplus phosphorus as phosphorus-deficiency indicators in Lake Erken. *Archiv für Hydrobiologie* 89: 54–87.

Pettersson, K. 1985. The availability of phosphorus and the species composition of the spring phytoplankton in Lake Erken. *Internationale Revue Gesamten Hydrobiologie* 70: 527–546.

Reichardt, W. and J. Overbeck. 1969. Zur enzymatischen Regulation der Phosphatmonoesterhydrolyse durch Cyanophyceenplankton. *Berichte Deutsches Botanische Gesellschaft* 81: 391–396.

SAS Institute Inc. *SAS User's Guide: Statistics*, Version 5 Edition. Cary, NC: SAS Institute Inc., 1985. 956 pp.

Siuda, W. 1984. Phosphatases and their role in organic phosphorus transformation in natural waters. A review. *Polskie Archiwum Hydrobiologii* 31: 207–233.

Stumm, W. and J.J. Morgan. 1981. *Aquatic Chemistry*. John Wiley and Sons, Inc., New York. 780 pp.

13

Phosphatase Activities in Lake Kinneret Phytoplankton

David Wynne, Bina Kaplan, and Thomas Berman

13.1 Introduction

Aquatic microorganisms (phytoplankton, bacteria) require a constant supply of nutrients in order to grow and divide and to carry out the multitude of metabolic processes occurring in the cell. The inorganic forms of macronutrients such as P and N are generally the only compounds that can be directly taken up and utilized by algae although, it has been shown that both organo-P and organo-N compounds can be used as indirect sources of nutrients (Antia et al., 1975; Cembella et al., 1984a; 1984b).

To metabolize organophosphoric ester compounds, algal cells have a variety of enzymes available, of which alkaline phosphatase (APase) has been the most widely studied. When external, ambient orthophosphate (soluble reactive phosphorus; SRP) concentrations are low (<2 μM), synthesis of APase is induced, enabling the hydrolysis of organophosphoric esters to be carried out (Figure 13.1, pathway A). The liberated orthophosphate is then available for further use by the cell.

Since APase activity increases at low ambient SRP concentrations (when P deficiency in the cells become more pronounced) it has been suggested that levels of this enzyme are a useful indicator of the P-status of algae (Fitzgerald and Nelson, 1966; Healey, 1978). Later studies, however, have suggested that APase activity may not completely reflect the degree of P limitation of phytoplankton. For example, even though high APase activities were observed in P-limited phytoplankton, addition of orthophosphate did not result in a concomitant drop in APase activity (Healey, 1973; Healey and Henzel, 1976). In addition, low APase activity can be found in algal cells, even at high ambient SRP concentrations (Wynne, 1981a). This latter activity may reflect that of a constitutive form of the enzyme, not involved in P nutrition and possibly utilizing

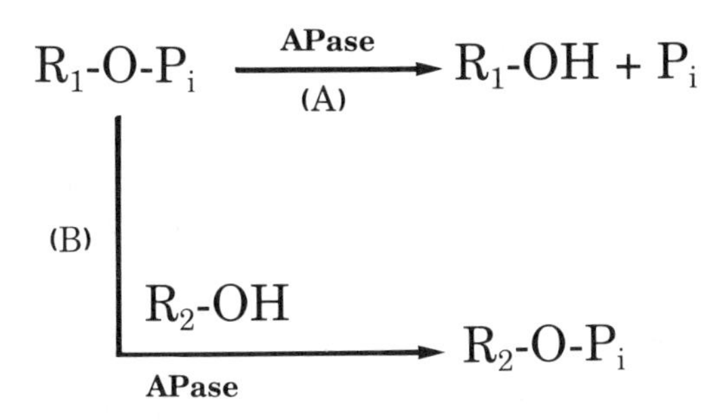

Figure 13.1 Schematic diagram showing the action of phosphatases on organophosphoric ester compounds. A = hydrolysis; B = phosphate transfer; R_1 and R_2 are any organic moiety.

the "phosphate transferase" ability of APase (Figure 13.1, pathway B; Barman, 1969) not usually considered in such studies. Furthermore, diurnal changes in cellular APase activity (Rivkin and Swift, 1979; Wynne, 1981a; 1981b) and various ecological factors, such as temperature and growth rate (Rhee, 1973); N:P ratio and spectral quality (Wynne and Rhee, 1986); and heavy metals (Wynne and Pieterse, unpublished data), all influence APase activity in phytoplankton, thereby making it difficult to draw conclusions on algal P status based solely on enzyme levels.

The Kinneret (the Biblical Sea of Galilee), situated in the northern part of Israel, is the only natural freshwater lake in the country. The chemistry, physics, and biology of Lake Kinneret have been extensively investigated over the past 20 years (Serruya, 1978; Berman, 1985). Concentrations of inorganic nutrients in the epilimnion of the lake can be characterized as follows (Berman et al., 1984): a peak of NH_4^+ during December-January; a rise in nitrite levels to a maximum in January; followed by a sharp increase in nitrate during early spring, largely as a result of runoff from agricultural areas during the rainy season. SRP concentrations in the epilimnion are always low (ranging from 0.03–0.3 μM) mainly due to the calcium precipitation of the incoming phosphate.

The phytoplankton population in the lake is dominated by the annual bloom of the freshwater dinoflagellate *Peridinium gatunense* (formerly *P. cinctum;* Hickel and Pollingher, 1988). Maximum biomass of *Peridinium* (>300 g m^{-2}) is reached in late spring approximately coinciding with high nitrate concentrations in the epilimnion. During the summer months, phytoplankton biomass drops to about 20 g m^{-2} and is mostly comprised of a highly productive population of nanoplankton (Pollingher, 1981).

Since *Peridinium* reaches such a high biomass in the lake despite low SRP concentrations, we initially concentrated on the phosphorus metabolism of this dinoflagellate. Results have shown that, whereas levels of *Peridinium* acid phosphatase are always high, alkaline phosphatase activity increases only towards the end of the bloom period, possibly coinciding with a switch from high to low N:P atomic ratios in the upper epilimnic waters (Wynne, 1977; 1981a; 1981b). These results, together with those from more recent studies, have suggested that *Peridinium* is not P limited in the Kinneret, at least not at the beginning of the bloom. In this paper we summarize some of our studies on the phosphatases of Kinneret phytoplankton populations and their role in P nutrition.

13.2 Sampling

Water samples were collected from a central or off-shore station in the Kinneret (Stations A and F, respectively) at depths of 1–5 m, mixed, filtered through a large-mesh (300-μm) net to remove zooplankton and large particles, and used as described below.

Phosphatase activity Phosphatase activity was determined using a modified filter method (Wynne, 1981a; Berman et al., 1990) as follows: Lake water was filtered through 47-mm (0.2-μm pore size) filters and the filter placed on the bottom of a 100-ml beaker. To measure activity in different size fractions of Lake Kinneret phytoplankton, 47-mm Nuclepore filters of various pore sizes were used. For the spectrophotometric phosphatase assay, 2.6 ml of buffer (for acid phosphatase: 0.1 M acetate, pH 5; for APase: 0.05 M Tris HCl containing 1 mM $MgCl_2$, pH 8.6) and 0.1 ml 10 mM *p*-nitrophenylphosphate (PNPP) were then added and the samples incubated at 37°C. After 1 hour of incubation 1 M NaOH (0.3 ml) was added to each beaker, the sample was filtered, and the yellow *p*-nitrophenol formed was measured at 410 nm. Concentrations were calculated using a molar extinction coefficient of 18,000.

For the fluorometric APase method, PNPP was replaced with 4-methylumbelliferyl phosphate (MUP; Pettersson, 1980). The reaction mixture comprised of 4 ml 0.05 Tris HCl buffer (pH 8.6, 1 mM $MgCl_2$) and 0.5 ml 0.1 mM MUP. After 1 hour, the reaction mixture was filtered (0.45 μm) and the fluorescence of the 4-methylumbelliferone (MUF) formed was measured on a Turner Model 430 spectrofluorometer using an excitation wavelength of 360 nm and an emission wavelength of 450 nm. Fluorescence was converted to concentration using a standard curve (0.4–40 nM) of MUF.

13.3 Results and Discussion

The filter method described above for determining phosphatase activities in natural phytoplankton populations was initially carried out with the addition of 0.2 ml of chloroform to help solubilize the cell membrane enabling access of the substrate to the enzyme (chloroform was chosen since it had been previously shown not to inhibit algal alkaline phosphatases; Berman, 1970). Various other membrane solubilizers were also tried and 1:1 mixture of chloroform and isobutanol was found to give optimal activities. The ability of this solvent mixture to enhance phosphatase activity was species specific, however and gave unpredictable results with natural lake samples (Table 13.1). We suggest therefore that phosphatase activities be measured both with and without solvent to determine which method gives highest activities with the particular sample under study.

By employing filters of suitable pore size, the filter phosphatase method could also be used to study enzymatic activity of various size fractions. Phosphatase activities in four size classes (>20 μm, 20–3 μm, 3–0.8 μm, and 0.8–0.2 μm) of Kinneret phytoplankton were measured. Significant activity was often found in the smallest size class, corresponding to picoplanktonic and bacterial populations (Berman et al., 1990) and in good agreement with previous investigations (e.g., Francko, 1983; Chróst and Overbeck, 1987).

During this investigation we compared the spectrophotometric and fluorometric algal phosphatase assays. For natural waters with low phytoplankton biomass and/or activities the fluorometric method required much less volume in order to obtain measurable results (10–20 ml, in contrast to 250–1,000 ml for the spectrophotometric method). Furthermore, the fluorometric determination of dissolved (extracellular) phosphatase activity could be carried out within 2–4 hours, instead of 2–4 days, using a spectrophotometric method (Berman, 1970). Similarly to observations by Pettersson and Jansson (1978), we found that absolute hydrolysis rates of PNPP were much higher than that of MUP, probably reflecting differences in binding of the two substrates to the enzyme. As expected, the size fractionation pattern of acid and alkaline phosphatase activity in natural samples from the lake water was similar (but not identical), irrespective of the assay method employed (Berman et al., 1990).

The fluorometric method was used to study the phosphatases present in filtered Kinneret water. In a preliminary experiment (Figure 13.2A), using a commercial bacterial (*E. coli*) alkaline phosphatase (Sigma), the enzyme required Mg^{2+} ions for maximum activity. When EDTA (1 mM), a known inhib-

Table 13.1. Effect of solubilizer on algal phosphatases[1]

Source	Ratio of phosphatase activity	
	Alkaline	Acid
Culture of:		
Pediastrum sp.	2.17	1.13
Scenedesmus sp.	0.92	0.92
Selenastrum sp.	0.98	1.04
Lake water:		
14 Nov. 1988	1.80	2.06
20 Nov. 1988	1.01	3.66
12 Dec. 1988	1.02	1.03
16 Jan. 1989	0.85	1.26
20 Feb. 1989	1.06	0.87

[1]Results are given as the ratio between activities with and without added solublizer (further explanation in text).

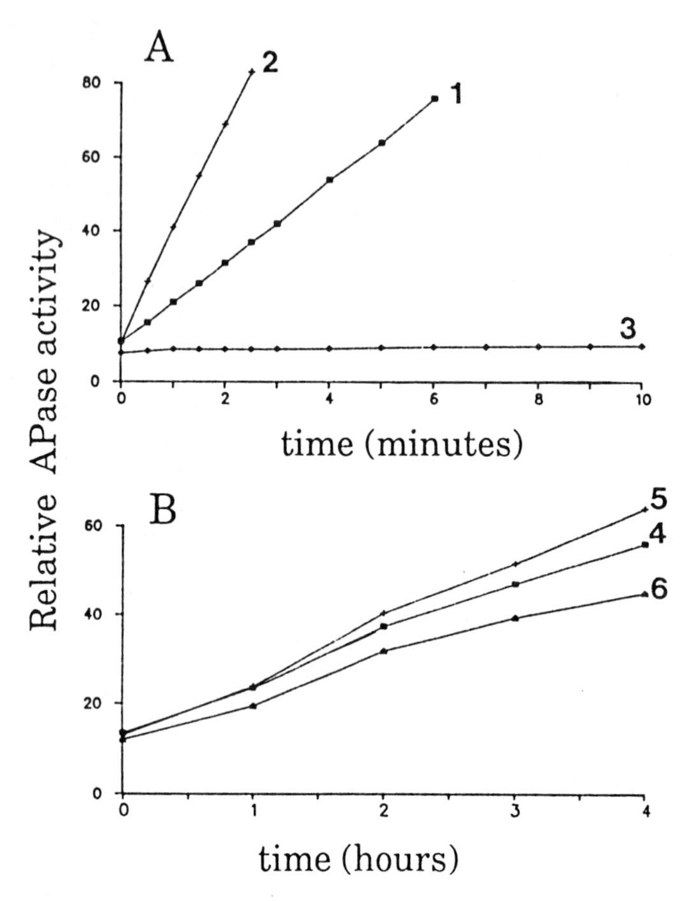

Figure 13.2 Fluorometric determination of alkaline phosphatase activity. (A) *E. coli* (Sigma): 1, no addition; 2, supplemented with 1 mM Mg^{2+} or 1 mM Mg^{2+} and 1 mM Ni^{2+}; 3, supplemented with 1 mM EDTA; (B) Filtered Kinneret water: 4, no addition or 1mM Ni^{2+}; 5, 1 mM Mg^{2+}; 6, 1 mM EDTA.

itor of alkaline phosphatase, was added to the reaction mixture, essentially all activity was abolished. On the other hand, addition of Ni^{2+} (1 mM), an inhibitor of 5′-nucleotidase (another type of phosphoesterase; Ammerman and Azam, 1985), had little effect on alkaline phosphatase activity. Using filtered Kinneret water as a source of enzyme, the result was more complex: Even though Mg^{2+} ions activated the alkaline phosphatase, EDTA did not abolish all activity, and Ni^{2+} had only a marginal effect on enzyme activity (Figure 13.2B).

For this investigation, we studied the activity of both acid and alkaline phosphatases in Lake Kinneret phytoplankton, even though only the latter is usually considered to respond to declining ambient SRP concentrations. Our results have shown that activities of both acid and alkaline phosphatases increased at lower SRP concentrations (Wynne, 1981). In addition, acid phosphatases (as well as alkaline phosphatases) could be detected in filtered lake water and in cell-free culture media. This finding suggests that acid phosphatases are also excreted by algae in spite of the localization of the enzyme in internal organelles, such as chloroplasts. (In our studies, filtration of phytoplankton was carried out under gentle vacuum only so that breakage of cells was unlikely).

It should be pointed out, that for some algae, the concept of "extracellular" enzymes may be difficult to define (see Chapter 3). For example, the APase that appears in cells towards the end of *Peridinium* bloom in Lake Kinneret is localized on the outer cell membrane (Wynne, 1977) and is very easily extracted by aqueous media (Carpene and Wynne, 1986). Therefore, for the Kinneret at least, APase found in filtered lake water during the bloom may not be truly extracellular but may simply reflect the poor degree of binding of the enzyme to the cell-membrane matrix. Furthermore, it was originally assumed that extracellular enzymes in the aquatic environment were produced by algae in response to various ecological triggers. However, the source of free enzymes in lake water is not necessarily restricted to phytoplankton; they could also originate from a variety of other organisms, ranging from bacteria and picocyanobacteria to zooplankton and fish. In these organisms, the release of enzymes (including phosphatases) may depend on a variety of factors, such as their physiological state and the influence of environmental conditions, without necessarily reflecting ambient nutrient concentrations. For example, zooplankton have been shown to release APase when they are under various forms of stress (Wynne and Gophen, 1981). Further investigations are necessary to elucidate these points.

References

Ammerman, J.W. and F. Azam. 1985. Bacterial 5'-nucleotidase in aquatic ecosystems. A novel mechanism of phosphorus regeneration. *Science* 227: 1338–1340.

Antia, N.J., Berland, B.R., Bonin, D.J. and S.Y. Maestrini. 1975. Comparative evaluation of certain organic and inorganic N sources for phytotrophic growth of marine microalgae. *Journal of Marine Biology Association (UK)* 55: 519–539.

Barman, T.E. 1969. *Enzyme Handbook*. Springer Verlag, Berlin. 928 pp.

Berman, T. 1970. Alkaline phosphatase and phosphorus availability in Lake Kinneret. *Limnology and Oceanography* 15: 663–674.

Berman, T. 1985. Lake Kinneret: Case history of conservation in a multifunctional warm lake. *Proceedings EWCA Congress, Lakes, Pollution and Recovery, Rome, Italy*, pp. 240–248.

Berman, T., Sherr, B.F., Sherr, E., Wynne, D. and J.J. McCarthy. 1984. The characteristic of ammonium and nitrate uptake by phytoplankton in Lake Kinneret. *Limnology and Oceanography* 29: 287–297.

Berman, T., Wynne, D. and B. Kaplan. 1990. Phosphatases revised: Analysis of particle associated enzyme activities in aquatic systems. *Hydrobiologia* 194: 235–245.

Carpene, E. and D. Wynne. 1986. Properties of an alkaline phosphatase from the dinoflagellate *Peridinium cinctum*. *Comparative Biochemistry and Physiology* 83B: 163–167.

Cembella, A., Antia, N.J. and P.J. Harrison. 1984a. The utilization of inorganic and organic phosphorus compounds as nutrients by eukaryotic microalgae: A multidisciplinary perspective. *CRC Critical Reviews in Microbiology* 10: 317–390.
Cembella, A., Antia, N.J. and P.J. Harrison. 1984b. The utilization of inorganic and organic phosphorus compounds as nutrients by eukaryotic microalgae: A multidisciplinary perspective. *CRC Critical Reviews in Microbiology* 11: 13–81.
Chróst, R.J. and J. Overbeck. 1987. Kinetics of alkaline phosphatase activity and phosphorus availability for phytoplankton and bacterioplankton in Lake Plußsee (North German eutrophic lake). *Microbial Ecology* 13: 229–248.
Fitzgerald, G.P. and T.C. Nelson. 1966. Extractable and enzymatic analyses for limiting or surplus phosphorus in algae. *Journal of Phycology* 2: 32–37.
Francko, D.A. 1983. Size fractionation of alkaline phosphatase activity in lake water by membrane filtration. *Journal of Freshwater Ecology* 2: 305–309.
Healey, F.P. 1973. Characteristics of phosphorus deficiency in *Anabaena*. *Journal of Phycology* 9: 383–394.
Healey, F.P. 1978. Physiological indicators of nutrient deficiency in algae. *Internationalen Vereinigung für Theoretische und Angewandte Limnologie Mittelungen* 21: 34–41.
Healey, F.P. and L.L. Henzel. 1976. Physiological changes during the course of blooms of *Aphanizomenon flos-aquae*. *Journal of Fisheries Research Board of Canada* 33: 36–41.
Hickel, B. and U. Pollingher. 1988. Identification of the bloom forming *Peridinium* from Lake Kinneret (Israel) as *P. gatunense* (Dinophyceae). *British Phycological Journal* 23: 115–119.
Pettersson, K. 1980. Alkaline phosphatase activity and algal surplus phosphorus as phosphorus-deficiency indicator in Lake Erken. *Archiv für Hydrobiologie* 89: 54–87.
Pettersson, K. and M. Jansson. 1978. Determination of phosphatase activity in lake water—a study of methods. *Verhandlungen der Internationalen Vereinigung für Theoretische und Angewandte Limnologie* 20: 1226–1230.
Pollingher, U. 1981. The structure and dynamics of the phytoplankton assemblages in Lake Kinneret. *Journal of Plankton Research* 3: 93–105.
Rhee, G.Y. 1973. A continuous culture study of phosphate uptake, growth rate and polyphosphates in *Scenedesmus* sp. *Journal of Phycology* 9: 495–506.
Rivkin, R.B. and E. Swift. 1979. Diel and vertical patterns of alkaline phosphatase activity in the oceanic dinoflagellate *Pyrocystis noctiluca*. *Limnology and Oceanography* 24: 107–116.
Serruya, C. 1978. *Lake Kinneret*. Dr. Junk Publ., The Hague. 501 pp.
Wynne, D. 1977. Alterations in activity of phosphatases during the *Peridinium* bloom in Lake Kinneret. *Physiologia Plantarum* 40: 219–224.
Wynne, D. 1981a. The role of phosphatases in the metabolism of *Peridinium cinctum* from Lake Kinneret. *Hydrobiologia* 83: 93–99.
Wynne, D. 1981b. Phosphorus, phosphatases and the *Peridinium* bloom in Lake Kinneret. *Verhandlungen der Internationalen Vereinigung für Theoretische und Angewandte Limnologie* 21: 523–527.
Wynne, D. and M. Gophen. 1981. Phosphatase activity in freshwater zooplankton. *Oikos* 37: 369–376.
Wynne, D. and G.Y. Rhee. 1986. Changes in alkaline phosphatase activity and phosphate uptake in P-limited phytoplankton, induced by light intensity and spectral quality. *Hydrobiologia* 160: 173–178.

14

Filtration and Buoyant Density Characterization of Algal Alkaline Phosphatase

David A. Francko

14.1 Introduction

Many planktonic algae and bacteria respond to acute orthophosphate limiting conditions by synthesizing alkaline phosphatase (APase). This class of enzymes permits biota to hydrolyze dissolved phosphomonoester substrates present in the water column, thereby providing an additional source of orthophosphate for biotic assimilation (Berman, 1970; Heath and Cooke, 1975; Francko and Heath, 1979; Pettersson, 1980; Wetzel, 1981; Heath, 1986; Francko, 1986a). As such, APase plays a central role in the overall process of phosphorus recycling and orthophosphate regeneration in epilimnetic waters (reviewed by Francko, 1986a).

APase in lake and marine waters is assessed by filter-fractionation techniques that resolve activity into particulate-associated and dissolved APase classes on the basis of apparent size. Particulate APase is generally assumed to be comprised largely of plasmalemma-bound, exoenzymatic activity of algae, an assumption that appears valid for some systems (cf., Heath and Cooke, 1975; Pettersson, 1980). Other authors have provided evidence that bacterial APase (Stewart and Wetzel, 1982) and APase capable of passing through small pore-size filters (usually 0.2–0.7 μm) may comprise the bulk of APase in some epilimnetic waters (Wetzel, 1981; Francko, 1983; 1986a). This latter category of APase is referred to as "dissolved" in the literature, although this definition is based strictly on filtration criteria. Further information on the speciation of APase in epilimnetic waters and laboratory cultures, especially putative extracellular forms, is needed in order to better understand the role of APase in overall P dynamics.

This chapter discusses the use of techniques for filtration and discontinuous sucrose-density-gradient centrifugation in order to fractionate APase synthesized in axenic algal cultures and cultures amended with additions of non-APase producing bacteria. Data were used to test the following theses: (1)

particulate and algal-associated APase are not functionally equivalent classes of enzyme, even in axenic cultures where algal APase synthesis represents the only source of enzyme; (2) extracellular APase is not necessarily equivalent to dissolved APase; and (3) the presence of non-APase producing particles influences the physical characteristics and potential ecological significance of exoenzymatic APase produced by algae.

14.2 Methods

Anabaena flos-aquae (G-R strain, originally isolated from Lake Erie), a filamentous cyanobacterium, and *Selenastrum capicornutum* (EPA strain, Corvallis, Oregon), a unicellular chlorophyte were grown in axenic batch culture (250 ml Moss medium [Moss 1972], pH 8.0) on a reciprocal shaker (60 μE m^{-2} s^{-1} illumination; 12:12 LD cycle; 22 $\pm$2°C). Axenicity of cell suspensions was assayed by inoculating aliquots onto Moss and nutrient agar plates (after 96-h incubation at 37°C) and by examination with phase contrast microscopy. Subcultures of a species of the bacterial genus *Pseudomonas* were isolated from contaminated cultures and maintained on nutrient agar plates. APase in cultures was induced by inoculating 15 ml of late-log-phase cells (algal density ca. 5 x 10^6 cells ml^{-1}) grown in normal medium into 235 ml of fresh Moss medium lacking phosphate. Cell transfers and subsequent APase analyses were conducted approximately at the mid-point of the light period. APase was induced within about 12 h. All APase assays were conducted using a fluorometric technique (Perry, 1972; Francko, 1983; *o*-methylfluorescein-phosphate, MFP, as substrate; Turner Model 110 equipped with a high-sensitivity sample holder).

Upon induction of APase, subsamples (100 ml) of cell cultures were withdrawn (cell density ca. 4 x 10^5 cells ml^{-1}). One 20-ml aliquot of culture water was set aside for estimation of total APase. Additional 20-ml aliquots were filtered through a double layer of Whatman No. 4 filter paper, or through Nuclepore polycarbonate filters with nominal pore sizes of 3.0, 0.6, and 0.2-μm. Filters were pre-rinsed with 15 ml of distilled, deionized water; a very gentle vacuum differential was used (0.1 atm); and filters were not sucked dry. Replicate subsamples of whole water and filtrates (4.0 ml each) were assayed for in vivo chlorophyll$_a$ content using a Turner 110 fluorometer equipped as described above and set at the highest excitation setting (Francko, 1986b). APase analyses were then conducted on these subsamples. In general, enzymatic activity was high enough (ca. 25 to 50 nmol MFP hydrolyzed l^{-1} min^{-1} in whole water samples) so that APase values could be determined using 1–2 min of incubation at room temperature.

The buoyant density of various APase classes in whole culture water and filtrates above was determined by subjecting portions of subsamples collected above to discontinuous sucrose-density-gradient centrifugation (Wollum and Miller, 1980; Heath and Francko, 1988). A 7-ml portion of 62% sucrose solution (w/w; buoyant density = 1.30 g ml^{-1}) was placed at the bottom of a 30-ml glass centrifuge tube. On top of this was placed 7 ml of a 36% sucrose solution (w/w; density = 1.15 g ml^{-1}), followed by 7 ml of a 13% sucrose solution (w/w; density = 1.05 g ml^{-1}). Centrifugation grade sucrose and sterile, distilled, deionized water were used to prepare sucrose solutions. Each solution was amended with Trisbase to a final concentration of 10 mM and the pH was adjusted to 8.0 prior to use.

A 4.0-ml portion of whole water or filtrate collected as described above was gently layered on top of the final sucrose solution. The tubes were then centrifuged for 25 min at 3,000 x g in a Sorvall swinging bucket rotor. Centrifugations were conducted at 4°C. Two-ml portions were collected from the top of tubes, placed in 2 ml of sterile Moss medium, and assayed for APase. Portions containing 1.30 g ml^{-1} sucrose were further diluted 1:7 with Moss medium prior to APase assay because the fluorescence of sucrose itself at high concentrations produced an unacceptably high background. A 4.0-ml sample of sterile Moss medium was fractionated as above, APase analyses were conducted on gradient subsamples, and values were used to correct experimental data for background fluorescence in the absence of APase. An additional sample of sterile Moss medium was centrifuged and subsamples were spiked with a known amount of calf intestinal mucosa APase prior to analysis. This procedure indicated that fluorescence yields per unit APase increased as the concentration of sucrose increased, and data from subsequent centrifugations of culture samples were corrected accordingly.

To establish the isopycnic position of true dissolved APase, a solution of purified calf intestinal mucosa alkaline phosphatase (Sigma) containing 12 μg protein ml^{-1} of sterile Moss medium, pH 8.0 was prepared. One 4-ml portion of this enzyme solution was centrifuged and assayed for APase directly as above. An additional portion of enzyme solution was filtered through a 0.2-μm Nuclepore filter and a 4.0-ml filtrate subsample was centrifuged and analyzed for APase. This procedure permitted determination of both the buoyant density and filter retention of dissolved APase.

In some procedures, axenic cultures of *Anabaena* were inoculated with a suspension of *Pseudomonas* prior to analyses of APase. Bacterial cells were collected via inoculating loop from petri dishes and resuspended in fresh Moss medium. Prior experimentation with these bacteria demonstrated that they represented the only nonalgal contaminant in algal cultures described in this paper and that they were apparently incapable of APase production, even when cells were subcultured in P-free Moss media containing glucose as a carbon source. Bacterial suspensions (A_{600nm} = 0.032) were prepared so that inocula (6 ml bacterial suspension plus 94 ml sterile algal culture) produced a final algal:bacterial biovolume ratio of greater than 100:1.

14.3 Results and Discussion

In demonstrably axenic cultures of *Anabaena* (Figure 14.1A), more than one-half of the total APase in P-starved cultures was capable of passing through small-

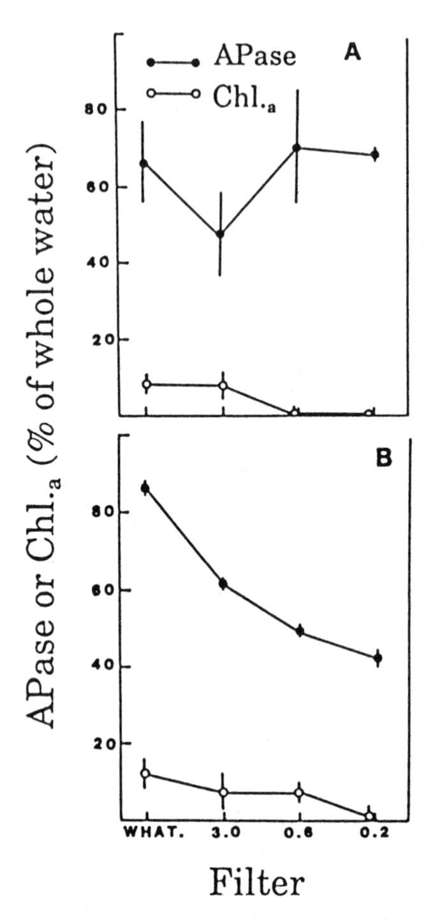

Figure 14.1 Alkaline phosphatase activity (APase) and chlorophyll$_a$ (Chl.$_a$) passing through Whatman filter paper (What.) and Nuclepore filters of decreasing pore size (3.0-, 0.6-, and 0.2-μm) from cultured *Anabaena*. (A) Data from axenic cultures. (B) Data from *Pseudomonas*-contaminated cultures. Data shown as mean ±SE (N = 2) of APase and of Chl.$_a$ passing through each kind of filter as a percentage of the values in whole, unfiltered culture samples.

pore-size Nuclepore filters as readily as through Whatman No. 4 filter paper (pore size generally >5 μm). In contrast, over 90% of the chlorophyll$_a$ could be removed from suspension by filtration through Whatman No. 4 filter paper. Since these data were derived from a culture sampled within 12 h after the induction of APase production, it appeared that much of the APase originally produced by *Anabaena* rapidly became part of the extracellular enzyme pool.

Anabaena suspensions contaminated with low (less than 1% bacterial biovolume as percent of total) titers of a *Pseudomonas* species exhibited a pattern of filtration chlorophyll$_a$ removal similar to that seen in axenic suspensions (Figure 14.1B). APase removal by succeedingly smaller-pore-size differed greatly from axenic variants. Only about 14% of total APase was removed by Whatman filters. Increasing amounts of APase were removed from solution by 3.0-μm and 0.6-μm filters, and about 40% of the total APase present in whole-culture water passed through 0.2-μm filters.

Pseudomonas cells were subcultured from contaminated *Anabaena* cultures. Microscopic examination and plate count assays demonstrated that this organism was the sole contaminant in APase-producing algal cultures. *Pseudomonas* cells inoculated into sterile Moss P-free medium did not produce APase, even after a one-week incubation period and in media supplemented with 1 g l^{-1} of glucose as a carbon source. These findings support the view that *Pseudomonas* contaminants could not have been responsible for APase production in non-axenic *Anabaena* cultures, but may have affected the apparent size distribution of non-*Anabaena*-associated exoenzymatic APase.

Filtration criteria give only limited information on the partitioning of particulate-associated and dissolved APase. Discontinuous density gradient centrifugation was used to differentiate the buoyant densities of algal-associated and dissolved forms of APase. Purified calf intestinal mucosa alkaline phosphatase behaved as expected of a dissolved moiety upon sedimentation to its isopycnic position in the sucrose gradient (Figure 14.2). Quantitative recovery of APase (98% of activity applied to the top of the gradient prior to centrifugation) occurred in the Moss medium fractions at the top of the gradient and at the interphase of the 1.05 g ml^{-1} sucrose solution. This behavior was not altered by filtration through a 0.2-μm filter, indicating that little if any aggregation or particle formation occurred in sterile enzyme solutions (93% recovery of added APase).

Preliminary experimentation indicated that the buoyant density of *Anabaena* cells as cultured was between 1.25 and 1.30 g ml^{-1}. Cells were clearly visible in, and $chlorophyll_a$ distribution was confined to, the two fractions immediately before and at the 1.30 g ml^{-1} sucrose cushion. The buoyant density distribution of APase in unfiltered and filtered culture samples is shown in Figure 14.3. When an unfiltered sample of axenic *Anabaena* culture was sedimented, only about 20% of the total APase appeared in these two fractions, i.e., associated with intact *Anabaena* cells. Approximately 55% remained at the top of the gradient, and the balance of APase sedimented at the bottom of the tube, the latter fraction behaving as particles at least as heavy as 1.30 g ml^{-1}.

In a sample passed through Whatman filter paper, most of the dissolved APase (activity in the first three fractions at the top of the tube) remained in the filtrate, whereas APase at the 1.30 g ml^{-1} interphase was reduced compared with the whole-water sample. This general pattern was repeated in samples passed through 3.0-, 0.6-, and 0.2-μm filters. Although much of the APase passing through these filters possessed a buoyant density associated with dissolved protein, approximately 30–32% of the filtrate activity had a buoyant density of about 1.30 g ml^{-1}.

Collectively, such data suggest that some of the APase behaving as a dissolved moiety through small pore-size filters nonetheless sedimented in a man-

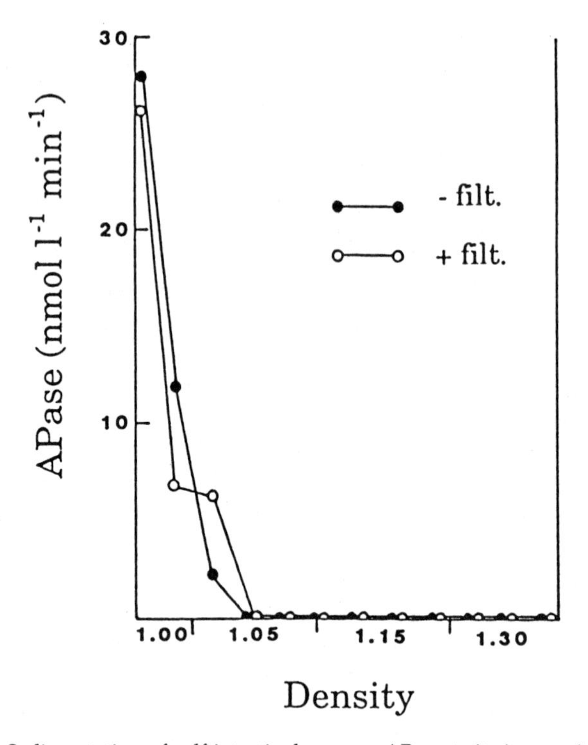

Figure 14.2 Sedimentation of calf intestinal mucosa APase to its isopycnic position in a discontinuous sucrose gradient. Subsamples were alternatively applied directly to the top of the sucrose gradient (- filt.) or filtered through a 0.2-μm Nuclepore filter prior to centrifugation (+ filt.).

ner consistent with a fairly dense particle or aggregate. In that no intact *Anabaena* or chlorophyll was capable of passing through the smaller filters, this dense fraction could not have been directly associated with algal cells. Rather, it appeared that APase originally produced by *Anabaena* and released as an extracellular enzyme was being sequestered by some particulate moiety present in the axenic culture medium. Since, as shown in Figure 14.2, purified calf intestinal mucosa APase dissolved in sterile media did not associate into dense aggregates or particles and was not appreciably retained on 0.2-μm filters, the data in Figure 14.3 suggest that high density, noncellular APase formation was biotically mediated.

Further experiments using a sterile *Anabaena* culture and a culture deliberately spiked with a known amount of non-APase producing *Pseudomonas* were used as a first test of the hypothesis that biotic particles, themselves incapable of APase production, sequester dissolved APase into a particulate-associated, dense form (Figure 14.4). In the axenic *Anabaena* culture, filter retention of APase followed the pattern previously shown in Fig. 14.1A; approximately the same amount of APase passed through Whatman filters and the smaller Nuclepore

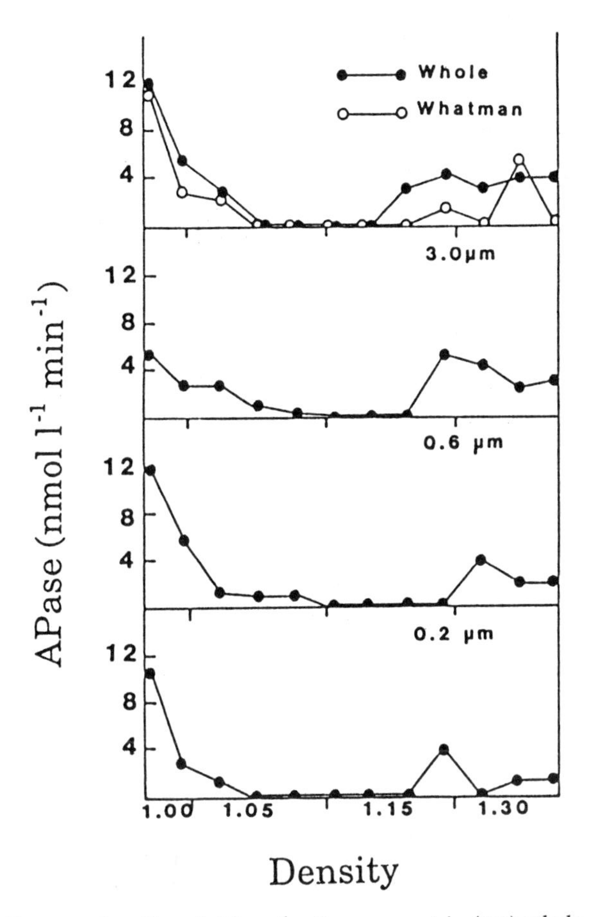

Figure 14.3 Buoyant densities of APase fractions present in (top) whole-culture water and Whatman filtrate and in three sizes of Nuclepore filtrates from axenic *Anabaena* suspensions, as determined by discontinous sucrose-density-gradient centrifugation.

filters. When bacteria were introduced at a low titer (*Anabaena*:bacterial biovolume at >100:1) and samples were incubated at room temperature for 3 h prior to APase analyses, the APase filtration profile more closely resembled that shown in Figure 14.1B. Significantly more APase was retained by 0.2-µm filters when bacteria were added as compared to the axenic variant.

That the change in filtration dynamics was directly due to *Pseudomonas* is supported by data shown in the inset to Figure 14.4. Most of the *Pseudomonas* cells were capable of passing through the Whatman and 3.0-µm filters. If the thesis that these bacterial cells were capable of sorbing extracellular APase produced by *Anabaena* is valid, proportionately more APase should have been present in the Whatman and 3.0-µm filtrates when bacteria were present. Similarly, dissolved APase putatively sorbed to bacteria should have been preferentially

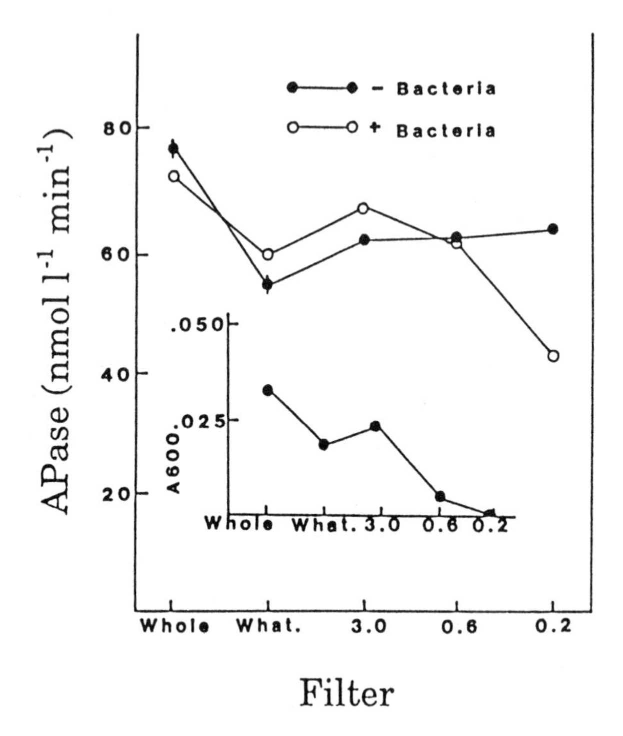

Figure 14.4 APase present in whole-culture water and passing through various filter types (Whatman and 3.0-, 0.6-, and 0.2-μm Nuclepore) in axenic *Anabaena* suspensions (- Bacteria) and algal suspensions inoculated with a low titer of *Pseudomonas* 3 h before filtration (+ Bacteria). Data shown as mean ±SE (N = 2). Standard errors smaller than the dimensions of the points are not shown. Inset shows the loss of *Pseudomonas* cells from the filtrate (absorbance at 600 nm) as a function of the Whatman and Nuclepore filter type.

retained by the 0.2-μm filters. As the data in Figure 14.4 demonstrate, these predictions were corroborated.

Aliquots of 0.2-μm filtrates from bacteria-free and bacteria-containing cultures above were sedimented in microfuge tubes containing 0.5 ml of 1.30 g ml^{-1} sucrose at the bottom and 0.5 ml of 0.2-μm filtrate at the top (Figure 14.5). Centrifugation and APase analyses were conducted as described in Section 14.2, except that 0.5-ml subsamples were added to Moss medium for APase analyses. Proportionately less APase was present at the top of the centrifuge tube in filtrates from bacteria-containing cultures, whereas more APase sedimented into the sucrose solution. In this experiment, the loss of APase from the low density fraction (about 10% of the total APase in unfiltered culture water) was exactly balanced by the increase in APase in the sucrose fraction. These results were also consistent with the view that *Pseudomonas* cells sequestered a portion of the previously dissolved APase into a cellular fraction removed by filtration

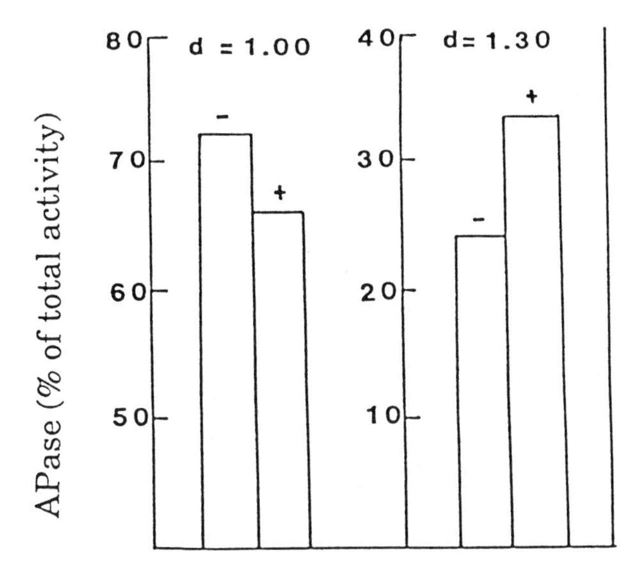

Figure 14.5 Relative amounts of (left) less dense and (right) more dense APase present in 0.2-μm filtrate subsamples from an axenic *Anabaena* culture (-) and an *Anabaena* culture inoculated with *Pseudomonas* (+). Cultures were incubated for 15 h prior to filtration and centrifugation. Values shown as the mean of replicate determinations as a percentage of total activity present in 0.2-μm filtrates (48.5 ±4 nmol l^{-1} min^{-1}).

and also enhanced the production of noncellular, dense APase capable of passing through 0.2-μm filters.

Although extensive experimentation was not conducted on other algal species, preliminary data suggest that the formation of a class of non-algal associated, particulate APase, even in axenic culture, may be taxonomically widespread. Axenic cultured *Selenastrum* cells displayed APase and $chlorophyll_a$ filtration dynamics similar to *Anabaena* (Figure 14.6). Like *Anabaena, Selenastrum* cells as cultured had a buoyant density of about 1.30 g ml^{-1}. Proportionately more APase in whole-water samples cosedimented with visible cells and $chlorophyll_a$ (i.e., one sample before and after the 1.30 density interphase) in *Selenastrum* centrifugation experiments compared with *Anabaena* (Figure 14.7). However, subsamples passed through 3.0-, 0.6-, and 0.2-μm filters each possessed similar amounts of high density APase. In each case, more than one-half of the APase behaving as dissolved activity in filtration analyses had a buoyant density of about 1.30 g ml^{-1}.

Taken together, the data support the following conclusions: (1) particulate APase and algal-associated APase are not functionally equivalent, even in axenic cultures where the original source of APase is algal; (2) APase that behaves as a dissolved moiety in filtration analyses is not necessarily dissolved, in that much of the activity passing through small pore-size filters has a much higher buoyant density than that of true dissolved APase; (3) biotic particles such as

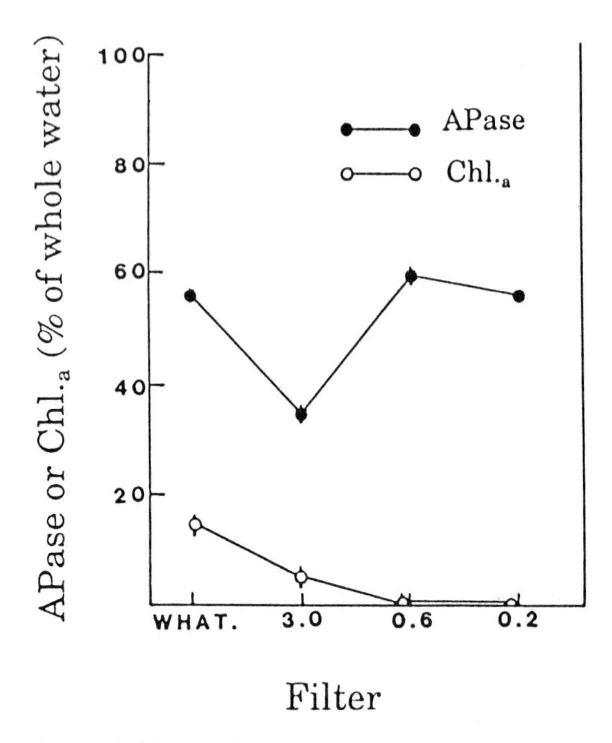

Figure 14.6 APase and chlorophyll$_a$ (Chl.$_a$) passing through Whatman filters and 3.0-, 0.6-, or 0.2-μm Nuclepore filters as the mean percentage (±SE; N = 2) of values in unfiltered water from axenic suspensions of *Selenastrum*. Standard errors smaller than the diameter of data points are not shown.

bacteria that themselves are incapable of APase production may sorb APase produced by algae; and (4) algae and bacteria apparently produce some extracellular substance that sorbs APase into a high density form capable of passing through small pore-size filters, thereby mimicking dissolved APase.

These conclusions have important implications for our understanding of the role of APase, especially exoenzymatic forms, in natural systems. For example, bacteria or algae incapable of producing their own APase but able to sorb APase produced by other species and released into their extracellular medium could introduce two ecologically interesting phenomena. First, filtration analyses of the kind typically conducted on natural waters would be incapable of separating true APase producers from those cells that merely sorb dissolved enzyme. Thus, literature data on the relative production of APase by various size-classes of phytoplankton and bacterioplankton may be incorrect. Second, organisms capable of sorbing extracellular APase may derive the same benefits from the enzymatic hydrolysis of phosphomonoesters occurring in close proximity to their surface that normal APase producers derive. The ecological consequences of such interactions for planktonic community dynamics in P-limited systems could be profound.

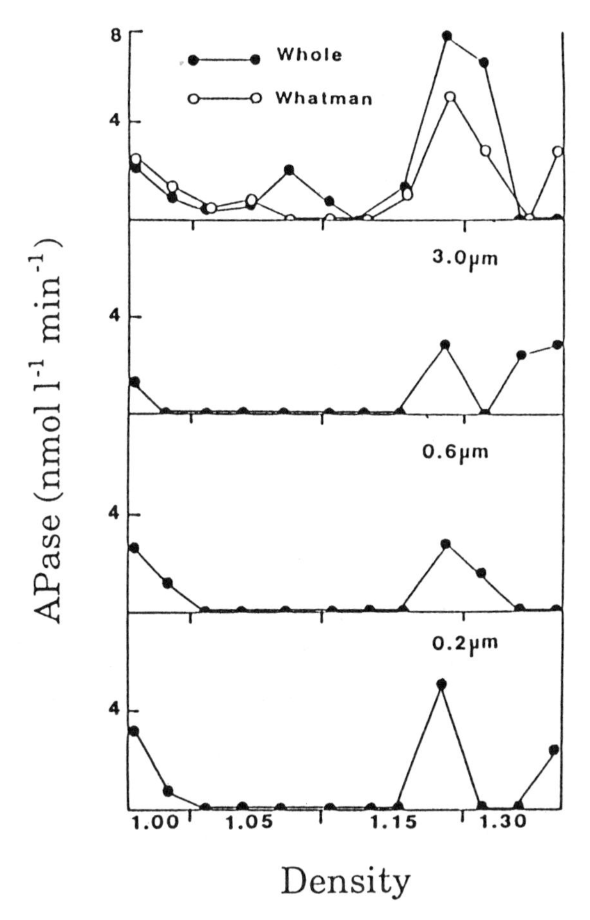

Figure 14.7 Buoyant densities of APase fractions present in an unfiltered axenic *Selenastrum* culture subsample or in Whatman and Nuclepore filtrates as determined by discontinuous sucrose density centrifugation.

These data are insufficient to speculate on the nature or mode of formation of high density, exoenzymatic APase capable of passing through small pore-size filters. Algae and bacteria are capable of producing many extracellular products (e.g., microfibrils) that may influence the partitioning of dissolved phosphorus in surface waters (reviewed by Cembella et al., 1984) and such products may be capable of aggregating or sorbing APase.

It does appear clear that significant quantities of lake, marine, and culture-water APase may occur in an exoenzymatic form that is functionally distinct from dissolved APase. It also appears that the partitioning of APase is more complex than previously known. The eventual fate and ecological importance of APase produced by one species of plankton and released into the water column represents a exciting and potentially important area of new research.

Acknowledgments I thank Dr. Safaa Al-Hamdani and Ms. Paula Cinnamon for technical support. Support was provided by the National Science Foundation (BSR-8907440), the Oklahoma State University Center for Water Research, and the Department of Botany, Miami University.

References

Berman, T. 1970. Alkaline phosphatases and phosphorus availability in Lake Kinneret. *Limnology and Oceanography* 15: 663–674.

Cembella, A.D., Antia, N.J. and P.J. Harrison. 1984. The utilization of inorganic and organic phosphorus compounds as nutrients by eukaryotic microalgae: a multidisciplinary perspective. *CRC Critical Reviews in Microbiology* 10: 317–391.

Francko, D.A. 1983. Size-fractionation of alkaline phosphatase activity in lake water by membrane filtration. *Journal of Freshwater Ecology* 2: 305–309.

Francko, D.A. 1986a. Epilimnetic phosphorus cycling: influence of humic materials and iron on co-existing major mechanisms. *Canadian Journal of Fisheries and Aquatic Sciences* 43: 302–310.

Francko, D.A. 1986b. Measurement of algal chlorophyll$_a$ and carbon assimilation by a tissue solubilizer method: A critical analysis. *Archiv für Hydrobiologie* 106: 327–335.

Francko, D.A. and R.T. Heath. 1979. Functionally distinct classes of complex phosphorus compounds in lake water. *Limnology and Oceanography* 24: 463–473.

Heath, R.T. 1986. Dissolved organic phosphorus compounds: do they satisfy planktonic phosphate demand in summer? *Canadian Journal of Fisheries and Aquatic Sciences* 43: 343–350.

Heath, R.T. and G.D. Cooke. 1975. The significance of alkaline phosphatase in a eutrophic lake. *Verhandlungen der Internationalen Vereinigung für Theoretische und Angewandte Limnologie* 19: 959–965.

Heath, R.T. and D.A. Francko. 1988. Comparison of phosphorus dynamics in two Oklahoma reservoirs and a natural lake varying in abiogenic turbidity. *Canadian Journal of Fisheries and Aquatic Sciences* 45: 1480–1486.

Moss, B. 1972. The influences of environmental factors on the distribution of freshwater algae: an experimental study. I. Introduction and the influence of calcium concentration. *Journal of Ecology* 60: 917.

Perry, M.H. 1972. Alkaline phosphatase activity in subtropical Central North Pacific waters using a sensitive fluorometric method. *Marine Biology* 15: 109–113.

Pettersson, K. 1980. Alkaline phosphatase activity and algal surplus phosphorus as phosphorus-deficiency indicator in Lake Erken. *Archiv für Hydrobiology* 89: 54–87.

Stewart, A.J. and R.G. Wetzel. 1982. Phytoplankton contribution to alkaline phosphatase activity. *Archiv für Hydrobiologie* 93: 265–271.

Wetzel, R.G. 1982. Longterm dissolved and particulate alkaline phosphatase activity in a hardwater lake in relation to lake stability and phosphorus enrichments. *Verhandlungen der Internationalen Vereinigung für Theoretische und Angewandte Limnologie* 21: 337–349.

Wollum, A.G. and R.H. Miller. 1980. Density centrifugation method for recovering *Rhizobium* spp. from soil for fluorescent-antibody studies. *Applied and Environmental Microbiology* 39: 466–469.

15

Cyclic Nucleotide Phosphodiesterase Activity in Epilimnetic Lake Water

Jon Barfield and David A. Francko

15.1 Introduction

Many correlative and perturbational studies conducted in situ and in the laboratory have demonstrated a putative mechanistic link between seasonal and diel variations in cyclic AMP (cAMP) concentrations and photosynthetic and heterotrophic carbon assimilation rates in algae and aquatic macrophytes (Francko, 1984a; 1984b; 1984c; 1986; 1987a; Francko and Wetzel, 1980; 1981b; 1982; 1984a; 1984b). Recent work on cultured *Selenastrum capricornutum*, a unicellular chlorophyte alga, suggests that fluxes in both extracellular (dissolved) and intracellular cAMP induce changes in the algal transmembrane electrogenic potential, with concomitant perturbation of photosynthetic carbon assimilation (PCA) rates (Francko, 1989a; 1989b). Alterations in both membrane potential and PCA rates could be induced in the laboratory using manipulations of dissolved and intracellular cAMP within the range of seasonal and diel variation occurring in freshwater and marine systems (a few pmol to over 0.1 μmol l^{-1} and ca. 10 pmol to over 1 μmol g^{-1} wet wt., respectively; Ammerman and Azam, 1981; 1982; Francko, 1983; 1987a).

The operation of a putative cAMP-mediated photosynthetic regulatory system in epilimnetic planktonic assemblages would be conditionally dependent both on the rates of cAMP production and release by biota (signal generation) and on the processes that remove the cAMP "signal" from cells and lake water. The above references contain data on cAMP production and release rates and demonstrate that biotic production by planktonic algae and bacteria, aquatic vascular plants, and attached epiphytes can account for most of the dissolved cAMP in lake water. To date, however, almost nothing is known about the mechanisms responsible for cAMP degradation in natural aquatic systems. This information is crucial for further elucidation of the ecophysiology of lacustrine cAMP metabolic pathways.

In bacterial and animal cells, cAMP degradation is enzymatically catalyzed by a class of cyclic nucleotide phosphodiesterases (cPDE). We are aware of only two definitive studies on cPDE in cultured algae, although numerous attempts to measure cPDE in terrestrial vascular plants have been attempted (reviewed by Amrhein, 1977; Brown and Newton, 1981; Francko, 1983). Fischer and Amrhein (1974) and Ownby and Kuenzi (1982) reported the presence of substantial cPDE activity in cultured *Chlamydomonas reinhardtii* and *Anabaena variabilis*, respectively. Like animal cell cPDE, enzymatic activity was strongly inhibited by the methylxanthine derivative theophylline and was highly specific for cAMP.

Circumstantial evidence for naturally occurring enzymatic hydrolysis of cAMP in freshwater systems was presented by Francko and Wetzel (1984a). Samples of littoral zone surface waters were filtered (0.6-μm filters) at time zero and incubated for 1–4 h in situ. Losses of cAMP in filtrates, presumably due to hydrolysis, ranged from 0.5–1.1 nmol l^{-1} h^{-1} in water samples containing 1–6 nmole of cAMP l^{-1}. Putative cAMP hydrolysis exhibited an optimum pH of 7.0 and a temperature optimum of 23°C. The capacity of lakewater filtrates to hydrolyze cAMP was irreversibly destroyed by heating water samples to 100°C for 5 minutes, but hydrolysis was not inhibited by 5 mM methyl-3-isobutylxanthine (MIX), a potent methylxanthine inhibitor of cPDE in animal cells (Amrhein, 1977).

In the above study, the presence of cAMP hydrolytic activity was inferred through measurements of cAMP loss rates in filtered water; direct measurements of cPDE activity were not made. In this study we employed a specific cPDE assay system to examine the amount and physicochemical characteristics of both particulate-associated and extracellular cPDE within epilimnetic waters from a eutrophic reservoir. We also examined the relationship between cPDE activity and changes in cAMP concentrations in lake water, permitting limited speculation on the ecological importance to cPDE.

15.2 Methods

Sampling Water samples were collected from the 0.1-m stratum of Sangre Isle Lake, (also known as Cedar Isle Lake), Payne County, Oklahoma, a small (3.2 ha), shallow (mean depth 2.0 m), eutrophic reservoir with limnological characteristics resembling those of a natural eutrophic lake (Francko, 1987b). Samples were collected at mid-day during May-October 1988, and during May of 1989 at an epilimnetic site located over the deepest portion of the reservoir. This system has been the site of previous studies on cAMP dynamics (Francko, 1983; 1984a; 1984b; 1984c). Water temperature measurements (Thermistor) were made in situ, whereas pH measurements were made in the laboratory.

Sample preparation Water samples were returned to the laboratory within 20 min of collection, and 150-ml portions were passed through 0.45-μm Millipore membrane filters using a vacuum differential of 0.5 atm. Particulate material was rinsed off the filters and resuspended in 20 ml of ice-cold 40 mM Tris-HCl, pH 7.6, containing 10 mM cysteine and 2 mM $MgSO_4$ in order to preserve the stability of putative cPDE activity (Ownby and Kuenzi, 1982).

Filtrate and particle-associated cPDE activity were measured by preparative and assay techniques previously developed for cultured algae (Ownby and Kuenzi, 1982). Resuspended particulate fractions were sonicated (Branson Model 185) on ice using four 10-s bursts on the highest power setting at 30-s intervals. The sonicated preparations were centrifuged (3,000 x g) for 5 min and resulting supernatants were assayed immediately. Control experiments indicated that additional cPDE activity was not released from particulate fractions by lengthier sonication. The filtrate fraction was collected directly from the filtration apparatus and kept on ice until the assay was performed. Experiments demonstrated that sonication did not enhance putative cPDE activity.

Procedure for cPDE assays Reaction tubes (1.5-ml polyethylene microfuge tubes) containing 100 μl of extract supernatant or filtrate (triplicate for each fraction) and 200 μl of sufficient Tris-HCl to yield a final concentration of 40 mM, pH 7.6, 2 mM $MgSO_4$, and 10 mM cysteine, were incubated at room temperature for 10 min. Additional replicate tubes were prepared as above and boiled for 15 min at time zero to serve as denatured controls. After incubation, 100 μl of a cAMP stock solution containing tracer quantities of $[^3H]$cAMP (40 nmol cold cAMP and 1.3 x 10^3 Bq ml^{-1} of $[^3H]$cAMP) were added to each tube. After 90 min of incubation at 32°C, the tubes were boiled for 1 min, chilled on ice, and centrifuged (3,000 x g). Following centrifugation, replicate aliquots (200 μl) from the supernatant in each tube were removed and added to fresh microfuge tubes containing 200 μl of an aqueous solution of *Ophiophagus hannah* venom (1 mg ml^{-1}) to hydrolyze to adenosine any 5′-AMP formed through putative cPDE activity. Venom-containing aliquots were incubated for 20 min at 32°C, boiled, chilled, and centrifuged as before. Aliquots (200 μl) were taken from the supernatant of each tube and mixed with 1.2 ml of an ethanolic slurry of anion exchange resin Dowex AG1-X8 (Londesbourough, 1976), shaken for 15 min, and centrifuged. $[^3H]$adenosine in sample supernatants was analyzed by removing 200 μl from the supernatant of each tube for liquid scintillation spectroscopy. These data were used to compute cPDE activity (nmol cAMP hydrolyzed per liter lakewater h^{-1}). Radioactivity in boiled controls was used to correct for nonenzymatic $[^3H]$cAMP breakdown and any unhydrolyzed $[^3H]$cAMP not removed by the anion-exchange resin prior to calculations of enzymatic activity.

Although cPDE assays were typically conducted in assay mixtures buffered at pH 7.6, we also measured the pH dependency of putative lakewater cPDE. Tris-maleate (pH 6.0 to 7.0) and Tris-HCl (pH 7.0 to 9.0) were used in this experiment. An additional series of experiments was conducted in water baths at nonstandard temperatures to evaluate the temperature dependence of cPDE activity. The substrate specificity of lakewater cPDE was evaluated by using cyclic GMP (cGMP), glucose-6-phosphate, and 5′-AMP as alternate substrates in the reaction mixture in addition to cAMP. In these experiments, 40 nmol of cAMP solution plus 40 nmol of alternative substrate were placed in reaction tubes. For the cAMP-alone control, an equivalent amount of assay buffer was added to correct for dilution effects. The methylxanthine sensitivity of lakewater cPDE activity was determined by incubating replicate particulate and filtrate samples amended by the addition of 10 mM theophylline (Francko and Wetzel, 1981a; Ownby and Kuenzi, 1982).

Chemical analyses Chlorophyll$_a$ concentrations in each lakewater sample analyzed were measured in replicate on particulate fractions retained by 0.45-μm Millipore membrane filters. Pigments were extracted in basic methanol and assayed fluorometrically (Wetzel and Likens, 1979). Portions of whole lake water were preserved with Lugol's iodine solution for later analysis of planktonic community composition.

Concentrations of naturally occurring cAMP in particulate matter and lakewater filtrates were simultaneously evaluated in replicate particulate and filtrate samples by preparative techniques and a competitive displacement assay previously published (Francko and Wetzel, 1980; 1982).

15.3 Results and Discussion

pH optimum Our data indicated that enzymatic activity consistent with known characteristics of cPDE from cultured algae occurred in both particulate and filtrate fractions from epilimnetic Sangre Isle Lake waters. We conducted pH optimum characterization experiments in May, June, and July 1988 and again in May 1989. Data for two sampling dates are shown in Figure 15.1, and indicate a broad pH optimum of about 7.0 and 7.5 in 1988 and 1989, respectively. Corresponding values for 4 June 1988 (pH 7.0) and 14 July 1988 (pH 8.0) were similar, although the range of one pH unit in the optimal values suggests that seasonally different isozymes of cPDE could exist. In all experiments, a precipitous drop in activity was noted when the pH was raised above 8.5. Since the pH of the lakewater samples themselves in the laboratory varied between 8.2 and 9.5, the above data suggest that cPDE activity values derived under standard assay pH conditions (pH 7.6) may have overestimated in situ activity.

Temperature optimum Temperature optimum experiments were conducted on 1 June and 6 June 1988. In each case, the temperature optimum was about 30–35°C (Figure 15.2) even though the in situ temperature of surface waters was about 25°C in early June. In each experiment, incubation of samples at 50°C for 15 min or more resulted in total loss of enzyme activity.

Substrates for cPDE Competition experiments indicated that lakewater cPDE preferentially utilized cAMP as a substrate (Table 15.1). Of the three alternate substrates we evaluated, glucose-6-phosphate and 5′-AMP exhibited no evi-

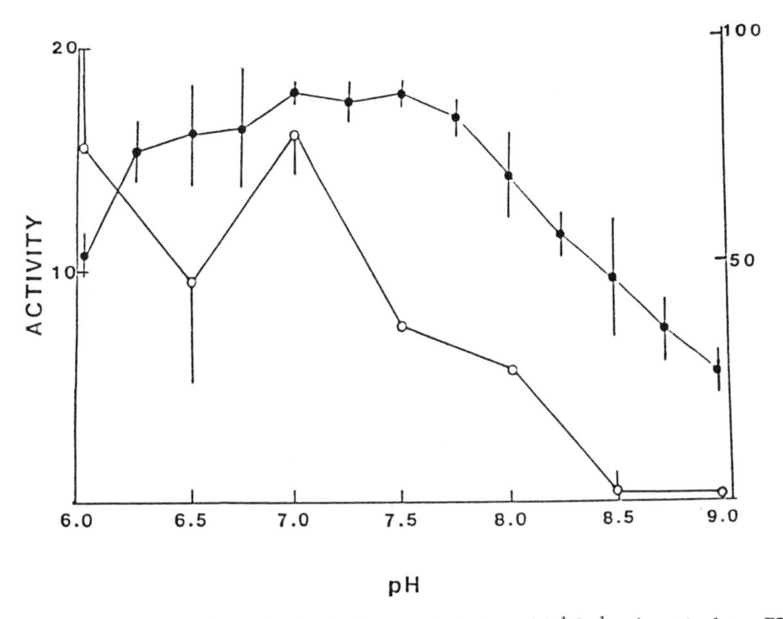

Figure 15.1 Effect of pH on the hydrolytic activity (nmol l^{-1} h^{-1}) of particulate cPDE sampled on 8 June 1988 (open circles) and 21 May 1989 (closed circles). Values on the left side of the ordinate refer to 1989 data whereas units on the right side correspond to 1988. Where error bars are not shown, ±SE (n = 3) was smaller than the diameter of points.

dence of competition. A slight depression in the calculated activity of the sample amended with cyclic GMP occurred, indicative of competition between the cyclic nucleotide substrates, but this compound reduced the activity toward cyclic AMP by only 10–15%. Particulate cPDE activity was reduced to undetectably low levels (<3 nmol l^{-1} h^{-1}) in samples incubated with 10 mM theophylline.

Taken together, the pH and temperature optima, substrate specificity, and methylxanthine sensitivity of cPDE activity in Sangre Isle Lake samples closely paralleled that of cultured algae described in the literature (Fischer and Amrhein, 1973; Ownby and Kuenzi, 1982).

Temporal dynamics in cPDE activity Temporal patterns of particulate-associated (PcPDE) and filtrate (dissolved; DcPDE) cyclic nucleotide phosphodiesterase activity for the period of study are shown in Figure 15.3 (top). Particulate activity ranged from about 15 to 60 nmol l^{-1} h^{-1} while dissolved activity (15 to 25 nmol l^{-1} min^{-1}) varied less during sampling period. The ratio of filtrate DcPDE to particulate-associated PcPDE per liter of lake water varied from 25 to 30, indicating that much more cPDE occurred in an extracellular form.

When particulate cPDE levels per unit of chlorophyll$_a$ were computed (Figure 15.3, top), a slightly different pattern emerged. Microscopic examination of

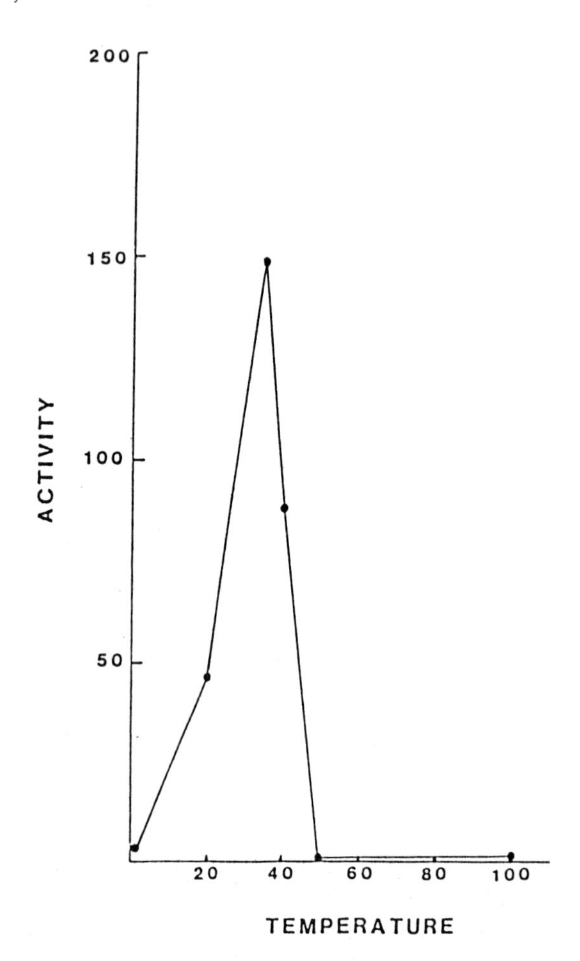

Figure 15.2 Effect of temperature on hydrolytic activity (nmol 1^{-1} h^{-1}) of particulate cPDE, 1 June 1988. All ±SE (n = 3) values were smaller than the diameter of points.

Table 15.1 Substrate specificity of cPDE activity in particulate fractions from epilimnetic water collected on 20 May 1988.

Substrate	Activity (nmol 1^{-1} h^{-1})[1]
cAMP	96.4 ±20.8
cGMP	82.2 ±26.4
5′-AMP	96.5 ±3.3
Glucose-6-P	89.9 ±0.8

[1]Based on three measurements; ±SE

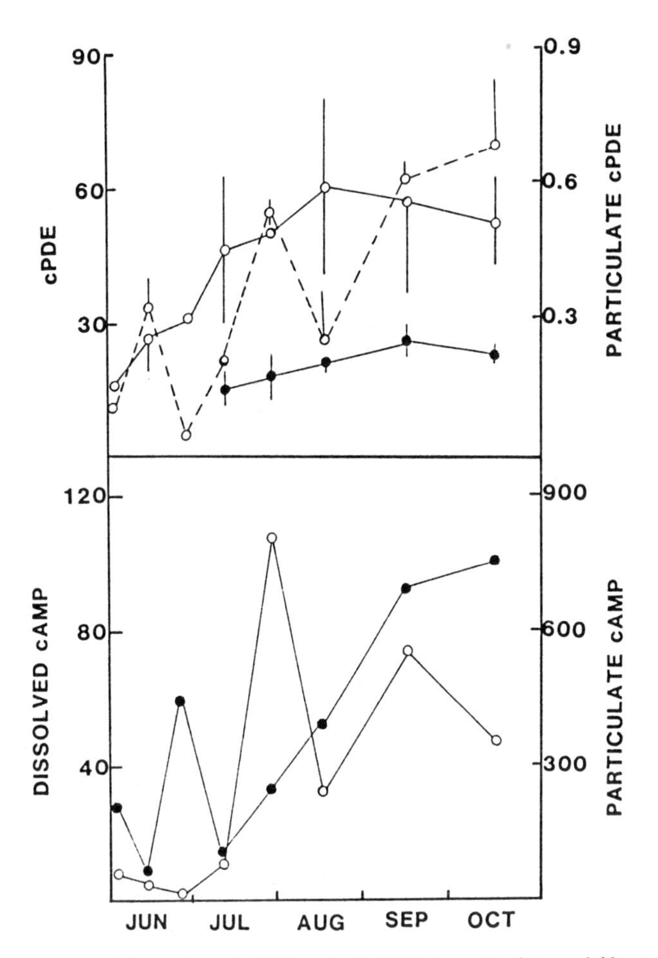

Figure 15.3 Seasonal dynamics of cPDE and of cAMP in particulate and filtrate fractions (PcPDE and DcPDE, and PcAMP and DcAMP, respectively) collected from the 0.1-m stratum of Sangre Isle Lake, 1988. (Top) PcPDE in nmol 1^{-1} h^{-1} (o-o; left ordinate) and nmol μg^{-1} Chl.a h^{-1} (o--o; right ordinate); DcPDE in nmol 1^{-1} min^{-1} (closed circles; left ordinate). Error bars denote ±SE (n = 3). (Bottom) PcAMP (open circles) and DcAMP (closed circles) in nmol μg^{-1} Chl.a and nmol l^{-1}, respectively. All ±SE (n = 2) values were smaller than the diameter of points.

Lugol's-preserved water samples indicated that two massive, nearly unialgal blooms of cyanobacteria comprised the bulk of particulate biomass during the summer period (*Anabaena* spp. between late July and early September; *Microcystis aeruginosa* between mid-June and late July). Chlorophyll$_a$-specific cPDE concentrations were highest during the early and late stages of development of both surface blooms and lowest during the mid-point of the bloom periods. Biomass-specific cPDE remained elevated after the crash of the *Anabaena* bloom in mid-September.

cAMP versus cPDE in lake water Filtrate (dissolved) and particulate-associated cAMP concentrations (Figure 15.3, bottom) were similarly dynamic. The highest concentrations of particulate cAMP occurred in late July and mid-September corresponding to the late stages of the *Microcystis* and *Anabaena* blooms, respectively.

Dissolved cAMP levels were high early in the *Microcystis* bloom period, dropped during the middle stage of the bloom, and then increased again during the development and senescence of the *Anabaena* bloom. Similar relationships between seasonal phytoplankton assemblages have been reported in Sangre Isle Lake water (Francko, 1983; 1987), where cAMP can comprise up to 8% of the dissolved phosphorus pool.

From early June until mid-July an inverse relationship between particulate cPDE activity (which behaved as an exoenzyme during this investigation) and dissolved and particulate cAMP existed, suggesting that this enzyme may have been important in regulating dissolved cAMP levels in situ during the study. This inverse correlation was not noted later in the study period. Based on cAMP and cPDE levels in the filtrate from mid-July to mid-October, the turnover time for dissolved cAMP in epilimnetic waters is much higher than in previously cited studies on cAMP "loss" rates in epilimnetic filtrates from Lawrence Lake (Francko and Wetzel, 1984a). As noted earlier, however, cPDE values reported here are likely overestimates of in situ hydrolytic rates since the ambient pH was always well above the measured pH optimum for naturally occurring cPDE in Sangre Isle Lake waters. Taken together, our evidence suggests that cPDE hydrolysis of cAMP may represent an important degradative pathway in epilimnetic waters.

Location and sources of cPDE Our evidence suggested that particulate-associated cPDE occurred largely as a surface-associated rather than as intracellular enzyme. Total cell protein measurements were made by a reagent dye binding assay (Bradford, 1976) on particulate-fraction subsamples prior to and following sonication. Sonicated aliquots contained several-fold more protein than nonsonicated samples, and a limit product was reached using the sonication procedure outlined in Section 15.2. In contrast, particulate cPDE values were statistically similar ($p < 0.05$; t-test, $n = 2$) in both nonsonicated and sonicated aliquots, supporting the view that the enzyme was accessible to cPDE assay reagents even without cell disruption.

This finding has important implications for the formation of extracellular cPDE as well as on the potential role of cPDE in cAMP dynamics. If particulate cPDE is surface-associated, the actual cAMP turnover time in the filtrate might have been considerably shorter than that described earlier. We did not attempt to determine the source of epilimnetic extracellular cPDE in this study. However, if particulate cPDE is loosely associated with the outer surface of cells it might be readily released from planktonic biota into a dissolved form. Alternatively, dissolved cPDE activity present in epilimnetic filtrates may in part have been imported from the extensive littoral zone that comprises the majority of the surface area of Sangre Isle Lake. Macerated tissue samples from *Ceratophyllum demersum*, the dominant macrophyte species in this lake, contain measurable cPDE activity (Neighbors and Francko, unpublished observations). On 12 July 1988, we made simultaneous measurements of dissolved cPDE levels at the epilimnetic sampling site and within an adjacent *Ceratophyllum* bed. The littoral filtrate contained about fivefold more cPDE than the epilimnetic filtrate (69 versus 15 nmol l^{-1} min^{-1}), further supporting the plausibility of littoral extracellular cPDE export.

The characteristics determined here for lakewater cPDE were similar to those of known cPDE isozymes from animal and algal systems, where this class of enzymes plays a central role in regulating cAMP-dependent physiological responses. Further characterization of lakewater cPDE and research on the ecological significance of cPDE activity appear warranted.

Acknowledgments We thank Ms. Paula Cinnamon for technical assistance. Support was provided by the National Science Foundation (BSR-8907440), the Oklahoma State University Center for Water Research and Department of Botany and Microbiology, and the Department of Botany at Miami University.

References

Ammerman, J.W. and F. Azam. 1981. Dissolved cyclic adenosine monophosphate (cAMP) in the sea and uptake by marine bacteria. *Marine Ecology Progress Series* 5: 85–89.

Ammerman, J.W. and F. Azam. 1982. Uptake of cyclic AMP by natural populations of marine bacteria. *Applied and Environmental Microbiology* 43: 869–76.

Amrhein, N. 1977. The current status of cyclic AMP in higher plants. *Annual Reviews in Plant Physiology* 238: 123–132.

Bradford, M.M. 1976. A rapid and sensitive method for the quantitation of microgram quantities of protein utilizing the principle of protein-dye binding. *Analytical Biochemistry* 72: 248–254.

Brown, E.G. and R.P. Newton. 1981. Cyclic AMP and higher plants. *Phytochemistry* 20: 2453–2463.

Fischer, R. and N. Amrhein. 1974. Cyclic nucleotide phosphodiesterase of *Chlamydomonas rheinhardtii. Biochimica et Biophysica Acta* 341: 412–420.

Francko, D.A. 1983. Cyclic AMP in photosynthetic organisms: Recent developments. *Advances in Cyclic Nucleotide Research* 15: 97–117.

Francko, D.A. 1984a. Phytoplankton metabolism and cyclic nucleotides. I. Nucleotide-induced perturbation of carbon assimilation. *Archiv für Hydrobiologie* 100: 341–354.

Francko, D.A. 1984b. Phytoplankton metabolism and cyclic nucleotides. II. Nucleotide-induced perturbations of alkaline phosphatase activity. *Archiv für Hydrobiologie* 100: 409–421.

Francko, D.A. 1984c. The significance of cyclic nucleotides in phytoplanktonic carbon metabolism: A current view. *Verhandlungen der Internationalen Vereinigung für Theoretische und Angewandte Limnologie* 22: 612–619.

Francko, D.A. 1986. Laboratory studies on *Nelumbo lutea*. (Willd.) Pers. II. Effect of pH on photosynthetic carbon assimilation. *Aquatic Botany* 26: 119–127.

Francko, D.A. 1987a. Size fractionation of phytoplanktonic C-assimilation: Effect of exogenously-applied cAMP. *Journal of Phycology Supplement* 23: 18.

Francko, D.A. 1987b. Limnological characteristics of Sangre Isle Lake, Oklahoma (U.S.A.). *Journal of Freshwater Ecology* 4: 53–60.
Francko, D.A. 1989a. Uptake, metabolism, and release of cyclic AMP in *Selenastrum capricornutum* (Chlorophyceae). *Journal of Phycology* 25: 300–304.
Francko, D.A. 1989b. Modulation of photosynthetic carbon assimilation in *Selenastrum capricornutum* (Chlorophyceae) by cAMP. An electrogenic mechanism? *Journal of Phycology* 25: 305–314.
Francko, D.A. and R.G. Wetzel. 1980. Cyclic adenosine 3′:5′-monophosphate: Production and extracellular release from green and blue-green algae. *Physiologia Plantarum* 49: 65–67.
Francko, D.A. and R.G. Wetzel. 1981a. Dynamics of cellular and extracellular cAMP in *Anabaena flos-aquae* (Cyanophyta): Correlation with metabolic variables. *Journal of Phycology* 17: 129–134.
Francko, D.A. and R.G. Wetzel. 1981b. Synthesis and release of cyclic adenosine 3′:5′-monophosphate by aquatic macrophytes. *Physiologia Plantarum* 52: 33–36.
Francko, D.A. and R.G. Wetzel. 1982. The isolation of cyclic adenosine 3′:5′-monophosphate from lakes of differing trophic status: Correlation with planktonic metabolic variables. *Limnology and Oceanography* 27: 27–38.
Francko, D.A. and R.G. Wetzel. 1984a. The physiological ecology of cyclic adenosine 3′:5′-monophosphate (cAMP) in vascular aquatic plants. I. In situ dynamics and relationship to primary production in a hardwater lake. *Aquatic Botany* 19: 13–22.
Francko, D.A. and R.G. Wetzel. 1984b. The physiological ecology of cyclic adenosine 3′:5′-monophosphate (cAMP) in vascular aquatic plants. II. cAMP as a modulator of carbon assimilation. *Aquatic Botany* 19: 23–35.
Londesborough, J. 1976. Quantitative estimation of 3′5′ cyclic AMP phosphodiesterase using anion exchange resin in a batch process. *Analytical Biochemistry* 71: 623–628.
Ownby, J. and F. Kuenzi. 1982. Changes in cyclic AMP phosphodiesterase activity in *Anabaena variabilis* during growth and nitrogen starvation. *Federation of European Microbiological Societies, Letters* 15: 243–247.
Wetzel, R.G. and G. Likens. 1979. *Limnological Analyses*. Saunders, Philadelphia. 357 pp.

16

Chitinase Activity in Estuarine Waters

Richard A. Smucker and Chi K. Kim

16.1 Introduction

Relatively few measurements of extracellular enzymes in aquatic environments have been reported (Khailov, 1968; Corpe and Winters, 1972; Verstraete et al., 1976; Wainswright, 1981) in spite of the fact that exoenzyme hydrolytic action on organic polymers has long been recognized as a prerequisite for biological membrane transport of component disaccharides and monomers (Pollock, 1962; Fogarty and Kelly, 1979; Keeney, 1983; Button, 1985; Takiguchi and Shimihara, 1988). Microbial growth is recognized to be nutrient-limited for microbes of the aquatic environment and that extracellular enzymatic hydrolysis is usually the rate-limiting step in rendering macromolecules available for cell membrane transport and metabolism (Button, 1985). Despite the universal recognition of the need for extracellular enzymatic hydrolysis of polymers, there are relatively few reports on its presence in aquatic systems. This is in contrast to the numerous citations for soil enzymology where over 50 different enzymes have been reported (for reviews see Skujins, 1976; 1978; and Burns, 1978).

Debate continues on whether and when individual cells export and release extracellular enzymes during active growth but it is clear that polymers with potential for metabolism must be hydrolyzed to yield fragments (usually disaccharides) before they can function as substrates for cell envelope bound (Hoppe, 1983; Somville, 1984; Chróst, 1989; 1990; Soto-Gil and Zyskind, 1989; Chapter 3) and cytosolic (Sims, 1984) disaccharases. The lack of data on extracellular enzymes that are relevant to ecological issues remains especially acute for the polymer chitin.

Chitin (poly-β-1,4-N-acetyl-D-glucosamine) is synthesized as a linear homopolymer but is subsequently modified, yielding a plethora of forms depending upon the producing organisms (Neville, 1975; Muzzarelli, 1977; Gooday, 1983; Cabib, 1987). Microbial cell cultures grown in the presence of appropriate substrate produce extracellular hydrolytic enzymes specific for chitin (Berger and Reynolds, 1958; Okutani, 1978; Ohtakara, 1982). In his review of chitinases, Jeuniaux (1966) described the two-step requirement for complete enzymatic degradation of chitin to yield free N-acetyl-D-glucosamine. In consecutive fashion chitinase (poly-β-1,4-(2-acetamido-2-deoxy)-D-glucoside glycanohydrolase; EC 3.2.1.14) hydrolyzes the polymers of N-acetyl-D-glucosamine, including tetramers and, to a lesser extent, trimers. Chitobiase (chitobiose acetamido-deoxyglucohydrolase; EC 3.2.1.29) hydrolyzes chitobiose (dimer of N-acetyl-D-glucosamine) and chitotriose to yield N-acetyl-D-glucosamine. The "social" function of extracellular hydrolysis can be seen in the report of Okutani (1978), where 100% of the seawater isolates could utilize N-acetyl-D-glucosamine (GlcNAc) but only 10% could hydrolyze chitin. At present, it is unknown whether chitinase-less microorganisms possess the enzyme chitobiase to hydrolyze the disaccharide chitobiose.

Kim and ZoBell (1972) found that high pressure had no effect on hydrolytic rates of introduced agarase, amylase, cellulase, and chitinase. Their data supports the concept of extracellular hydrolytic activity in aquatic systems even under deep-sea conditions. However, direct evidence for extant aquatic chitinase activity was unavailable until use of radiolabeled substrates permitted appropriately specific and sensitive analysis in natural waters (Smucker, 1982; 1986). Chitinase activity was demonstrated as inducible in naturally compact anaerobic muds amended with crab exoskeletal remains and incubated in aerobic flowing seawater (Smucker and Kim, 1987).

Two major constraints on the popularity of specific exoenzyme assessment in aquatic systems are the recognition of the significance of the individual polymer and the availability of suitable methods. Chitin is well recognized as a major organic component of arthropod exoskeletons (Hackman and Goldberg, 1981), molluscs (Jeuniaux et al., 1982), fungi and ascarides (Hackett and Chen, 1978). In all of these cases, the chitin polymer is covalently linked in varying degrees with amino acids and peptides (Brine and Austin, 1981). The plankton *Euphasia superba* alone produces 20–300 x 10^9 tons of chitin per year (Anderson et al., 1978). These high levels of chitin in the context of all other marine arthropod production has generated considerable interest in chitin turnover in aquatic systems (Seki and Taga, 1963; Kim and ZoBell, 1972; Goodrich and Morita, 1977; Warnes and Rux, 1982; Poulicek and Jeuniaux, 1982; Donderski, 1984; Pel and Gottschal, 1986; Herwig et al., 1988).

A complete appraisal of the possibilities for the ecological roles of chitinases must however include the recent disclosure of chitin as a primary product of pelagic photosynthesis (Smucker and Dawson, 1986). Previous measurements of carbon partitioning in primary products of phytoplankton-fixed ^{14}C have been confused due to the presence of chitin in the "protein" fraction which has operationally been defined as the trichloracetic acid precipitate (Olive et al., 1969; Morris et al., 1974). Since the chitin produced by diatoms is in the β form (parallel molecular arrangement), is 100% N-acetylated, is not conjugated with other glycans or with proteins and is not calcified, this highly crystalline form of chitin has been labeled "chitan" (McLachlan, 1965).

Enzymatic attack of the crystalline diatom chitan is unique with respect to hydrolysis of colloidal chitin from crabs (Smucker and Warnes, 1989). Chitinase from *Streptomyces griseus* processively releases the chitinase-definitive end product chitobiose (the disaccharide N,N′-diacetylchitobiose) from diatom chitan while carrying out an endolytic attack on the α-chained (antiparallel molecules) colloidal chitin. Endolytic attack of colloidal chitin releases a suite of high-order oligomers before yielding the disaccharide. The buoyant nature of chitan fibers imparts a high degree of buoyancy to phytoplankton. Other functions of chitan for diatoms are unknown. Undoubtedly, global chitin production estimates will increase dramatically when phytoplankton chitan production is included.

The principle of cell-free enzyme hydrolysis of chitin is employed in chitin agar plate counts for determining the numbers of chitin-hydrolyzing microorganisms (Hsu and Lockwood, 1975; Smucker and Kim, 1984). The method is based upon enzymatic clearing of the agar-suspended particles. Another method for demonstrating a potential for polymer degradation is monitoring weight loss of appropriate purified particles entrained in fine-mesh containers (Warnes and Rux, 1982). These "plaiting and baiting" methods are useful for determining the degradation potential of given organisms or potential of a given aquatic site, but they do not permit assessment of instantaneous extant enzyme activity. Furthermore, these methods are essentially limited to microbial activities; excluding activities of filter-feeding arthropods and of suspension feeding lamellibranchs (Jørgensen, 1973; Mayasich and Smucker, 1987).

An example of the difficulty in using colony clearing as evidence for normal exoenzyme production is in interpretation of colony clearing by in *Streptomyces* spp. during sporulation and germination. It is commonly recognized that most *Streptomyces* spp. can produce zones of clearing (zones of chitin hydrolysis) on chitin agar overlay plates (Hsu and Lockwood, 1972; Smucker and Kim, 1984). In contrast, it is not well known that during sporulation, *Streptomyces* synthesize their own chitin which remains intact, enveloping mature spores (Smucker and Pfister, 1978; Smucker and Morin, 1986). Consequences of sporulation chitin production is that even spores produced following serial growth on sucrose and then glucose (which strongly represses chitinase production) release chitinase during germination in the presence of chloramphenicol (Smucker and Morin, 1986). Although this germination chitinase release may contribute to total cell-free chitinase production in culture and in environmental samples it does not relate teleologically to vegetative extracellular chitinase status. Chitinase released during germination is a consequence of localized spore envelope autolysis. In a similar fashion chitinases from many sources (including molting processes of arthropods) could contribute to the biogeochemical pool of cell-free chitinase activities. It is clear therefore that substrate utilization methods are not the best for assessing holistic extant environmental functions.

Several methods previously proposed for chitinase analysis are cumbersome, insensitive, not sufficiently specific, or have restricted application. Methylumbelliferyl (MUF) substrates (Yang and Hamaguchi, 1980; Hoppe, 1983; Meyer-Reil, 1986; Robbins et al., 1988) and *p*-nitrophenol substrates (Verstraete et al.,

1976; Bowers et al., 1980) have been used for estimating hydrolytic activity including presumptive chitinase assessment. The conjugated substrates offer ease of use and the interpretation of kinetics is somewhat straightforward but even the trimer- and tetramer conjugates yield results which cannot necessarily be interpreted as chitinase activity. An important point is that the aglycone (chromophore) is not detected until it is cleaved from the synthetic oligomer (Yang and Hamaguchi, 1980). True chitinase action by definition yields the disaccharide, chitobiose, but little, if any, monomer production.

In the present study, extant environmental cell-free chitinase activity was determined as a function of geographical location and as a function of microbial viable counts in selected sites of the Chesapeake Bay system. Radiolabeled chitin provided an effective and selective tool for assessing extant chitinase activity.

16.2 Study Location and Methods

Study locations Water and sediment samples were retrieved from sites within the Patuxent River (Figure 16.1), Mill Creek (Figure 16.2) and the Chesapeake Bay. Sites within the lower Patuxent River were immediately adjacent to oyster bars. Patuxent River and a few sites in the Chesapeake Bay were selected to reflect a profile of salinity. Mill Creek, a subcomponent of the Patuxent River drainage area, was chosen for expediency for weekly profile sampling. Water samples were obtained from appropriate depths with submersible pumps. Sediment was retrieved using a Peterson sampler from which only the top 10-mm of sediment was sampled. All samples were placed on ice until processing, which occurred within 12-h post-sampling. Salinity and temperature were determined with a Hach salinometer and turbidity (NTU) with a Hach Turbidimeter, Model 2100A.

Microorganisms *Penicillium islandicum* was grown on chitosan (N-deacetylated chitin) medium in order to test the resistance of radiolabeled colloidal chitin substrate to *P. islandicum* extracellular chitosanase (Monaghan, 1975; Smucker and Wright, 1986). *Streptomyces griseus* chitinase was purified from commercial (Sigma) chitinase or from cultures of *S. griseus* OSU#433 using ion exchange, size exclusion (Smucker et al., 1986) and hydrophobic interaction chromatography (Goheen and Engelhorn, 1984).

Radiolabeled chitin substrate In a properly designed chitinase assay (Smucker and Wright, 1984), the degree of N-acetylation is important in order to minimize competing enzyme hydrolysis. The method of Hirano et al. (1976) was used as modified by Molano et al. (1977). Purified chitosan derived from blue crab chitin (Sigma) having approximately 40% free N-groups, was dissolved in dilute acetic acid, and then reacetylated in methanolic acetic acid with radiolabeled acetic anhydride (New England Nuclear or ICN) at room temperature.

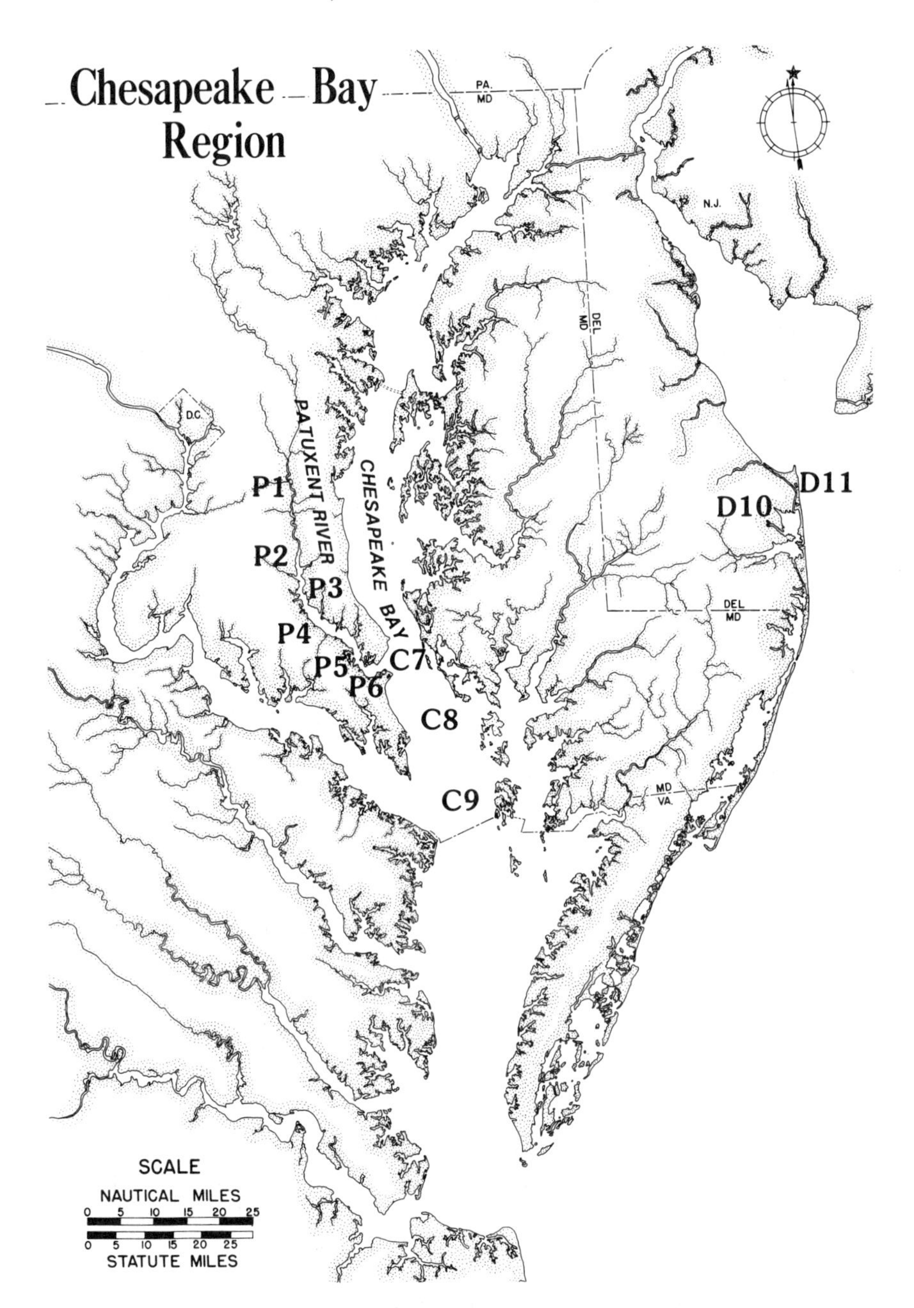

Figure 16.1 Patuxent River and Chesapeake Bay Site map. Site codes and descriptions: P1, Upper Marlboro (dominated by marsh); P2, Lower Marlboro (dominated by marsh); P3, Teague Point (silted oyster bar); P4, Jack Bay (oyster bar; sediment samples were difficult to obtain directly on oyster bars, therefore sediment was obtained proximal to a given bar); P5, Helen's Point (oyster bar); P6, Hog Island (oyster bar); C7, PR Buoy (channel margin, soft bottom); C8, near Hooper Island (soft bottom); C9, Bloodsworth Island site (soft bottom); D10, Rehoboth Bay (near-shore sandy beach); D11, Rehoboth, Delaware (near-shore ocean beach).

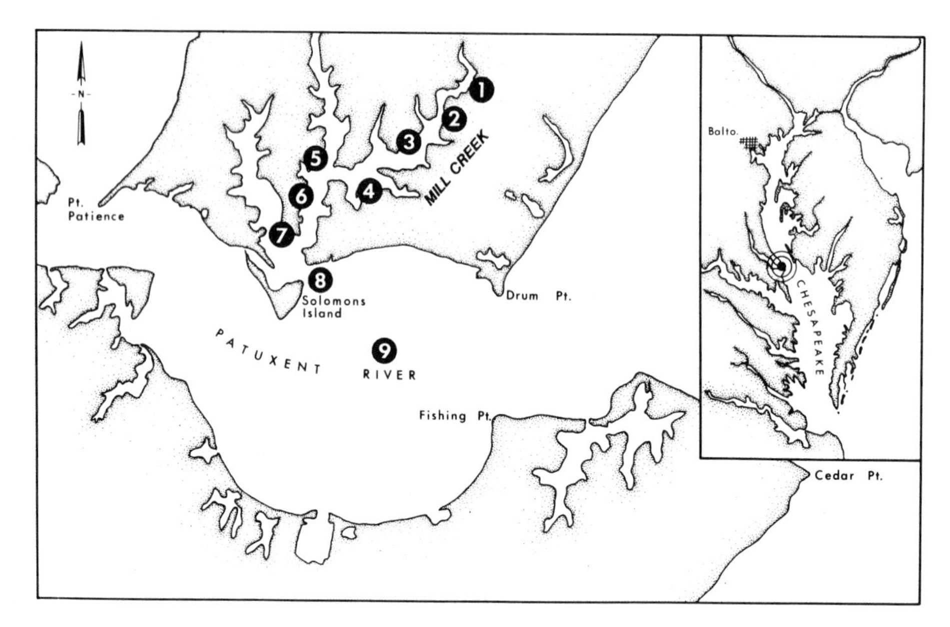

Figure 16.2 Mill Creek site map. 1, Marina area; 2, Brook's Cove; 3, Old House Cove; 4, mouth of Leason Cove; 5, Lusby Point; 6, mouth of Bow Cove; 7, Ship Point; 8, entrance to Solomon's Island Harbor; 9, Middle Ground "Red" no. 4. Sites 1 through 7 are in areas of shoreline homes with individual septic systems. Numerous housing starts in the area produce surface runoff. Shoreline erosion is affected by moderate use of power boats during the summer months.

Background radioactive solubles were removed by exhaustive washing by alternate homogenization in a low volume blender and then vacuum filtration over medium porosity scintered glass. Once the soluble background was low enough, the colloidal dispersion of regenerated chitin was stored at 5°C in the presence of 0.02% unbuffered sodium azide. The expected degree of N-acetylation was confirmed by C:N ratios and by infrared spectral analyses (Sannan et al., 1978) and was supported by resistance to hydrolysis by *Penicillium islandicum* chitosanase, which hydrolyzes only adjacent free amino- and N-acetylated β-1,4-linked residues (Monaghan, 1975). Carbon/nitrogen ratios determined by C:H:N analyses ranged from 7.9 to 8.15 in parallel preparations of unlabeled product. Theoretical C:N ratio of fully N-acetylated chitin is 8.0. Total amino sugars were determined by the method of Ohno et al. (1985) using N-acetyl-D-glucosamine (GlcNAc) as standard. The colloidal dispersions were essentially pure with typical yields of 99.9% expected GlcNAc recovery when fully hydrolyzed after treatment with 6N HCl for 24 h at 100°C.

General conditions for chitinase assay Chitinase (EC 3.2.1.14) determinations were based upon monitoring release of 2.8% trichloracetic acid-soluble oligomers. While it is possible that chitobiase participates in the reaction system, all chitinolytic activity is entirely dependent upon the initial endolytic action of chitinase (reviewed by Cabib, 1987). Essential features of the typical assay were reported by Smucker and Kim (1987): 0.5 ml assay sample containing 0.02% sodium azide in a 10 mm x 75 mm test tube; to which was added 30 μl radiolabeled chitin (variances and specific masses are outlined where indicated). The triplicate samples were incubated at indicated times and temperatures. The reactions were terminated by the addition of 200 μl 10% trichloracetic acid (TCA). The TCA treatment denatured enzymes and coprecipitated undigested chitin along with proteins. Solubilized radiolabeled chitin oligomers were separated from the residual particulate chitin substrate by vacuum filtration over 25 mm Gelman A/E glass fiber filters. Each tube was then rinsed three times with 200 μl distilled water. Filtrates were collected directly into 20 ml liquid scintillation vials containing 9 ml of Aquasol (NEN) and counted for a total of 10,000 counts or 10 min, whichever came first. In all assays, the potential for nonenzymatic tritium exchange was considered in both the design and the interpretation of results (Oldham, 1968; Waterfield et al., 1968)

Chitinolytic bacterial plate counts Colloidal chitin overlay plates were used for determinations of chitinoclastic bacteria (Smucker and Kim, 1984) and additionally scored for total bacteria. Colonies which exhibited zones of colloidal chitin hydrolysis were scored as chitinoclasts. Those colonies which showed no zones of hydrolysis were scored as nonchitinoclastic; but were included in the total viable count bacteria estimator. Agar plates were adjusted with NaCl to mimic one-third-strength sea water. Plates were incubated at ambient laboratory temperatures and scored after two weeks. Very few fungal colonies were visible on any of the plates. Independent registry of fungi, eubacteria, and actinomycete populations will be reported elsewhere. Only total bacteria (including actinomycetes) and chitinoclastic bacteria will be reported and discussed in this chapter.

16.3 Chitinase Assay

Assay development The effect of various ways of water sample processing on chitinase activity is shown in Table 16.1. Removal of bacteria by centrifugation yielded the most activity during the four hour assay period. The reason for the lower value in the intact water sample was presumably due to bacterial uptake of the chitinase-released oligomer chitobiose. The apparently low value for the 0.4-μm-pore-size filtered water sample may be explained by the inefficiency of 0.4-μm filters in removing bacteria from environmental samples. Filtration with 0.2-μm-pore-size filters would remove most of the bacteria and yet permit passage of most cell-free enzymes. These results support the hypothesis that chitinase activity is predominantly in the extracellular phase and are consistent with the classic notion of exoenzyme action (Pollock, 1962). The remainder of assays and results for chitinase activity reported in this work were

Table 16.1 The effect of various water treatments on chitinase activity

Treatment	Chitinase activity (cpm)[1]			
	Background (time 0)			
	Water only[2]	All components[3]	4-h assay	Net observed
None	16.8	488	513	25.2
Boiled	20.6	481	474	−6.5
Filtered:				
0.4-μm pore	17.4	486	495	8.2
0.2-μm pore	15.6	480	580	40.1
Centrifuged (10,000 × g; 10 min)	15.0	427	483	56.3

[1]These assays were run without toluene or other inhibitor. ^{3}H-CPM (counts per minute) are given in order to demonstrate the level of counting sensitivity.
[2]"Water only" refers to the sample background activity immediately after the treatment listed.
[3]"All components" include the treated 0.5-ml water sample plus 30 μl (560 μmoles) of [^{3}H]chitin, 200 μl 10% TCA, three 200-μl water washes.

Table 16.2 Substrate saturation for chitinase determinations in the Patuxent River water column[1]

Regenerated colloidal chitin assay (μmoles GlcNAc equivalents/0.5 ml)	Enzyme solubilized (^{3}H)-CPM
186	254
373	10,146
560	11,819
746	11,690

[1]Trichloroacetic acid control values were substracted independently for each substrate concentration. The river water samples were unfiltered and were not centrifuged; with toluene added.

performed in the presence of transport-limiting levels of toluene added to intact water samples or extracts of sediment. Toluene was added several minutes before starting the assays by addition of radiolabeled substrate.

Minimum colloidal chitin substrate concentration to achieve substrate saturation was determined for representative estuarine water (Table 16.2) and sediment chitinase determinations. While there were small adsorptive losses of solubilized oligomers to the filtration/processing system (Table 16.3) these losses appeared to be consistent and not significantly affected by additions of toluene.

Presence of kaolin (a dominant clay in the Patuxent River) depressed the kinetics of partially purified *Streptomyces griseus* chitinase activity (Figure 16.3). The temperature profile of the control and the kaolin treatments were quite similar, indicating that the clay interfered with chitinase by a relatively straightforward diffusion-limiting mechanism. Judging by the temperature response

Table 16.3 Adsorptive losses of chitinase-solubilized oligomers and GlcNAc to the filtration processing system

Sample treatment	^{3}H-DPM recovered (n = 3)	Percent change
1. 50 μl direct[1,2] (no final filtration)	60,985 (±98)[2]	—
2. 50 μl in 0.5 ml sterile sea water[3]	58,277 (±1,643)	4.4 loss
3–5 Same as no. 2 plus toluene		
3. + 10 μl toluene	63,219 (±1,199)	3.7 gain
4. + 20 μl toluene	58,718 (±3,326)	3.8 loss
5. + 30 μl toluene	63,736 (±1,678)	4.5 gain

[1]A stock solution of soluble ^{3}H-labeled GlcNAc and oligomers was established by filtering enzymatically released products from ^{3}H-chitin through a Gelman A/E glass fiber filter; before additions of toluene or 10% TCA. All treatments were run in triplicate.
[2]This sample received no final filtration but was added directly into the liquid scintillation cocktail.
[3]Treatments no. 2–5 consisted of a 50 μl addition of stock solution to 450 μl of sterile prefiltered sea water (15 ppt salinity) placed in a test tube as for all other enzyme assays. Adsorptive losses to the filtration system are noted in the "Percent change" column.

profiles, kaolin provided little thermal protection for the chitinase. In one experiment, ^{14}C-chitin was used in order to investigate chitinase activity in sediment slurry (Figure 16.4). Using dilute, but otherwise intact, sediment, boiled controls showed higher activities at time zero. Results of such an experiment are difficult to correctly interpret in the earliest stages of a time course, since autoclaving or heating produced a less porous sediment matrix, resulting in higher values at time zero (but not later in the assay) for the heated controls. An explanation for this is the additive filtration effect of the sediments accumulating on the glass fiber filters. Heating by boiling or autoclaving denatures many organic compounds resulting in a more open matrix and permitting noticeably higher filtration flux rates. This matrix treatment effect was also observed in some water samples at time zero (Table 16.1). Of course, future work on this topic should include alternatives to the terminal filtration technique. However, the present studies were all based upon filtration separation of TCA-soluble and -insoluble radiolabelled products.

For sediments from two different sites, 1.8 mM GlcNAc was sufficient to provide enzyme saturation under the conditions of the assay (Figure 16.5). It is important to keep in mind that the time period of incubation can significantly alter the observed K_m of chitinase reactions (Smucker and Wright, 1986) due to progressive enzymatically developed exposure of new sites within colloidal chitin particulates.

Standard chitinase assay conditions for water samples Except where noted otherwise, 0.5-ml water samples were dispensed into 10 x 70 mm test tubes. Controls for each treatment were treated at time zero with 200 μl aqueous 10%

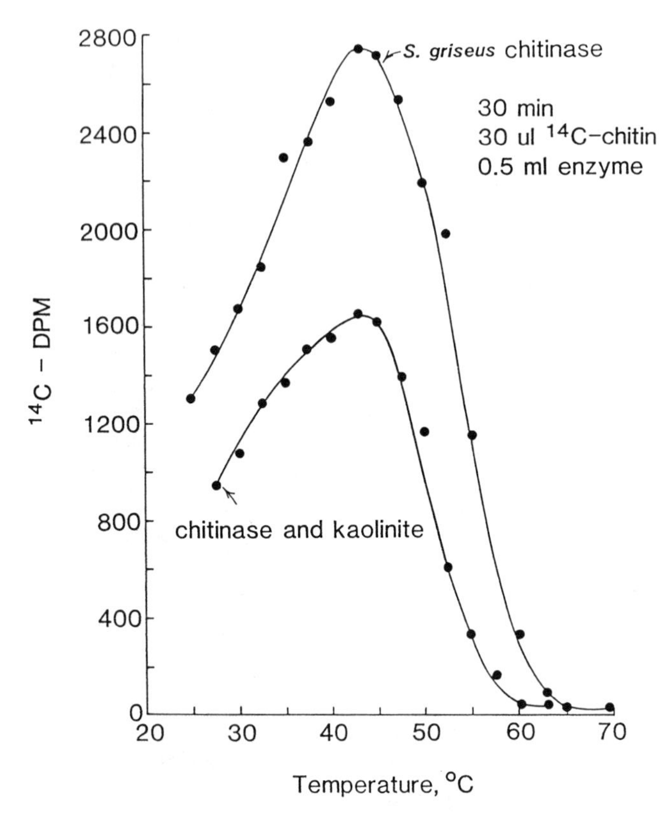

Figure 16.3 Kinetics of chitinase with and without kaolin as a function of incubation temperature.

trichloracetic acid. Controls and experimental replicates were then subjected to additions added in the following sequence: 10 μl toluene, 30 μl ^{3}H-chitin (560 μmole). Several minutes lapsed between the addition of toluene and the addition of radiolabeled substrate, in order to permit toluene interaction with cell membranes. Tubes were placed vertically in a tube rack and placed in an air incubator. Since the major objective of this work was to assess changes in chitinase levels, all incubations were done at a standardized 30°C in a reciprocating platform incubator. There was negligible water loss due to evaporation within the four hour incubation period.

Standard chitinase reaction conditions for sediment samples For comparative studies of sediment chitinase activity, sediment samples were obtained from the top cm of several subsites of a pair of sediment grabs. In the laboratory, a given sample was mixed with a stainless steel scoop and 20 g of the mix was transferred to 60 ml of autoclaved and filtered water (5,000 MW, Amicon hollow fiber) appropriate for the sample salinity. This mix was homoge-

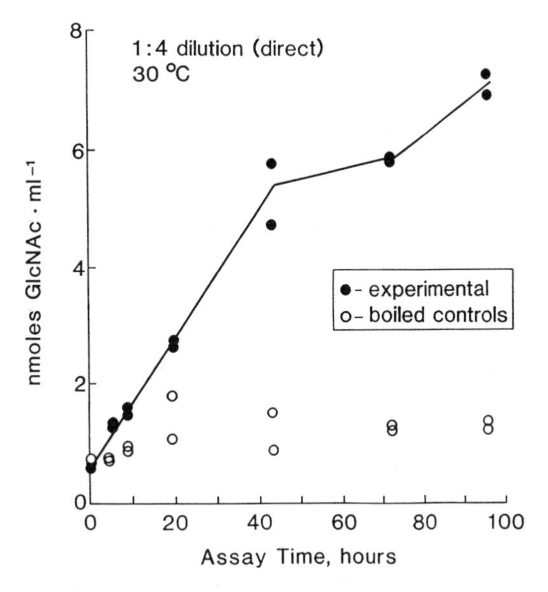

Figure 16.4 Kinetics of sediment chitinase activity as a function of time. ^{14}C-chitin substrate was used since the unextracted total sediment matrix would have caused problems with ^{3}H-chitin and non-catalytic ^{3}H exchange.

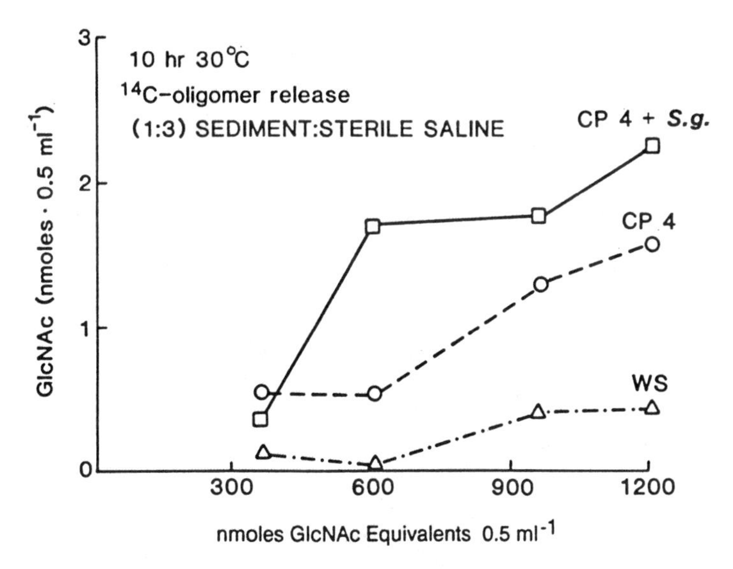

Figure 16.5 Chitin saturation of sediment chitinase activity (CP4, Cove Point site in central Chesapeake Bay; WS, Western Shore site at mid-Bay) *S. griseus* chitinase was added to the CP4 sample, in one case. These sediments were not taken directly at sites indicated in Figure 16.1, but are indicative of the Central Bay sites.

nized for 60 sec in a Virtis open-blade homogenizer. An aliquot was removed for viable microbial plate counts and the remainder was centrifuged for 20 min at 10,000 times g in a Sorval SS-34 rotor. The supernate was dispensed in 500 μl aliquots in test tubes and 30 μl toluene was added several minutes before radiolabeled substrate addition. For these extracted sediments, ^{3}H-chitin was used in order to achieve appropriate sensitivity within the desired 4-h assay period. Boiled samples with time zero trichloracetic acid blanks were processed in the same way as the water column samples.

16.4 Chitinase Activity in Water and Sediments

Chitinase in a Patuxent River/Chesapeake Bay profile Chitinase activity for the upper water column paralleled that of the lower water column (Table 16.4 and Figure 16.6B). Stations P3 and P4, which are the brackish-water stations, exhibited the highest water column activities. For all of the downstream sites in the Patuxent River and in the three Chesapeake Bay sites, activity was lower. For sediment, the pattern was more dramatic; and in the reverse direction. Sediment in the downstream sites had remarkably higher chitinase activities. This effect in sediment may be partially explained by the increased depth of the water column towards the mouth of the river and into the bay. Integrated water column chitinase activity would include the depth factor, partially reflecting the increases in sediment activity. However, the activities expressed for sediments are listed as uncorrected for the fourfold sediment dilution in order to simplify graphic presentation. One can easily see that even vertically integrated water column values may not explain the high values of sediment activities at the sites with greater depth. Another unexpected result of the sediment data is the inverse correlation between chitinase activity and chitinolytic bacterial populations (Fig. 16.6C). That is, the lack of correlation is unexpected if bacteria are the presumed dominant force for chitin degradation. This phenomenon is discussed below.

Chitinase time series variability at the Mill Creek Site All water samples (except M9b taken at 20 feet, Table 16.5) used for this portion of the study were taken from the upper photic zone (see Figure 16.2), most samples were taken from one meter subsurface (Table 16.5). Total water column depth was inversely related to microbial plate count data and chitinase (Fig. 16.7). There was an apparent positive correlation of chitinase activity with chitinolytic bacteria and with total numbers of viable-count bacteria. Since rainfall was remarkably low during the summer of 1983, negligible allochthonous input would account for the pattern observed. However, it is conceivable that tidal action affected proportionally more resuspension in the more shallow regions of the Mill Creek study.

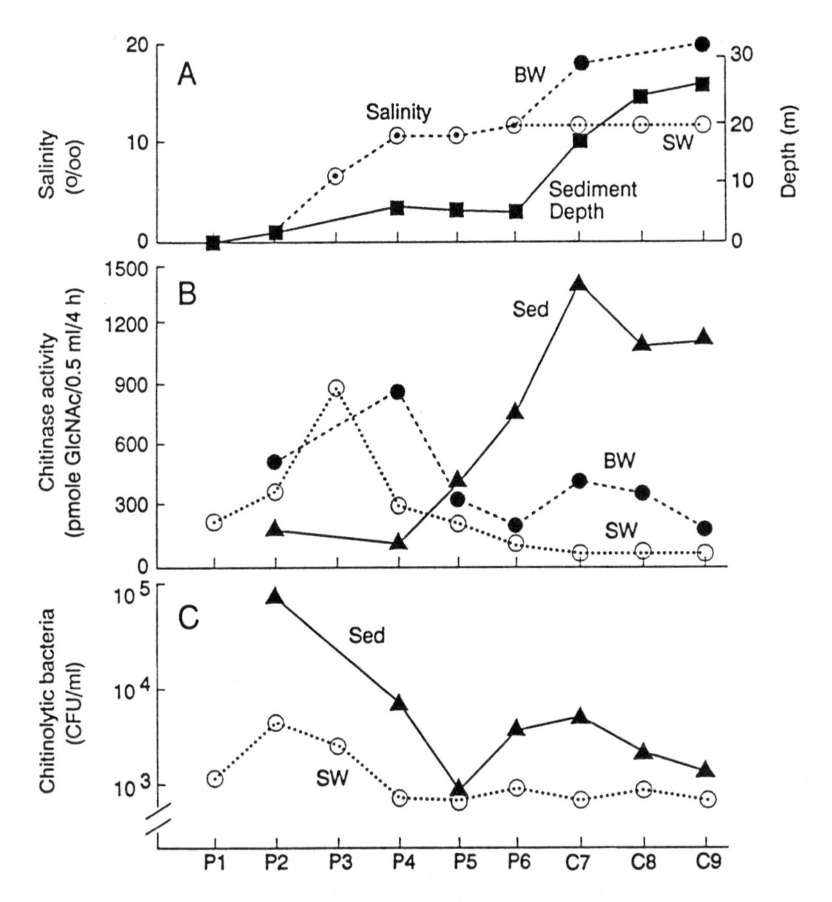

Figure 16.6 Chitinase activity in Patuxent River water and sediment, July, 1983 cruise (BW, water taken one meter off-bottom; SW, water taken one meter sub-surface; SED, extracts of sediment diluted four fold). A, salinity and depth; B, chitinase activity; C, chitinolytic bacteria.

Chitinase activity correlation with bacterial populations Data summarized in Figure 16.7 show that 3-month means of total plate counts and chitinolytic microbial plate counts share a similar trend with chitinase activity. The data in this figure are 3-month means with a total of 12 dates with duplications for each data point. Correlation of bacterial populations with chitinase activity is expected if one assumes that bacteria are the dominant force for turnover of chitin.

The important comparison between the Mill Creek data and that of the Patuxent River is the lack of correlative trend in the Patuxent River data for chitinase with chitinoclastic bacterial populations (Figure 16.6). One explanation for this may be the presence of oyster bed communities in the Patuxent River. We have previously shown that oysters produce a bacterially indepen-

Table 16.4 Chitinase activity, chitinolytic bacteria, and other characteristics in the Patuxent River and Chesapeake Bay

		Physical parameter					Biological parameter			
Site code	Samples type	Depth (m)	Temperature (½C)	pH	Salinity (‰)	Conductivity ($\mu S\ cm^{-1}$)	Dissolved oxygen ($mg\ O_2\ l^{-1}$)	Turbidity (rel. units)	Chitinase[3] activity	Total bacteria ($10^2\ CFU/ml^{-1}$)
P1a[3]	S.W.[1]	—	28.5	7.24	—	—	—	30.0	221	00
P1b[3]	S.W.	—	28.5	7.16	—	—	—	25.0	32	230
	Sed.	—	—	—	—	—	—	—	234	35
	S.W.	—	28.2	6.71	1.4	2.6	7.3	16.0	382	260
P2	B.W.	2.0	28.2	6.71	1.3	2.7	6.4	60.0	521	—
	Sed.	—	—	6.55	—	—	—	—	190	810
P3	S.W.	—	29.5	7.01	7.1	13.1	—	3.6	881	120
	S.W.	—	27.7	7.52	10.4	17.9	7.2	7.6	305	45
P4	B.W.	7.0	27.7	7.01	11.1	19.2	5.1	4.5	878	—
	Sed.	—	—	—	—	—	—	—	110	45
	S.W.	—	28.3	8.21	10.9	18.6	—	3.5	233	25
P5	B.W.	5.5	27.7	7.58	11.3	19.6	7.1	2.5	323	—
	Sed.	—	—	8.26	—	—	—	—	404	25
	S.W.	—	26.9	8.00	11.8	20.1	—	2.0	158	15
P6	B.W.	4.5	26.9	8.01	11.8	19.9	—	2.4	152	—
	Sed.	—	—	7.67	—	—	—	—	756[4]	30
	S.W.	—	26.4	8.15	11.9	19.9	—	1.6	39	10
C7	B.W.	16.5	25.4	7.31	18.5	29.5	—	2.0	442	—
	Sed.	—	—	8.29	—	—	—	—	1,386[4]	200
	S.W.	—	26.4	8.10	12.1	20.3	—	1.5	68	75
C8	B.W.	25.0	26.1	7.58	13.7	22.7	—	1.3	368	—
	Sed.	—	—	7.83	—	—	—	—	1,080	70
	S.W.	—	26.9	8.15	11.9	20.1	—	1.1	56	33
C9	B.W.	27.0	25.5	7.44	20.2	32.3	—	1.2	184	—
	Sed.	—	—	8.35	—	—	—	—	1,148	150
D10	S.W.	—	29.5	7.86	—	—	—	9.5	322	220
D11	S.W.	—	24.5	7.71	—	—	—	45.0	176	320

[1]S.W., 1 m below surface water, B.W., bottom water 1 m above sediment, Sed., sediment.
[2]Chitinase activity determined as described in the text. Data expressed as picomoles GlcNAc/0.5 ml/4h. Sediment chitinase activity the centrifuged supernatant of the fourfold sediment dilution.
All the samples were taken on 26 July 1983. Site P1a was a shallow marsh-feeder stream; P1b was mid-channel Patuxent River at Upper Marlboro.
The samples of surface and bottom waters were taken from 1 m beneath the water surface and 1 m above the sediment surface, respectively.

Table 16.5 Chitinase activity, chitinolytic bacteria, and other characteristics in Mill Creek[1]

		Physical parameters				Biological parameters			
Site code	Sampling depth (ft)	Depth (ft)	Temperature (°C)	pH	Turbidity (NTU)	Chitinase activity[2]	Total bacteria (10^2 CFU/ml^{-1})	Chitinolytic bacteria (10^2 CFU/ml^{-1})	Ratio of Chitinolytic total bacteria (%)
M1	2	3	28.8	7.73	5.38	503(108)[3]	168(40)[3]	50(14)[3]	36(10)[3]
M2	3.5	7	28.6	7.61	4.42	347(109)	111(34)	20(10)	25(11)
M3	3.5	7	28.6	7.81	3.76	440(100)	114(40)	26(10)	29(12)
M4	4	10	28.6	8.04	2.92	344 (75)	120(34)	24 (7)	24 (9)
M5	4	11	28.5	8.09	3.04	405(100)	193(76)	27 (6)	19(11)
M6	4	14	28.3	8.21	3.36	278 (67)	114(32)	16 (6)	19(10)
M7	4	15	28.1	8.25	4.28	291 (69)	108(51)	11 (4)	13 (5)
M8	4	20	27.2	8.14	3.02	298 (52)	270(36)	27 (8)	19 (6)
M9a[3]	4	32	26.7	8.05	2.28	242 (48)	105(10)	21 (7)	20 (5)
M9b[3]	20	32	26.7	7.64	2.58	222 (63)	96(30)	15 (6)	15 (4)

[1]Mean values for 12 weekly samples taken June through August, 1983.
[2]Chitinase activity determinations as described in text. Data expressed as picomoles GlcNAc/0.5 ml/4 h.
[3]9a, surface; 9b, 20 feet.

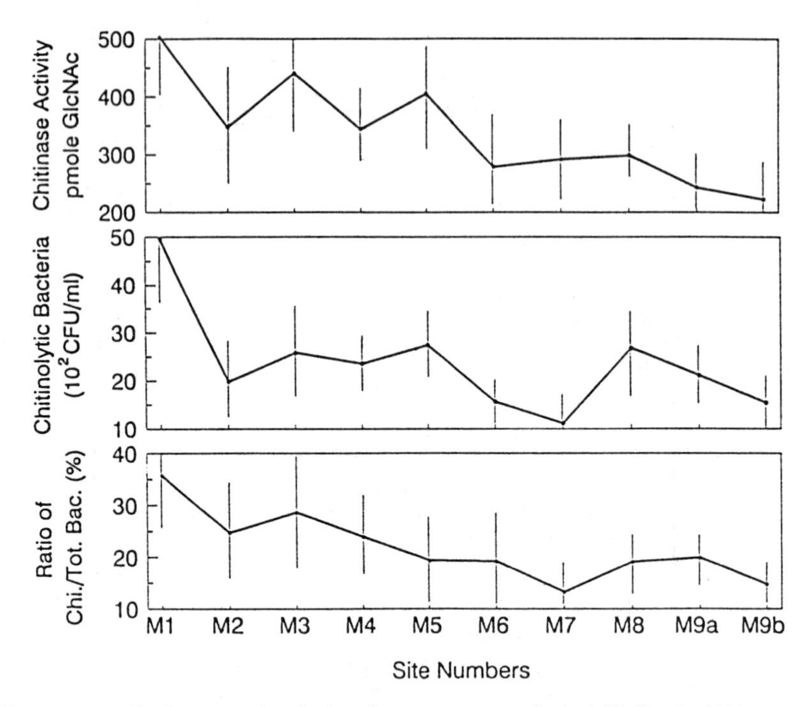

Figure 16.7 Chitinase activity during the summer months in Mill Creek, 1983.

dent constitutive suite of chitinases (Mayasich and Smucker, 1987) with a quantitatively important role in respective membrane transport enzymology (Mayasich and Smucker, 1986) and biodeposits (Smucker and Kim, unpublished observations). These and other invertebrate feeding activities may contribute to the total extracellular activity of chitinases, especially, as observed from samples of the sediment surface.

16.5 Significance of Chitinolytic Activity in Aquatic Environments—Conclusions

Data has been presented which supports the hypothesis that extracellular chitinase activity plays a dynamic role in estuarine waters. Chitinase activity in the water column and in intact sediments exhibited typical protein enzymatic properties such as substrate saturation, acid denaturation, and heat denaturation. Previously, we have also demonstrated induction of chitinase activity in estuarine anaerobic mud dosed with crab exoskeleton and incubated aerobically in flowing natural estuarine water (Smucker and Kim, 1987). These data demonstrate properties expected of microbial enzyme production and enzymatic activity in general. The general concept of extracellular endolytic action of polysaccharases is supported by direct evidence of cell-free disaccharides in estuarine waters (R. Dawson, personal communication). Identified disaccharides included sucrose, gentiobiose, maltose, and cellobiose. They were separated by strong anion exchange chromatography and measured by pulsed amperometric detection (Hardy et al., 1988; Townsend et al., 1988). Identification of the chitobiose disaccharide is under investigation in our laboratories.

The present work demonstrates the potential utility for development of a chitin and chitinase role in pelagic and benthic components of the estuary. Trends in chitinase activity in the 1983 Patuxent River and Mill Creek studies indicate the usefulness of including this enzyme system in ecosystem characterizations. Microbial populations do not adequately explain observed chitinase in the main stem of the Patuxent River.

Measurements of chitan primary production (Smucker and Dawson, 1986) were not incorporated into the experimental design of the 1983 study. This means that chlorophylla data and primary production data were not specifically available. Due to the patchiness of phytoplankton populations, the data from other investigations made during the same time period were not directly applicable to this work. However, one can conclude from the available data that the downstream region of the river is more strongly dominated by autochthonous primary production (Sigleo et al., 1982). With this assumption and because of the increased depth of the vertical photic zone, increased chitan input from the downstream water column's primary production may occur. There is no reason to limit this sort of investigation to estuarine systems, since there is a large potential for primary production of chitan in oceanic systems (Smucker and Dawson, 1986). Of course, it is also well known that micro- and macrozooplankton of the arthropoda are prolific, exhibiting high production rates of exuvial chitin which is susceptible to the same chitinase action as is diatom chitan (Smucker and Dawson, 1986).

Future work in this area should include primary production data, chitinoclastic bacterial populations, and phytoplankton genera scoring, along with the conventionally measured parameters.

Acknowledgements The authors recognize the valuable collegial support of Carl E. Warnes, Rodger Dawson and Robert Ulanowicz. Many thanks to Fran Younger for designing and creating the figures and to Linda Gott for her assistance with the manuscript. This work was done at the Center for Environmental and Estuarine Studies, The University of Maryland System, Solomons, Maryland, USA.

References

Anderson, C.G., Depablo, N. and C.R. Romo. 1978. Antarctic krill (*Euphasia superba* Dana) as a source of chitin and chitosan. pp. 5–10 in Muzzarelli, R.A.A. and Pariser, E.R. (editors), *Proceedings of the First International Conference on Chitin and Chitosan*. Massachusetts Institute of Technology, Cambridge.

Berger, L.R. and D.M. Reynolds. 1958. The chitinase system of a strain of *Streptomyces griseus*. *Biochimica et Biophysica Acta* 29: 522–534.

Bowers, G.N., Jr., McComb, R.B., Christensen, R.G. and R. Schaffer. 1980. High purity 4-nitrophenol: purification, characterization and specifications for use as a spectrophotometric reference material. *Clinical Chemistry* 26: 724–729.

Brine, C.J. and P.R. Austin. 1981. Chitin isolates: species variation in residual amino acids. *Comparative Biochemistry and Physiology* 70B: 173–178.

Burns, R.G. 1978. *Soil Enzymes*. Academic Press, San Francisco. 339 pp.

Button, D.K. 1985. Kinetics of nutrient-limited transport and microbial growth. *Microbiological Reviews* 49: 270–297.

Cabib, E. 1987. The synthesis and degradation of chitin. pp. 59–101 in Meister, A. (editor), *Advances in Enzymology and Related Areas of Molecular Biology*, vol. 59. John Wiley and Sons, New York.

Chróst, R.J. 1989. Characterization and significance of β-glucosidase activity in lake water. *Limnology and Oceanography* 34: 660–672.

Chróst, R.J. 1990. Microbial ectoenzymes in aquatic environments. pp. 47–78 in Overbeck, J. and Chróst, R.J. (editors), *Aquatic Microbial Ecology: Biochemical and Molecular Approaches*. Springer Verlag, New York.

Corpe, W.A. and H. Winters. 1972. Hydrolytic enzymes of some periphytic marine bacteria. *Canadian Journal of Microbiology* 18: 1483–1490.

Donderski, W. 1984. Chitinolytic bacteria in water and bottom sediments of two lakes of different trophy. *Acta Microbiologica Polonica* 33: 163–170.

Fogarty, W.M. and C.T. Kelly. 1979. Developments in microbial extracellular enzymes. pp. 45–102 in Wiseman, A. (editor), *Topics in Enzyme and Fermentation Biotechnology*. Halsted Press, New York.

Goheen, S.C. and S.C. Engelhorn. 1984. Hydrophobic interaction high performance liquid-chromatography. *Journal of Chromatography* 317: 55–65.

Gooday. G.W. 1983. The microbial synthesis of cellulose, chitin and chitosan. pp. 85–127 in Bushell, M.E. (editor), *Progess in Industrial Microbiology*, vol. 18. Elsevier, New York.

Goodrich, T.D. and R.Y. Morita. 1977. Bacterial chitinase in the stomachs of marine fishes from Yaquina Bay, Oregon, USA. *Marine Biology* 41: 355–360.

Hackett, C.J. and K.C. Chen. 1978. Quantitative isolation of native chitin from resistant structures of *Sordaria* and *Ascaris* species. *Analytical Biochemistry* 89: 487–500.

Hackman, R.H. and M. Goldberg. 1981. A method for determination of microgram amounts of chitin in arthropod cuticles. *Analytical Biochemistry* 110: 277–280.

Hardy, M.R., Townsend, R.R. and Y.C. Lee. 1988. Monosaccharide analysis of glycoconjugates by anion exchange chromatography with pulsed amperometeric detection. *Analytical Biochemistry* 170: 54–62.

Herwig, R.P., Pellerin, N.B., Irgens, R.L., Maki, J.S. and J.T. Staley. 1988. Chitinolytic bacteria and chitin mineralization in the marine waters and sediments along the Antarctic Peninsula. *Federation of European Microbiological Societies, Microbiology Ecology* 53: 101–112.

Hirano, S., Ohe, Y. and H. Ono. 1976. Selective n-acylation of chitosan. *Carbohydrate Research* 47: 315–320.

Hoppe, H.G. 1983. Significance of exoenzymatic activities in the ecology of brackish water: Measurements by means of extracted methylumbelliferyl substrates. *Marine Ecology Progress Series* 11: 299–308.

Hsu, S.C. and J.L. Lockwood. 1975. Powdered chitin agar as a selective medium for enumeration of actinomycetes in water and soil. *Applied Microbiology* 29: 422–426.

Jeuniaux, C. 1966. Chitinases. pp. 644–650 in Colowick, S.P. and Kaplan, N.O. (editors). *Methods in Enzymology*, vol. 8. Academic Press, New York.

Jeuniaux, C., Voss-Foucart, M.F. and J.C. Bussers. 1982. Preliminary results of chitin biomass in some benthic marine biocenoses. pp. 200–204 in Hirano, S. and Tokura, S. (editors), *Chitin and Chitosan*. Tottori University, Tottori.

Jørgensen, C.B. 1966. *Biology of Suspension Feeding*. Pergamon Press, Oxford. 357 pp.

Keeney, D.R. 1983. Principles of microbial processes of chemical degradation, assimilation, and accumulation. pp. 153–164 in D.W. Nelson, D.W., Elrick, D.W. and Tanji, K.K. (editors), *Chemical Mobility and Reactivity in Soil Systems*. American Society for Agronomy, Madison, Wisconsin.

Khailov. K.M. 1968. Extracellular microbial hydrolysis of polysaccharides dissolved in sea water. *Mikrobiologiya* 37: 518–522 (English translation 424–427.).

Kim, J. and C.E. ZoBell. 1972. Agarase, amylase, cellulase and chitinase activity at deep-sea pressures. *Journal of Oceanography Society Japan* 28: 131–137.

Mayasich, S.A. and R.A. Smucker. 1986. Glycosidases in the American oyster, *Crassostrea virginica* Gmelin, digestive tract. *Journal of Experimental Marine Biology and Ecology* 95: 95–98.

Mayasich, S.A. and R.A. Smucker. 1987. Role of *Cristispira* sp. and other bacteria in the chitinase and chitobiase activities of the crystalline style of *Crassostrea virginica* (Gmelin). *Microbial Ecology* 14: 157–166.

McLachlan, J., McInnes, A.G. and M. Falk. 1965. Studies on the chitan (chitin: poly-N-acetylglucosamine) fibers of the diatom *Thalassiosira fluviatilis* Hustedt. I. Production and isolation of chitan fibers. *Canadian Journal of Botany* 43: 707–713.

Meyer-Reil, L.-A. 1986. Measurement of hydrolytic activity and incorporation of dissolved organic substrates by microorganisms in marine sediments. *Marine Ecology Progress Series* 31: 143–149.

Molano, J., Duran, A. and E. Cabib. 1977 A rapid and sensitive assay for chitinase using tritiated chitin. *Analytical Biochemistry* 83: 648–656.

Monaghan, R.L. 1975. The discovery, distribution and utilization of chitosanase. *Ph.D. dissertation, Rutgers University*.

Morris, I., Glover, H.E. and C.S. Yentsch. 1974. Products photosynthesis by marine phytoplankton: the effect of environmental factors on the relative rates of protein synthesis. *Marine Biology* 27: 1–9.

Muzzarelli, R.A.A. 1977. *Chitin*. Pergamon Press, New York.

Neville, A.C. 1975. *Biology of the Arthropod Cuticle*. Springer Verlag, New York.

Ohno, N., Suzuki, I. and T. Yadomae. 1985. A method for monitoring elution profiles during the chromatography of amino sugar-containing oligo- and polysaccharides. *Carbohydrate Research* 137: 239–243.

Ohtakara, A., Mitsutomi, M. and E. Nakamae. 1982. Mode of hydrolysis of chito-oligosaccharides with *Pycnoporus cinnabarinus* β-N-acetylhexosaminidase: application of high-performance liquid chromatography. *Agriculture Biological Chemistry* 46: 293–295.

Okutani, K. 1978. Chitin and N-acetylglucosamine decomposing bacteria in the sea. pp. 582–586 in Muzzarelli, R.A.A. and Pariser, E.R. (editors), *Proceedings of First International Conference on Chitin and Chitosan*. Massachusetts Institute of Technology, Cambridge.

Oldham, K.G. 1968. Spurious isotope effects in enzymatic reactions with tritium labelled substrates. *Journal of Labeled Compounds* 4: 127–133.

Olive, J.H., Benton, D.M. and J. Kishler. 1969. Distribution of ^{14}C in products of photosynthesis and its relationship to phytoplankton composition and rate of photosynthesis. *Ecology* 50: 380–386.

Pel, R. and J.C. Gottschal. 1986. Mesophilic chitin-degrading anaerobes isolated from an estuarine environment. *Federation of European Microbiological Societies, Microbiology Ecology* 38: 39–49.

Pollock, M.R. 1962. Exoenzymes. pp. 121–178 in Gunsalus, I.C. and Stanier, R.Y. (editors), *The Bacteria*, vol. 4. Academic Press, New York.

Poulicek, M. and C. Jeauniaux. 1982. Biomass and biodegradation of mollusk shell chitin in some marine sediments. pp. 200–204 in Tokura, S. (editor), *Chitin and Chitosan*. Tottori University, Tottori.

Robbins, P. W., Albright, C. and B. Benfield. 1988. Cloning and expression of a *Strep-*

tomyces plicatus chitinase (chitinase-63) in *Escherichia coli*. *Journal of Biological Chemistry* 263: 443–447.

Sannan, T., Kurita, K., Ogura, K. and Y. Iwakura. 1978. Studies on chitin: 7. I.R. Spectroscopic determination of degree of acetylation. *Polymer* 19: 458–459.

Seki, H. and N. Taga. 1963. Microbiological studies on the decomposition of chitin in marine environment. IV. Disinfecting effect of antibacterial agents on the chitinoclastic bacteria. *Journal of Oceanographic Society of Japan* 19: 152–157

Sigleo, A. C., Hoering, T.C. and G.R. Helz. 1982. Composition of estuarine colloidal material: organic components. *Geochemica et Cosmologica Acta* 46: 1619–1626.

Sims, A.P., Kopetzki, E., Schulz, B. and J.A. Barnett. 1984. The use of phenolic glycosides for studying the aerobic or anaerobic transport of disaccharides into yeasts. *Journal of General Microbiology* 130: 1933–1940.

Skujins, J. 1976. Extracellular enzymes in soil. *CRC Critical Reviews in Microbiology* 4: 383–421.

Skujins, J. 1978. History of abiotic soil enzyme research. pp. 1–49 in Burns, R.G. (editor), *Soil Enzymes*. Academic Press, London.

Smucker, R.A. 1982. Determination of chitinase hydrolytic potential in an estuary. pp. 135–139 in Hirano, S. and Tokura, S. (editors), *Chitin and Chitosan*. Japanese Society of Chitin and Chitosan, Sapporo.

Smucker, R.A. 1986. Important developments in biotechnology. pp. 173–178 in *Leading Edges and Major Developments and New Applications in Membrane Technology*. Business Communications Co., Norwalk.

Smucker, R.A. and R. Dawson. 1986. Products of photosynthesis by marine phytoplankton: chitin in TCA "protein" precipitates. *Journal of Experimental Marine Biology and Ecology* 104: 143–152.

Smucker, R.A. and C.K. Kim. 1984. Effects of phosphate on *Streptomyces griseus* chitinase production. pp. 397–405 in Zikakis, J. (editor), *Chitin, Chitosan and Related Enzymes*. Academic Press, Orlando.

Smucker, R.A. and C.K. Kim. 1987. Chitinase induction in an estuarine system. pp. 347–355 in Llewellyn, G.C. and O'Rear, C.E. (editors), *Biodeterioration Research 1*. Plenum Press, New York.

Smucker, R.A. and L.G. Morin. 1986. *Streptomyces* sp. chitin development during sporulation. pp. 397–405 in Szabo, G., Biron, S. and Goodfellow, M. (editors), *Biological, Biochemical and Biomedical Aspects of Actinomycetes*, Part A. Akademiai Kiado, Budapest.

Smucker, R.A. and R.M. Pfister. 1978. Characteristics of *Streptomyces coelicolor* A3(2) aerial spore rodlet mosaic. *Canadian Journal of Microbiology* 24: 397–408.

Smucker, R.A. and C.E. Warnes. 1989. Differential mode of chitinase hydrolysis of chitin and chitan. pp. 75–80 in Llewellyn, G.C. and O'Rear, C.E. (editors), *Biodeterioration Research* 2. Plenum Press, New York.

Smucker, R.A., Warnes, C.E. and C.J. Haviland. 1986. Chitinase production by a freshwater pseudomonad. pp. 549–553 in Llewellyn, G.C. and O'Rear, C.E. (editors), *Biodeterioration 6*. Farnham House, Slough.

Smucker, R.A. and D.A. Wright. 1986. Characteristics of *Crassostrea virginica* crystalline style chitin digestion. *Comparative Biochemistry and Physiology* 83A: 489–493.

Somville, M. 1984. Measurement and study of substrate specificity of exoglucosidase activity in eutrophic water. *Applied and Environmental Microbiology* 48: 1181–1185.

Soto-Gil, R.W. and J.W. Zyskind. 1989. N,N'-Diacetylchitobiase of *Vibrio harveyi*: Primary structure, processing and evolutionary relationships. *Journal of Biological Chemistry* 264: 14778–14783.

Takiguchi, Y. and Shimahara, K. 1988. N,N'-diacetylchitobiose production from chitin by *Vibrio anguillarum* strain E-383a. *Letters Applied Microbiology* 6: 129–131.

Townsend, R.R., Hardy, M.R., Hindsgaul, O. and Y. Chuan Lee. 1988. High-performance anion-exchange chromatography of oligosaccharides using pellicular resins and pulsed amperometric detection. *Analytical Biochemistry* 174: 459–470.

Verstraete, W., Voets, J.P. and P. van Lancker. 1976. Evaluation of some enzymatic methods to measure the bio-activity of aquatic environments. *Hydrobiologia* 49: 257–266.

Wainwright, M. 1981. Enzyme activity in intertidal sands and salt-marsh soils. *Plant and Soil* 59: 357–363.

Warnes, C.E. and T.P. Rux. 1982. Chitin mineralization in a freshwater habitat. pp. 191–195 in Hirano, S. and Tokura, S. (editors), *Chitin and Chitosan*. Tottori University, Tottori.

Waterfield, W.R., Spanner, J.A. and F.G. Standford. 1968. Tritium exchange from compounds in dilute aqueous solutions. *Nature* 218: 472–473.

Yang, Y. and K. Hamaguchi. 1980. Hydrolysis of 4-methylumbelliferyl N-acetyl-chitotetraoside catalyzed by hen lysozyme. *Journal of Biochemistry* 88: 829–836.

17

Significance of Extracellular Enzymes for Organic Matter Degradation and Nutrient Regeneration in Small Streams

Jürgen Marxsen and Karl-Paul Witzel

17.1 Introduction

Extracellular enzymatic hydrolysis is the first step in the microbial degradation of macromolecular organic matter. Investigations into this process in lotic systems are scarce (e.g., Duddridge and Wainwright, 1982; Boon, 1989; Marxsen and Witzel, 1990), even though it may limit microbial substrate uptake and production. The first publications on extracellular enzymes in aquatic systems appeared more than 20 years ago (Overbeck, 1961; Reichardt et al., 1967; Chapter 1), but for a long time afterwards there were few investigations into this subject. One reason was probably the lack of an appropriate method. However, in recent years artificial substrates that release colored (Meyer-Reil, 1981; 1983; 1984; Hoppe et al., 1983) or fluorescent (Pettersson and Jansson, 1978; Hoppe, 1983; Somville and Billen, 1983; Rego et al., 1985; Chróst, 1990) compounds after enzymatic hydrolysis have been introduced into the field of aquatic microbial ecology. In particular, the fluorogenic model substrates have made sensitive, simple, and rapid measurements of enzyme activity possible.

Most recently, 4-methylumbelliferyl-(MUF-)compounds have been used for the investigation of limnic and marine sediment and water samples (e.g., Hoppe, 1983; Somville, 1984; Niesslbeck et al., 1985, Chróst and Krambeck, 1986; King, 1986; Meyer-Reil, 1986; 1987; Chróst and Overbeck, 1987; Holzapfel-Pschorn et al., 1987; Hoppe et al., 1988; Jacobsen and Rai, 1988; Chróst, 1989; Chróst et al., 1989; Münster et al., 1989). The subject has been reviewed recently by Chróst (1990). However, investigations on running waters are still scarce (Marxsen and Witzel, 1990; Freeman et al., 1990), even though the extracellular enzymatic hydrolysis of macromolecular organic matter is likely to play an important role in the regeneration of nutrients, as well as the decomposition of organic matter, in small unpolluted streams with large allochthonous inputs. In such systems, low-molecular-weight organic compounds capable of being taken up directly by bacteria and fungi amount to only about 20% of total dissolved organic matter (Thurman, 1985). Their quantitative significance is even less if particulate material is included.

This paper reports first on methodological investigations into the suitability of MUF-compounds for studying enzymatic activity in differing streams and stream habitats, after which some initial results obtained with this technique are presented. Sandy sediments were mainly considered because, on a priori grounds, microbial activity in this habitat was expected to be the more important in the streams under investigation. Applying the fluorometric technique to the determination of extracellular enzymatic activity of sandy sediments is more problematic than with water samples. To overcome this, a more complicated and time-consuming procedure than usual was necessary.

The main substrates used so far were MUF-β-D-glucoside (MUF-β-Glc) and MUF-phosphate (MUF-P). Cellulose is the most common organic compound in the small upland streams investigated here. It is composed of glucose molecules linked by β-D-glycosidic bonds. These are the same bonds as the bonds between 4-methylumbelliferon (MUF) and the glucose molecule of MUF-β-Glc. Among the several enzymes involved in the degradation of cellulose is β-glucosidase (EC 3.2.1.21), which hydrolyses cellobiose as well as MUF-β-Glc.

Soluble reactive phosphorus occurred at relatively low concentrations in the streams investigated, whereas total phosphorus was abundant in the streambed sediments. Thus extracellular phosphatases could be important for the supply of phosphorus to the heterotrophic microorganisms in the sediment, as well as for the autotrophic algae which may develop in large numbers, especially on the surfaces of fine sediments.

In this chapter, the term " extracellular enzymes" refers to those enzymes which function outside the cell, without considering whether they are free or cell bound. In experiments using sediment samples, it is almost impossible to distinguish between "free enzymes" (which have been termed "extracellular enzymes" elsewhere; Chróst, 1990), defined as enzymes dissolved in water or adsorbed to surfaces other than those of their producers, and "ectoenzymes", defined as enzymes being in contact with their producers and with a polymeric substrate located outside the cell (Chróst, 1990; see also Chapter 3). A technique to distinguish these differing enzymes would be much more complicated and time-consuming than is already the case with the current method. There would also be an increased risk of introducing experimental artifacts, negating any possible advantages of the extra information accrued.

17.2 Methods and Experimental Design

Characteristics of the streams investigated The three streams under investigation are situated between 200 and 400 m above sea level, near the town of Schlitz, some 100 km northeast of Frankfurt/Main, Germany. 1) The Breitenbach catchment is mainly forested, but in its middle and lower parts, the stream flows through a predominance of grassland, some 50 m from the forest edge. The upper reach is flanked by trees on at least one side. The stream water has a low alkalinity, with pH values near to neutral. More detailed descriptions can be found in Marxsen (1980a; 1988). The sediments, which originate from red sandstone, are very heterogeneous in size. This heterogeneity may fluctuate considerably with time. Two types of deposits dominate: gravel of approximately 1- to 4-cm diameter, mixed with stones up to about 20 cm or more, occurs where higher flow predominates, and during seasons with higher water levels (winter and spring). Secondly, sand, with a mean grain size of about 0.5 mm or less, is found mainly in slow-flowing areas and during seasons with lower water levels (summer and autumn). 2) The investigated section of the Rohrwiesenbach flows through forest composed mainly of beech (Marxsen, 1980a). This approaches the natural status of headwaters in Central Europe. In contrast to most streams in the Schlitz area, the Rohrwiesenbach is rather calcareous, with pH about 8. The sediments are relatively fine, and are mostly composed of sand mixed with silt. 3) The third stream under investigation is the spring headwater of the river Jossa. This is another woodland stream, but the catchment forests are almost exclusively coniferous. There are many small tributaries, the discharges of which are highly variable during the year, depending on the rainfall. For most of the year the pH is around 4, although sometimes lower. The sediments are similar to those of the Breitenbach.

Sampling Sediment cores were taken with a plastic cylinder pressed into the sandy deposits. The core was removed after supporting the lower end of the cylinder with a thin piece of plastic or aluminum. If different layers were to be investigated, the core was divided into the appropriate layers immediately after its removal. Sandy sediment samples were transported to the laboratory in plastic bags, and experimentation began no later than 2 hours after sampling.

Enzyme assays The substrates: 4-methylumbelliferyl-β-D-glucopyranoside (MUF-β-Glc), 4-methylumbelliferyl phosphate (MUF-P), L-leucine-4-methylcoumarinyl-7-amide (Leu-MCA), and 4-methylumbelliferone (MUF) for calibration, were obtained from Serva Feinbiochemica, Heidelberg (Germany). They were dissolved in 2 ml of methylcellosolve (Hoppe, 1983) before stock solutions (100 ml) with concentrations of 2 mM were prepared by dilution with ultrapure water (Seralpur PRO 90C). The stock solutions were frozen at −20°C. After thawing, they were diluted to the appropriate concentrations.

Experiments to determine the extracellular enzymatic activity associated with sandy sediments used naturally occurring deposits. One ml of sediment was diluted with 8 ml of stream water that had previously been filtered (Whatman GF/C) and boiled for 30 minutes. Samples were incubated at the ambient stream temperature in a shaking water bath. At first 3, and later 4, replicates were used to measure the initial fluorescence at between 10 and 20 min after substrate addition. Another 3 or 4 replicates were used to measure the final fluorescence at the end of the experiments. After fixation by boiling for 3 min, and subsequent cooling to room temperature, 1 ml of alkaline glycine-ammonium hydroxide quench solution (0.2 M ammonium hydroxide + 0.05 M glycine, adjusted to pH 10.5 with 5 M NaOH) was added to convert MUF to its more fluorescent anion form (Daniels and Glew, 1984). Following centrifugation (15 minutes at 10,000 rpm) the fluorescence of the supernatant was determined. Deviations from this procedure while developing the method are mentioned, as appropriate.

Artificial substrata were used for investigating activity on coarser streambed sediments. These were 1.5-mm-diameter glass beads strung on nylon monofilament line supported on a plexiglass plate. They were incubated in 20-cm-diameter opaque tubes anchored to the streambed (Lock and Ford, 1985; Ford and Lock, 1987; Freeman et al., 1990). In this way, a flow of water over the beads was maintained for 8 weeks prior to the experiments. During this time, a heterotrophic biofilm similar to the natural biofilms occurring on streambed gravel and stones developed.

Extracellular enzymatic activities of the slime coated beads were determined by adding 40 beads to 6 tubes containing 8 ml of boiled stream water. After transferring them to a shaking water bath at ambient stream temperature, 0.4 ml of fluorogenic substrate (MUF-β-Glc and MUF-P) was added to a final concentration of 48 $\mu mol\ l^{-1}$. After 15 min, 3 of the tubes were removed and fixed by boiling. The remaining 3 were incubated for a further 4 hours and then fixed as described. Following cooling, 1 ml of the quench solution was added and fluorescence was measured without prior centrifugation.

MUF fluorescence measurements Fluorescence was measured using a Kontron SFM 25 spectrofluorometer with a 10 nm slit. The initial investigations were done at an excitation wavelength setting of 330 nm and an emission setting of 450 nm. However, detailed investigations on the fluorescence of free MUF and different MUF-substrates under the experimental conditions used have shown that the optimum excitation wavelength setting with the addition of alkaline glycine-ammonium hydroxide solution was 365 nm. The maximum fluorescence of free methylumbelliferone was found at this setting, whereas only a weak fluorescence of the MUF-compounds studied was measured (Figure 17.1). However, there were marked differences among the background fluorescence patterns of the non-hydrolyzed MUF-compounds. While only weak fluorescence was observed for MUF-β-Glc, much higher values were found for the intact MUF-P and Leu-MCA molecules. For the latter two compounds, the excitation spectra indicated that the excitation minimum had not yet been reached

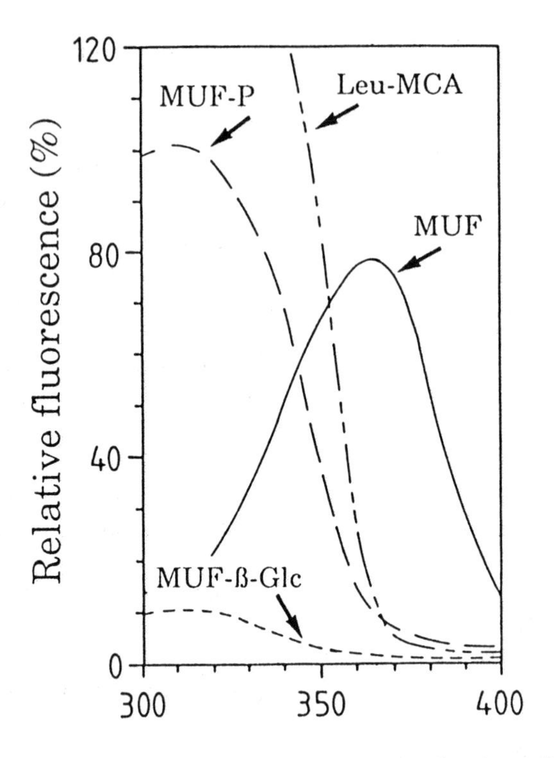

Figure 17.1 Fluorescence of MUF, MUF-β-Glc, MUF-P, and Leu-MCA at different excitation wavelengths (300–400 nm) and constant emission wavelength (450 nm). All compounds were dissolved in ultrapure water, to which 10% alkaline glycine-ammonium hydroxide solution was added; concentrations were 1 μM for MUF and 50 μM for the MUF-substrates.

at 365 nm. This would have been reached at around 380–390 nm, which would be far from the maximum fluorescence of free MUF.

The variability in background fluorescence between different MUF-compounds, which in some cases is rather high, is an important reason why extracellular enzymatic activity was measured by determining the increase in fluorescence between two points in time. If the fluorescence of added MUF-compounds is assumed to be zero and the fluorescence of the liberated MUF is only measured at the end of the experiment, then the enzymatic activity would probably be overestimated.

Another problem arising was interference from the innately high fluorescence of dissolved organic matter in the studied samples. This varied considerably between different sediments. Similar observations have been reported by Münster et al. (1989) for polyhumic lake water. It was therefore always necessary to perform special calibrations for each sample, including blanks and standards.

Determination of particle surface area The true surface area of sandy sediments was calculated using the formula of Fenchel (1969), which was modified from Prenant (1960). Briefly, the procedure was as follows. The number of quartz grains in 1 g of sediment is calculated from the relationship:

$$\log n = -0.14225 - (2.5 \times \log d_1) - (0.5 \times \log d_2)$$

where n is the number of grains retained by a sieve, and d_1 and d_2 are the mesh sizes in cm of 2 different sieves where $d_1 > d_2$.

The average radius of grains retained is assumed to be:

$$r = \sqrt{\frac{d_1 d_2}{4}}$$

Using the formula $4 \times \pi \times r^2 \times n$, the surface area of each grain size fraction can be calculated. By considering the relative proportions of the various fractions, the total surface area of the sample can be derived by addition.

17.3 Results and Discussion

Effects of experimental disruption of sediments Experimental investigations using sediments are inherently at risk from experimental artifacts. Homogenization techniques can destroy natural gradients and the fine structure of microenvironments. They can also disrupt microbial aggregates and the biofilms on sediment particles. Such disturbances could result in an increase in microbial activity. For example, by way of analogy, Meyer-Reil (1986) reported an average 1-order-of-magnitude increase in extracellular enzymatic hydrolytic activity in suspended brackish water sediments, when compared with undisturbed cores investigated using a core injection technique (see also Chapter 5).

Despite the apparent advantage of using core injection techniques, homogenizing the sediments does at least result in all the microorganisms being exposed to the same external substrate concentrations (Craven and Karl, 1984; Karl and Novitsky, 1988). This is important if reliable results for microbial activity measurements, which are concentration dependent, are to be obtained. Karl and Novitsky (1988) concluded from their experiments that their sediment-seawater slurry technique did not bias the measurement of the microbial uptake and assimilation of adenine. They state that these results may be unique for the Kahana Bay (Hawaii) sediments investigated, but probably not for adenine metabolism.

In this study, similar experiments revealed no significant effect on the rates of extracellular enzymatic hydrolytic activity using an experimental treatment with sandy sediments from the upland streams in the Schlitz area. In the first set of experiments, no increase, or only a slight increase (maximum, 1.5-fold) in enzymatic activity was observed when sediment dilution with stream water was increased stepwise from 1:2 (sediment volume:water volume) up to 1:32

(Table 17.1). Thus, it was shown that extensive dilution of the sediments did not appear to stimulate extracellular enzymatic hydrolysis, or, at the most, the effect was only very slight. However, it was not possible to assess any effects of the initial disruption on the basis of these experiments.

An indirect evaluation of any such effects was made with a second set of experiments in which extracellular enzymatic activity was observed for up to 24 hours (Table 17.2, Figure 17.2). If disruption of the sediments had resulted in either a more or a less favorable environment, then a systematic increase or decrease in the activity of the microbial community would have been observed with time (Karl and Novitsky, 1988). In our experiments, the enzymatic activity rates of the sediment microbial communities did not differ significantly for up to 6 h (Figure 17.2) or even more (Table 17.2). Thus, we concluded that the incubation technique used in this study had little effect on the measured activity parameter. Nevertheless, we are currently attempting to improve the method by using relatively undisturbed sediments in through-flow cores similar to those of Wallis (1981) and Fiebig (1988). If this work does suggest that

Table 17.1 Effects of diluting Breitenbach surface sediments with stream water on MUF-β-Glc hydrolytic rates[1]

	Rate of MUG-β-Glc hydrolysis ($nmol\ ml^{-1}\ h^{-1}$)	
Sediment volume:water volume	28 October 1987	1 June 1988
1:2	5.9 ±2.7	9.4 ±0.3
1:4	6.4 ±1.4	8.6 ±3.9
1:8	5.9 ±2.8	12.6 ±0.9
1:16	5.6 ±2.0	15.2 ±1.4
1:32	7.1 ±3.0	14.4 ±0.7

[1]Between 2 to 32 ml of filtered and boiled stream water was added to 1 ml sediment. The data given are values of MUF (±95% confidence limits) liberated from MUF-β-Glc. Experiments were conducted at 10.0°C for 2 hours. For routine experiments, 1 ml sediment was diluted with 8 ml of stream water.

Table 17.2 Effects of incubation time on rates of MUF-β-Glc hydrolysis following sediment sample dilution[1]

Time after dilution (h)	Rate of MUF-β-Glc hydrolysis ($nmol\ ml^{-1}$ sediment h^{-1})
2	2.52 ±0.46
6	2.47 ±0.28
12	2.65 ±0.23
18	2.79 ±0.17
24	2.51 ±0.09

[1]Data, ±95% confidence limits. 26 August 1987, 12°C.

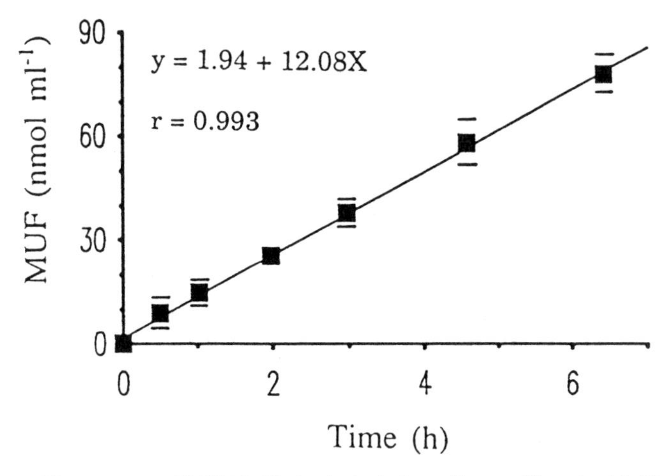

Figure 17.2 Time course of MUF-β-Glc hydrolysis in sediment diluted with filtered and boiled stream water. Sediment:water = 1:20 vol/vol; 5 May 1987, 10.0°C incubation temperature. Data are given in nmol MUF ml^{-1} sediment ±95 % confidence limits.

disrupting the sediments did bias our results, the results will, in any case, remain comparable relative to each other.

Substrate concentration response Sediment samples from small streams are usually heterogeneous. Even after careful mixing, homogenous subsamples are difficult to obtain. This would explain the often rather high confidence limits of the experimental results (see Table 17.1). Despite this, we did manage to obtain good agreement between replicates in most experiments. It was even possible to derive substrate saturation curves (Figure 17.3). The enzymatic activities were usually in good accordance with saturation kinetics when the concentrations were not too high. However, varying degrees of substrate inhibition often began to be observed between 50 and 100 μM (Marxsen and Witzel, 1990). This meant that the original intention of characterizing the maximum enzymatic activity by applying substrate concentrations in excess of saturation, which avoids time-consuming complete kinetic experiments, was not possible. Because of this inhibition effect, there was only a narrow range of concentrations between about 50 and 70 μM in which experiments using only one concentration gave results comparable to the maximum enzyme activity, Vmax. Concentrations within this narrow range were therefore used in routine measurements. Thus the data obtained represent potential hydrolytic rates, from which it is not possible to deduce information on the actual rates of hydrolysis.

Extracellular enzymatic activity in stream water and sediment Not surprisingly, a very low level of hydrolytic activity for MUF-β-Glc and MUF-P was observed in the water column, as compared with the sandy sediments (Figure

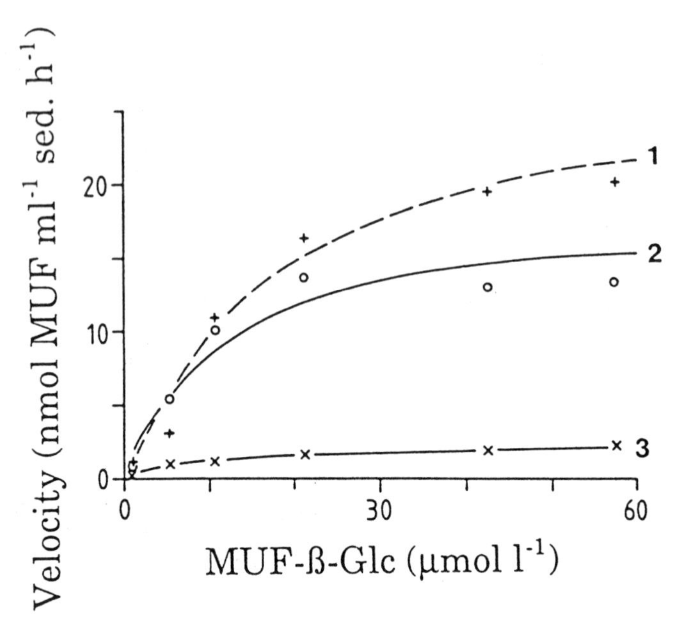

Figure 17.3 Substrate saturation curves for MUF-β-D-glucoside hydrolytic activity in sediment samples from different depths in the middle reach of the Breitenbach. The three lines correspond to depths of: 1, 0–4 cm (V_{max} = 28.4 nmol ml^{-1} h^{-1}, K_m = 19.8 μmol l^{-1}); 2, 4–7.5 cm (V_{max} = 16.0 nmol ml^{-1} h^{-1}, K_m = 7.6 μmol l^{-1}); 3, 7.5–11 cm (V_{max} = 2.5 nmol ml^{-1} h^{-1}, K_m = 10.9 μmol l^{-1}). (19 May 1987; 2 h incubation time; 8.0 °C incubation temperature).

17.4). Nevertheless, as a function of bacterial biomass, the activities were rather high.

Phosphatase activity values in stream water were similar to those in lake water samples from northern Germany (Chróst and Overbeck, 1987; Chróst et al., 1989), Finland (Münster et al., 1989), England (Jones, 1972), and Poland (Halemejko and Chróst, 1984), and were in the lower range of values for Baltic Sea coastal waters (Hoppe, 1983). They were lower than values reported for Australian rivers and billabongs (Boon, 1989), which were determined with a different method (photometrically with *p*-nitrophenol phosphate).

β-glucosidase activity in stream water was on the same order of magnitude as that reported by Chróst (1990), Chróst et al. (1989), and Münster et al. (1989) for lake water, and by Hoppe (1983) for Baltic Sea coastal waters. It was higher than values reported for North Sea coastal waters, but lower than those from the Scheldt Estuary and from eutrophic pond water (Somville, 1984). Activity in the sandy deposits corresponded to that reported by Meyer-Reil (1986) for marine sediments.

Extracellular enzymatic activity in different types of deposits and in streams of different pH β-glucosidase activity on sand was about 10 to 20% of that on coarser grains for the circumneutral Breitenbach and the calcareous Rohr-

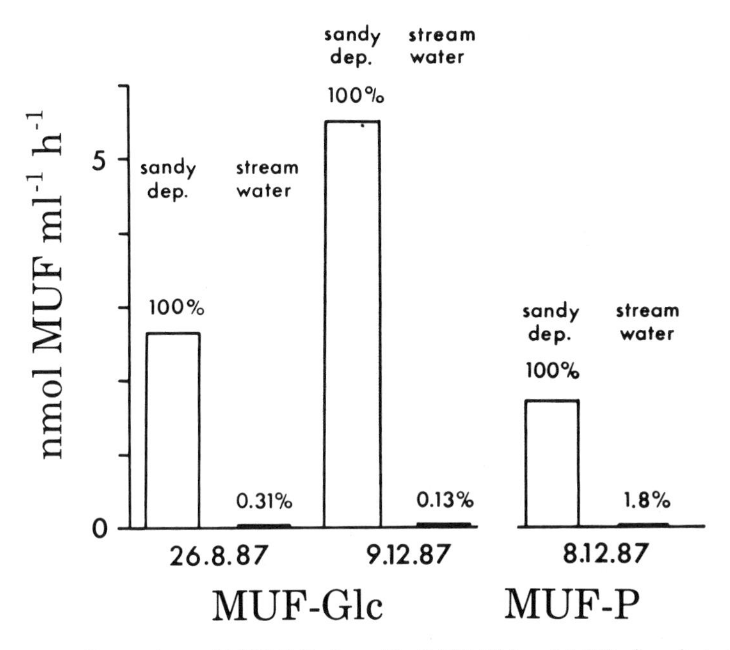

Figure 17.4 Comparison of MUF-β-D-glucoside (MUF-Glc) and MUF-phosphate (MUF-P) hydrolysis in Breitenbach sandy sediments (0–3 cm depth) and stream water. Experiments were conducted at initial saturation concentrations.

wiesenbach, when compared on the basis of grain surface area (Figure 17.5). Similar results were found for phosphatase activity. For the streambed as a whole, this means that by far the greater proportion of extracellular enzymatic activity occurs in the sandy deposits, since their total surface area was much higher than that of the gravel and stones. This is also true for the acidic Jossa headwater, where values per unit grain surface area were similar for the two types of deposits investigated. MUF-β-Glc hydrolyzing activity in the sandy deposits was similar in all streams (Figure 17.5, Table 17.3). This was not surprising for the Breitenbach and the Rohrwiesenbach, since earlier investigations into bacterial biomass and microbial activity (using other techniques) gave similar results for these two streams (Marxsen, 1980b; 1980c; Freeman et al., 1990). The similarly high β-glucosidase activity measured in the Jossa sediment was surprising. Results from previous experiments had shown that glucose uptake in the water column of this stream was only about 10 to 20% of that in the other two streams. The high level of hydrolytic activity is presumed not to be the result of similarly high quantities of enzymes occurring in the sediments. Instead, it is more likely a consequence of the lower pH. The pH of the Jossa headwater is close to the pH optimum of β-glucosidase activity. We found that purified β-glucosidases had only 14% of their hydrolytic activity in natural Breitenbach water, compared with that at pH 5.

β-glucosidase activity on glass beads, the model substrata for coarser sur-

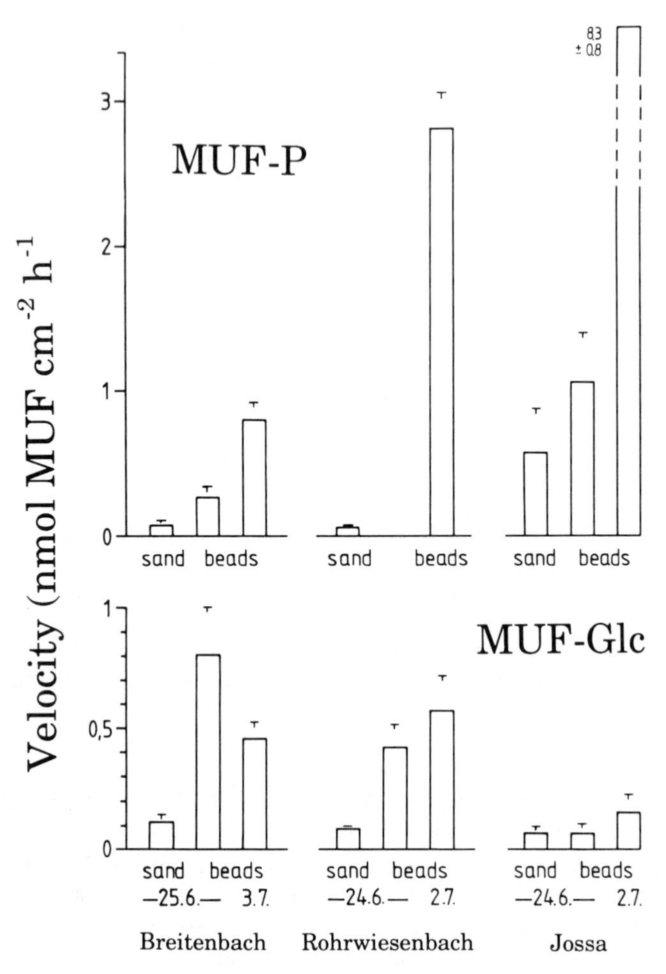

Figure 17.5 MUF-phosphate (MUF-P, upper panel) and MUF-β-D-glucoside (MUF-Glc, lower panel) hydrolysis on the surface of sandy sediments (0–3 cm depth) and glass beads (artificial substratum on which heterotrophic biofilms similar to natural biofilms on coarse sediments developed; data from different days). The values are given with ±95% confidence limits. 1987; 48 μM substrate concentration; 2–6 h incubation time; 12.0°C incubation temperature.

face sediments, was similar in the Breitenbach and the Rohrwiesenbach, but much lower in the acidic Jossa headwater (between 2 and 33% of the values obtained in the other two streams). These results agree well with those from earlier experiments, where a similar trend for glucose uptake in the water column was demonstrated (Table 17.3).

Different trends were seen with phosphatase activity: The highest activity was measured in the acidic Jossa, both in the sandy deposits and on the glass beads. Activity on glass beads from the Rohrwiesenbach lay between that mea-

Table 17.3 Chemical parameters and enzymatic activity in (A) stream waters, and (B) sandy sediments of the three streams[1]

Parameter	Breitenbach	Rohrwiesenbach	Jossa
DOC (mg C l^{-1})	2.1±0.1	13.2±1.1	8.2±0.4
pH	6.7	7.9	4.2
PO_4^{3-} (μg P l^{-1})	36	24	6
V_{max} (nmol l^{-1} h^{-1})	4.1	3.6	0.52
POM (mg ml^{-1} sediment)	4.4±0.4	9.8±1.0	25.7±3.0
P_{tot} (μg P ml^{-1} sediment)	39±4	52±13	20±2
β-Glucosidase activity (nmol ml^{-1} sediment h^{-1})	6.8±1.8	6.1±1.0	5.1±2.0
Phosphatase activity (nmol ml^{-1} sediment h^{-1})	4.8±1.7	4.6±0.7	44±24

[1]DOC, dissolved organic carbon; PO_4^{3-}, soluble reactive phosphorus; V_{max}, maximum glucose uptake velocity, 1985 annual mean; POM, total particulate organic matter; P_{tot}, total phosphorus. Data (and if available ±95% confidence limits) are from June 1987, except for V_{max}.

sured in the other two streams. These results concur with the differences in soluble reactive phosphorus concentrations in the stream waters. This was very low in the Jossa, and highest in the Breitenbach.

Phosphorus and phosphatase activity in sandy sediments of the Breitenbach
High phosphatase activities (between 1 and 7 nmol ml^{-1} sediment h^{-1}) were observed in the surface layers of the sandy Breitenbach deposits. These values are much higher than have been found previously with this method in lake, sea, and stream-water samples (Hoppe, 1983; Siuda and Chróst, 1987; Chróst and Overbeck, 1987; Chróst et al., 1989; Münster et al., 1989). The streambed sediments of the Breitenbach contained high levels of nonreactive phosphorus (Figure 17.6), sometimes more than 100 mg P l^{-1} sediment. The soluble reactive phosphorus concentrations were factors of 10 lower. Even total dissolved phosphorus concentrations in the interstitial waters always remained far below 1 mg P l^{-1}, even under anaerobic conditions (H.H. Schmidt, personal communication). Also, considerably less than 1 mg of reactive phosphorus is experimentally soluble from 1 liter of sediment in (aerobic) stream water.

For three cores taken between May and December 1987 (down to 11 cm below sediment surface), the C:P ratios for total organic carbon and soluble reactive phosphorus were higher than 10,000. However, if total phosphorus is considered, the C:P ratios were between 14 and 62. These data suggest that the pool of particulate phosphorus in the sandy deposits could be a valuable P source for the microorganisms if it is made available to them. The composition of the total phosphorus pool is still unknown. However, the high bacterial biomass and production occurring in the sandy Breitenbach deposits (Marxsen, 1988), and the high level of phosphatase activity suggest that there is a considerable flux of phosphorus in these habitats, perhaps by rapid close-

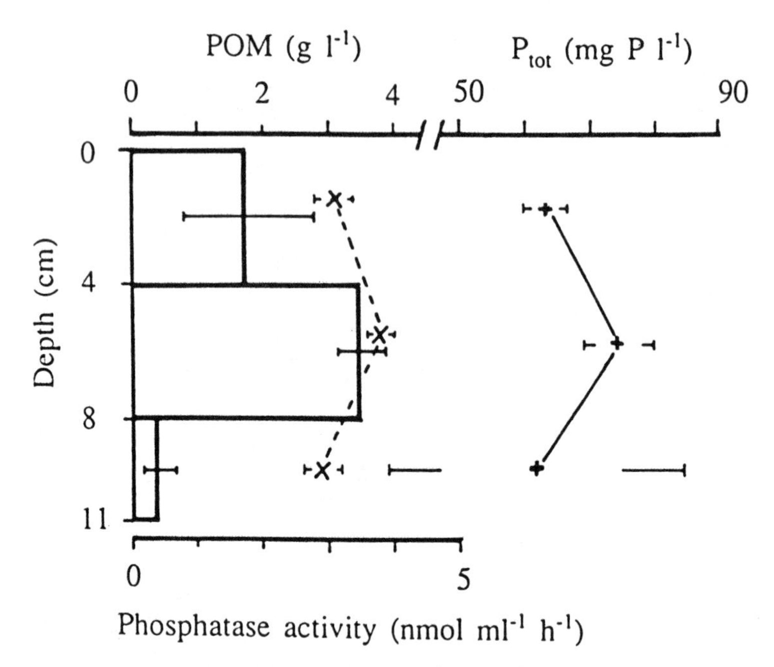

Figure 17.6 Depth profile of: bars, MUF-phosphate hydrolytic activity; +, total phosphorus, P_{tot}; and x, particulate organic matter, POM, from sandy sediment in the middle reach of the Breitenbach. 8 December 1987; 53 μM substrate concentration; 4.5 h incubation time; 8.0°C incubation temperature.

circuit cycling. The phosphorus supplied from the hydrolytic activity of extracellular phosphatases is probably not only important for the heterotrophic microorganisms in the sediment, but also for the autotrophic algae which may develop in particularly high numbers on the surfaces of fine sediments (Cox, 1990).

Conclusions and perspectives Despite some persisting methodological problems, there can be little doubt that extracellular enzymes are important for the flux of organic matter and nutrients in the small streams investigated in this study. However, our research is in the early stages, and we are still some way from a detailed knowledge of the occurrence and biochemical activity of microbial extracellular enzymes in these systems.

Using 4-methylumbelliferyl-compounds, the fluorometric technique can be an important tool for enhancing our knowledge of these important aspects of stream ecology. If the single-substrate approach is used, the method can reveal the maximum hydrolytic activity of free and bound extracellular enzymes. The kinetic approach enables a more detailed characterization of the enzymes, including their regulation. If, in addition, the natural substrate concentrations are measured, then the actual hydrolytic rates in the ecosystem can be determined.

Future work should be directed at determining the natural substrate pools (total organic matter and specific compounds, e.g., different phosphorus compounds, proteins, cellulose), as well as at investigating the utilization of the hydrolysis products (e.g., microbial uptake of glucose and amino acids). It will also be useful to evaluate the regulation of enzymatic activity by substrates and products, and the influence of physical and chemical parameters (e.g., temperature, oxygen, pH).

Acknowledgments The authors are indebted to B. Knöfel for technical assistance and for drawing the figures, and to Dr. D.M. Fiebig for revising the English text.

References

Boon, P.I. 1989. Organic matter degradation and nutrient regeneration in Australian freshwaters: I. Methods for exoenzyme assays in turbid aquatic environments. *Archiv für Hydrobiologie* 115: 339–359.

Chróst, R.J. 1989. Characterization and significance of β-glucosidase activity in lake water. *Limnology and Oceanography* 34: 660–672.

Chróst, R.J. 1990. Microbial ectoenzymes in aquatic environments. pp. 47–78 in Overbeck, J. and Chróst, R.J. (editors), *Aquatic Microbial Ecology: Biochemical and Molecular Approaches*. Springer Verlag, New York.

Chróst, R.J. and H.J. Krambeck. 1986. Fluorescence correction for measurements of enzyme activity in natural waters using methylumbelliferyl-substrates. *Archiv für Hydrobiologie* 106: 79–90.

Chróst, R., Münster, U., Rai, H., Albrecht, D., Witzel, K.P. and J. Overbeck. 1989. Photosynthetic production and exoenzymatic degradation of organic matter in the euphotic zone of a eutrophic lake. *Journal of Plankton Research* 11: 223–242.

Chróst, R.J. and J. Overbeck. 1987. Kinetics of alkaline phosphatase activity and phosphorus availability for phytoplankton and bacterioplankton in Lake Plußsee (north German eutrophic lake). *Microbial Ecology* 13: 229–248.

Cox, E. 1990. Studies on the algae of a small softwater stream. II. Algal standing crop (measured by chlorophyll-a) on soft and hard substrata. *Archiv für Hydrobiologie, Supplement* 83: 553–566.

Craven, D.B. and D.M. Karl. 1984. Microbial RNA and DNA synthesis in marine sediments. *Marine Biology* 83: 129–139.

Daniels, L.B. and R.H. Glew. 1984. 2.6 β-D-glucosidases in tissue (glucocerebrosidase, β-glucosidase). pp. 217–226 in Bergmeyer, H.U., Bergmeyer, J. and Graßl, M. (editors), *Methods of Enzymatic Analysis*, vol. 4, 3rd edition. Verlag Chemie, Weinheim.

Duddridge, J.E. and M. Wainwright. 1982. Enzyme activity and kinetics in substrate-amended river sediments. *Water Research* 16: 329–334.

Fenchel, T. 1969. The ecology of marine microbenthos. IV. Structure and function of the benthic ecosystem, its chemical and physical factors and the microfauna communities with special reference to the ciliated Protozoa. *Ophelia* 6: 1–182.

Fiebig, D.M. 1986. A study of riparian zone and stream water chemistries, and organic matter immobilization at the stream-bed interface. Ph.D. thesis, University of Wales, Bangor.

Ford, T.E. and M.A. Lock. 1987. Epilithic metabolism of dissolved organic carbon in boreal forest rivers. *Federation of European Microbiological Societies, Microbiology Ecology* 45: 89–97.

Freeman, C., Lock, M.A., Marxsen J. and S. Jones. 1990. Inhibitory effects of high molecular weight dissolved organic matter upon metabolic processes in biofilms from contrasted rivers and streams. *Freshwater Biology* 24: 159–166.
Hałemejko, G.Z. and R.J. Chróst. 1984. The role of phosphatases in phosphorus mineralization during decomposition of lake phytoplankton blooms. *Archiv für Hydrobiologie* 101: 489–502.
Holzapfel-Pschorn, A., Obst, U. and K. Haberer. 1987. Sensitive methods for the determination of microbial activities in water samples using fluorogenic substrates. *Fresenius Zeitschrift für Analytische Chemie* 327: 521–523.
Hoppe, H.G. 1983. Significance of exoenzymatic activities in the ecology of brackish water: measurements by means of methylumbelliferyl-substrates. *Marine Ecology Progress Series* 11: 299–308.
Hoppe, H.G., Gocke, K., Zamorano, D. and R. Zimmermann. 1983. Degradation of macromolecular organic compounds in a tropical lagoon (Cienaga Grande, Colombia) and its ecological significance. *Internationale Revue der gesamten Hydrobiologie* 68: 811–824.
Hoppe, H.G., Kim, S.J. and K. Gocke. 1988. Microbial decomposition in aquatic environments: Combined process of extracellular enzyme activity and substrate uptake. *Applied and Environmental Microbiology* 54: 784–790.
Jacobsen, T.R. and H. Rai. 1988. Determination of aminopeptidase activity in lake-water by a short term kinetic assay and its application in two lakes of different eutrophication. *Archiv für Hydrobiologie* 113: 359–370.
Jones, J.G. 1972. Studies on freshwater microorganisms: Phosphatase activity in lakes of different degrees of eutrophication. *Journal of Ecology* 60: 777–791.
Karl, D.M. and J.A. Novitsky. 1988. Dynamics of microbial growth in surface layers of a coastal marine sediment ecosystem. *Marine Ecology Progress Series* 50: 169–176.
King, G.M. 1986. Characterization of β-glucosidase activity in intertidal marine sediments. *Applied and Environmental Microbiology* 51: 373–380.
Lock, M.A. and T.E. Ford. 1985. Microcalorimetric approach to determine relationships between energy supply and metabolism in river epilithon. *Applied and Environmental Microbiology* 49: 408–412.
Marxsen, J. 1980a. Untersuchungen zur Ökologie der Bakterien in der fließenden Welle von Bächen. I. Chemismus, Primärproduktion, CO_2-Dunkelfixierung und Eintrag von partikulärem organischen Material. *Archiv für Hydrobiologie, Supplement* 57: 461–533.
Marxsen, J. 1980b. Untersuchungen zur Ökologie der Bakterien in der fließenden Welle von Bächen. II. Die Zahl der Bakterien im Jahreslauf. *Archiv für Hydrobiologie, Supplement* 58: 26–55.
Marxsen, J. 1980c. Untersuchungen zur Ökologie der Bakterien in der fließenden Welle von Bächen. III. Aufnahme gelöster organischer Substanzen. *Archiv für Hydrobiologie, Supplement* 58: 207–272.
Marxsen, J. 1988. Evaluation of the importance of bacteria in the carbon flow of a small open grassland stream, the Breitenbach. *Archiv für Hydrobiologie* 111: 339–350.
Marxsen, J. and K.-P. Witzel. 1990. Measurement of exoenzymatic activity in streambed sediments using methylumbelliferyl-substrates. *Archiv für Hydrobiologie Beihefte Ergebnisse Limnologie* 34: 21–28.
Meyer-Reil, L.-A. 1981. Enzymatic decomposition of protein and carbohydrates in sediments: Methodology and field observations during spring. *Kieler Meeresforschungen Sonderheft* 5: 311–317.
Meyer-Reil, L.-A. 1983. Benthic response to sedimentation events during autumn to spring at a shallow water station in the western Kiel Bight. II. Analysis of benthic bacterial populations. *Marine Biology* 77: 247–256.
Meyer-Reil, L.-A. 1984. Seasonal variations in bacterial biomass and decomposition of particulate organic material in marine sediments. *Archiv für Hydrobiologie Beihefte Ergebnisse Limnologie* 19: 201–206.
Meyer-Reil, L.-A. 1986. Measurement of hydrolytic activity and incorporation of dis-

solved organic substrates by microorganisms in marine sediments. *Marine Ecology Progress Series* 31: 143–149.

Meyer-Reil, L.-A. 1987. Seasonal and spatial distribution of extracellular enzymatic activities and microbial incorporation of dissolved organic substrates in marine sediments. *Applied and Environmental Microbiology* 53: 1748–1755.

Münster, U., Einiö, P. and J. Nurminen. 1989. Evaluation of the measurements of extracellular enzyme activities in a polyhumic lake by means of studies with 4-methylumbelliferyl-substrates. *Archiv für Hydrobiologie* 115: 321–337.

Niesslbeck, P., Voigt, M., Kim, S.J., Bolms, G. and H.G. Hoppe. 1985. Auswirkungen von Salzgehalts- und Temperaturänderungen auf die extrazelluläre Enzymaktivität marin-pelagischer Mikroorganismen. *Berichte des Instituts für Meereskunde, Christian-Albrechts-Universität Kiel 145, Kiel.* 55 pp.

Overbeck, J. 1961. Die Phosphatasen von *Scenedesmus quadricauda* und ihre ökologische Bedeutung. *Verhandlungen der Internationalen Vereinigung für Theoretische und Angewandte Limnologie* 14: 226–231.

Pettersson, K. and M. Jansson. 1978. Determination of phosphatase activity in lake water—a study of methods. *Verhandlungen der Internationalen Vereinigung für Theoretische und Angewandte Limnologie* 20: 1226–1230.

Prenant, A. 1960. Études écologiques sur les sables intercotidaux. I. Questions de méthode granulométrique. Application à trois anses de la Baie de Quiberon. *Cahiers de Biologie Marine* 1: 295–340.

Rego, J.V., Billen, G., Fontigny, A. and M. Somville. 1985. Free and attached proteolytic activity in water environments. *Marine Ecology Progress Series* 21: 245–249.

Reichardt, W., Overbeck, J. and L. Steubing. 1967. Free dissolved enzymes in lake waters. *Nature* 216: 1345–1347.

Siuda, W. and R.J. Chróst. 1987. Alkaline phosphatase activity (APA) and phosphorus availability for microplankton in lakes. *Acta Microbiologica Polonica* 36: 207–233.

Somville, M. 1984. Measurement and study of substrate specificity of exoglucosidase activity in eutrophic water. *Applied and Environmental Microbiology* 48: 1181–1185.

Somville, M. and G. Billen. 1983. A method for determining exoproteolytic activity in natural waters. *Limnology and Oceanography* 28: 190–193.

Thurman, E.M. 1985. *Organic Geochemistry of Natural Waters.* Martinus Nijhoff/Dr W. Junk Publishers, Dordrecht. 497 pp.

Wallis, P.M. 1981. The uptake of dissolved organic matter in groundwater by stream sediments—a case study. pp. 97–111 in Lock, M.A. and Williams, D.D. (editors), *Perspectives in Running Water Ecology.* Plenum Press, New York.

18

Enzyme Activities in Billabongs of Southeastern Australia

Paul I. Boon

18.1 Introduction

Williams (1988) in a recent review noted that most limnologists live and work in the Northern-Hemisphere. The inland waters of Australia are markedly different from those of Europe or North America (Lake et al., 1985; Williams, 1988), so it is not clear whether concepts developed with Northern-Hemisphere freshwaters are applicable to the freshwaters of Australia. For instance, the rivers of the Australian mainland are often turbid and saline (Hart and McKelvie, 1986; Mackay et al., 1988), and because of the flat terrain and widely variable rainfall, their discharge is generally low, but occasionally is high enough to cause widespread flooding (Walker, 1979). The flat terrain also results in the formation of extensive floodplains on which billabongs are a common and characteristic feature. Billabong is the aboriginal name given to a river branch that forms a backwater or stagnant pool fed mainly during floods, or to an ox-bow lake cutoff from the main flow of the river.

Billabongs have been the topic of very little research in the past, although it is evident that they differ greatly from their parent river in terms of flow regime, nutrient concentrations, inputs of autochthonous and allochthonous organic matter, and populations of bacteria, phytoplankton, and invertebrates (Walker and Hillman, 1977; Hillman, 1986; Boon, 1990). Just as Australian rivers are dissimilar in many ways to their counterparts in the Northern Hemisphere, billabongs are likely to be quite different aquatic environments from rivers.

In this chapter, I summarize some recent findings on the breakdown of organic matter and the regeneration of nutrients in the water column of various rivers and billabongs in the Murray-Darling Basin of southeastern Australia. The major aims of the research were to determine the range of enzymes active in billabongs, to determine whether there were consistent differences in enzyme activity in billabongs, rivers, and lakes, and to examine the environmental factors that most strongly influenced enzyme activity. It also proved possible to compare rates of enzyme activity in Australian freshwaters with those in the more commonly studied waters of the Northern Hemisphere.

18.2 Methods and Experimental Design

Field sites A total of 18 field sites was used to examine differences in enzyme activity in different freshwater systems (Figure 18.1). These included six billabongs and six river sites spread along the length of the River Murray, and sites on four other rivers, and at two lakes. The river distance between the least proximate sites (Site 1, on the Darling River, and Site 18, on the Mitta Mitta River) was greater than 1,800 km. Features of some sites are summarized

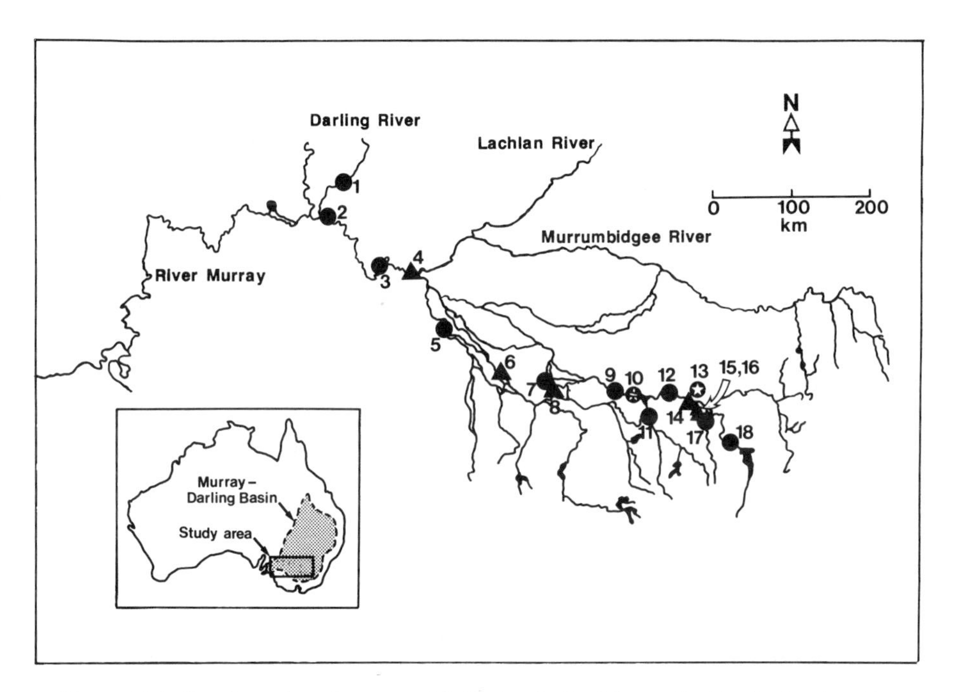

Figure 18.1 Location of study area and field sites in the Murray-Darling Basin of southeastern Australia. Billabongs are indicated by filled triangles; river sites by filled circles; lakes by stars. 1, Darling River at Burtundy; 2, River Murray at Wentworth; 3, River Murray at Robinvale; 4, Boundary Bend Billabong; 5, River Murray at Swan Hill; 6, Kate Malone Billabong; 7, River Murray at Stewart's Kitchen (Barmah Forest); 8, Red Tank Billabong (Barmah Forest); 9, River Murray at Chinaman's Bend; 10, Lake Mulwala; 11, Ovens River; 12, River Murray at Corowa; 13, Golfcourse Lake; 14, Horseshoe Lagoon; 15, Ryans 1 Billabong; 16, Ryans 2 Billabong; 17, Kiewa River at Killara; 18, Mitta Mitta River.

in Walker and Hillman (1977), Hillman (1986), Mackay et al. (1988), and Boon (1989), but a number remain unstudied.

Rates of alkaline phosphatase and aminopeptidase activity were determined from May 1987 to September 1988, although not all sites were visited on each occasion. The activity of other enzymes was determined for Ryans 2 Billabong, also. To determine the environmental conditions that most strongly influenced enzyme activity, the Kiewa River and Ryans 2 Billabong were sampled weekly or fortnightly over the period November 1987 to October 1988. Four enzymes (alkaline phosphatase, aminopeptidase, lipase, and α-D-glucosidase) were routinely examined, as well as the environmental parameters outlined below. Details of the sampling procedures are given in Boon (1990).

Enzyme assays Enzyme activity was determined with derivatives of *p*-nitrophenol and *p*-nitroaniline, according to Boon (1989; 1990). Substrate concentration was (with the exception of studies on kinetics) 200 μM, incubation temperature 20°C, and pH either 7.5 or 9.0, according to site and substrate. In the studies on kinetics, substrate concentrations from 10 to 2,000 μM were used (Boon, 1989). Unfiltered surface waters were used for all assays, except when samples were size-fractionated. For these studies, samples were gently filtered through Whatman GF/C filters then through Nuclepore polycarbonate membrane filters (0.2-μm pore size) prior to the assay. More finely resolved fractionation was obtained by using continuous centrifugation and tangential flow ultrafiltration with Millipore filters.

Environmental conditions Water samples from the Kiewa River and Ryans 2 Billabong were analyzed for nutrient concentrations ($NH4^+$; NO_2^- + NO_3^-; total dissolved primary amines, TDPA; Kjeldahl nitrogen; soluble reactive phosphorus, SRP; and total phosphorus) and other key environmental parameters (turbidity, suspended solids, temperature, etc). Analyses generally followed APHA (1976), with the exceptions noted in Boon (1990). Microbial biomass indicators (chlorophyll$_a$ and bacterial numbers) were quantified as outlined in Boon (1990).

Statistical analysis Between four and six replicates were used for enzyme assays, and three replicates for other water analyses. Experimental design was such that Analysis of Variance (ANOVA) could be used for data interpretation, but in most cases ANOVA was unnecessary or impossible (Boon, 1990). Two types of cluster analysis (*K*-means, and Czekanowski's Index with similarity values hierarchically linked with average-linkage clustering) were used to examine patterns of enzyme activity in the different sites (Hellawell, 1978; Wilkinson, 1988). Relationships between enzyme activity and environmental conditions were examined with linear regressions and product-moment correlation coefficients (Wilkinson, 1988) after the data were plotted.

18.3 Results

Range of enzyme activity The activity of eight enzymes in Ryans 2 Billabong was measured on five occasions throughout 1987–1988 (Table 18.1). Aminopeptidase was consistently the most active enzyme (93–1,015 μmol l^{-1} d^{-1}), followed by alkaline phosphatase (18–125 μmol l^{-1} d^{-1}). Gluco- and galactosidases were much less active, with rates always <3.5 μmol l^{-1} d^{-1}. There was detectable activity of N-acetyl-β-D-glucosaminidase, with rates varying from 2.4–5.8 μmol l^{-1} d^{-1}. Lipases were also detected, but activities were extremely variable.

Differences in enzyme activity among billabongs, rivers, and lakes Monthly variations in aminopeptidase activity in Ryans 2 Billabong, the Kiewa River, and Lake Mulwala are shown in Figure 18.2. Activity (means ± standard errors) in the billabong, river, and lake ranged from 24 (±1) to 1,134 (±17), 5 (±0.2) to 27 (±0.5), and 7 (±0.2) to 14 (±0.2) μmol l^{-1} d^{-1}, respectively. Hence, aminopeptidase activity was consistently greater in the billabong than the river or lake, and much more variable in time. This pattern was maintained when other sites in the Murray-Darling Basin were examined for alkaline phosphatase and aminopeptidase activity (Table 18.2). Aminopeptidase activity in the various rivers ranged from 1.8–32.1 μmol l^{-1} d^{-1}, and in lakes from 2.9–45 μmol l^{-1} d^{-1}. Activity was generally greater in Ryans 2 Billabong (57–896 μmol l^{-1} d^{-1}) than in other billabongs (9–353 μmol l^{-1} d^{-1}), so values for this site are given separately. Alkaline phosphatase activity was also greater in billabongs than in rivers or lakes. It should be noted that rates shown for the same sites in Tables 18.1 and 18.2 and Figure 18.2 are different because sampling was not always concurrent.

Results of the cluster analysis of data on spatial variations in enzyme activity are shown in Figure 18.3. Analysis by *K*-means and Czekanowski's Index gave similar conclusions (Boon, 1990), so only results of the former analysis are shown. Five billabongs were amenable to *K*-means analysis (data from Red Tank Billabong were unsuitable, because the site was sometimes flooded). On

Table 18.1 Activity of eight enzymes in the water column of Ryans 2 Billabong

Enzyme	Range of activity (μmol l^{-1} d^{-1})[1]
Alkaline phosphatase	18–125
Aminopeptidase	93–1,015
Endopeptidase	0.8–3.1
Lipase	1.0–12
α-D-Glucosidase	0.1–3.5
β-D-Glucosidase	0.1–3.4
β-D-Galactosidase	0.2–2.9
N-Acetyl-β-D-glucosaminidase	2.4–5.8

[1]Values shown are the range of rates, measured five times during 1987–1988.

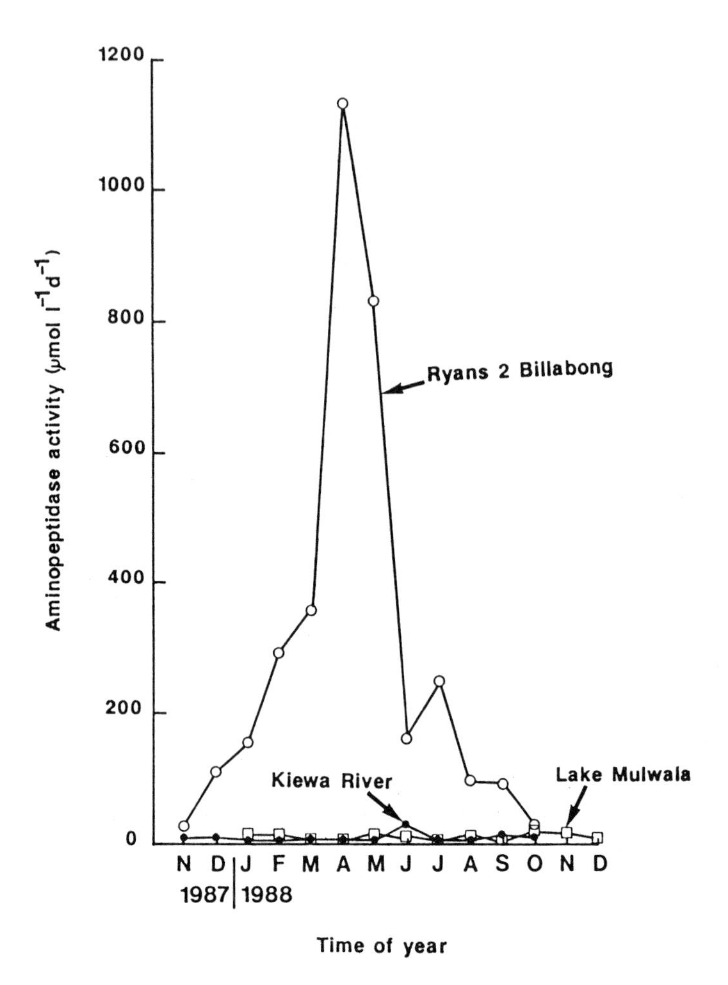

Figure 18.2 Aminopeptidase activity in the water column of the Kiewa River, Lake Mulwala, and Ryans 2 Billabong. Means (n = 4 or 6) are shown. Standard errors are smaller than the symbols used.

the basis of aminopeptidase activity, the billabongs separated completely from the river sites, which formed a single cluster. In terms of alkaline phosphatase activity, the billabongs again separated cleanly, except Ryans 1 Billabong, which grouped closely with the Darling River. All other rivers formed a single cluster. Golfcourse Lake was idiosyncratic, probably because it received treated waste from a paper mill and some storm-water runoff.

Enzyme activity in billabongs and rivers was also compared by examining the heterogeneity of kinetic parameters. The effect of increasing substrate concentrations on alkaline phosphatase, aminopeptidase, lipase, and α-D-glucosidase activity was determined for the Kiewa River, Ryans 2 Billabong, Horseshoe Lagoon, and the River Murray at Chinaman's Bend. Because of problems

Table 18.2 Activity of alkaline phosphatase and aminopeptidase in the water column of various billabongs, rivers, and lakes of southeastern Australia

Site	Range of activity (μmol l^{-1} d^{-1})[1]	
	Alkaline phosphatase	Aminopeptidase
Billabongs:		
Ryans 2 Billabong	27–222	57–896
Other billabongs	5.8–58	9–353
Rivers:		
River Murray	0.8–4.1	3.7–17.7
Darling River	4.1–8.2	8.5–13.4
Other rivers	1.3–7.3	1.8–32.1
Lakes:		
Lake Mulwala	0.9–4.0	6.4–14
Golfcourse Lake	5.1–41	2.9–45

[1]Values shown are the range of rates, measured nine times during 1988.

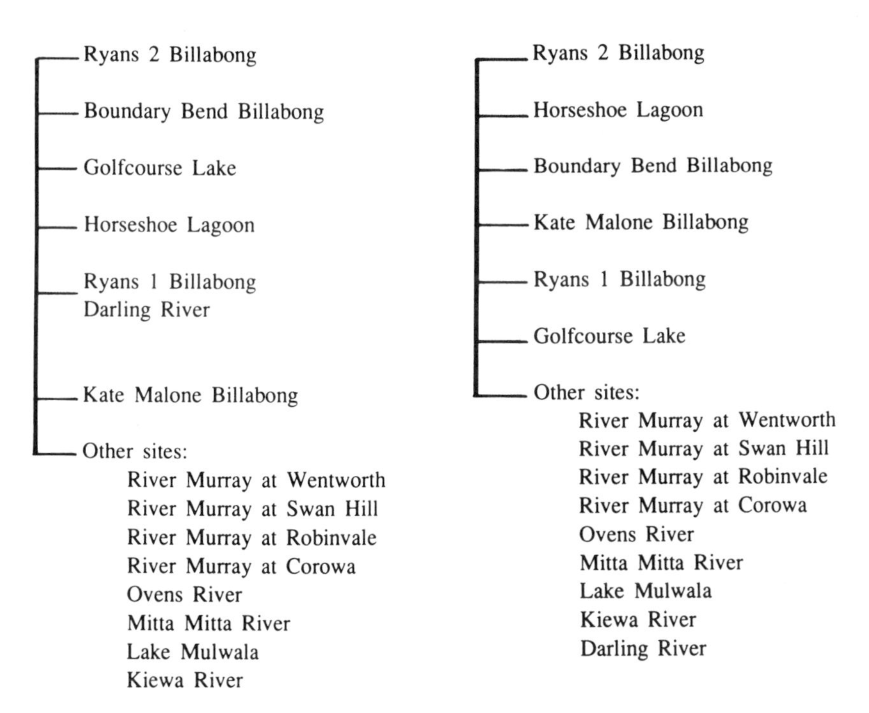

Figure 18.3 Cluster analysis (*K*-means) of alkaline phosphatase and aminopeptidase activity in the water column of various billabongs, rivers and lakes in the Murray-Darling Basin.

with the solubility of some substrates and very low α-D-glucosidase activity in the river sites, not all the 16 possible site-enzyme combinations were amenable to analysis. Of the systems that could be analyzed, six gave linear Lineweaver-Burk plots, but in the majority of cases the plots showed a marked tendency to curve towards the origin at low values of S^{-1} (Boon, 1989). This is indicative of extreme heterogeneity of enzyme activity. Of the various enzymes, aminopeptidase kinetics conformed most often to the Michaelis-Menten model. Of the various sites, conformity was most common with Horseshoe Lagoon. V_{max} and K_m ranged from 14–81 μmol l^{-1} d^{-1} and 62–83 μM, respectively.

Relationship with environmental conditions The magnitude of microbial populations and the load of suspended solids were the two factors most frequently and most highly correlated with enzyme activity. Aminopeptidase activity in Ryans 2 Billabong was strongly correlated with $chlorophyll_a$ contents ($r = 0.81$, $n = 33$) and bacterial numbers ($r = 0.84$, $n = 33$). Figure 18.4 shows the close relationship existing among phytoplankton and bacterial biomass indicators, aminopeptidase activity, and the regeneration of amino acids and ammonium in the water column of this billabong. Phytoplankton populations peaked in March-April, and were followed quickly by increases in bacterial numbers. Subsequently, a short-lived peak in aminopeptidase activity was followed by the release of amino acids (measured as total dissolved primary amines), and later by ammonium. The pattern of α-D-glucosidase activity closely followed that of aminopeptidase, but with a broader peak. However, there was no clear relationship between lipase activity and algal or bacterial populations, nor with any other parameter excepting suspended solids. Alkaline phosphatase activity in the billabong was also significantly correlated with phytoplankton and bacterial numbers ($r = 0.53$ and 0.51, respectively, $n = 33$), but correlations were weaker than those found for aminopeptidase, and there were no clear patterns evident when the rate of alkaline phosphatase activity was plotted against either microbial parameter. Alkaline phosphatase activity was not significantly correlated with concentrations of soluble reactive phosphorus or total phosphorus.

The activity of all four enzymes routinely examined in Ryans 2 Billabong was highly correlated with the load of suspended solids ($r = 0.56$ for alkaline phosphatase, $n = 30$; $r = 0.86$ for aminopeptidase, $n = 30$; $r = 0.39$ for lipase, $n = 18$; $r = 0.93$ for α-D-glucosidase, $n = 13$). Similarly, for the Kiewa River there was a high correlation between enzyme activity and suspended solids ($r = 0.85$ for alkaline phosphatase, $n = 31$; $r = 0.89$ for aminopeptidase, $n = 31$). In both sites, generally, there was little enzyme activity in samples filtered through 0.2-μm-pore-size filters prior to the assay (Table 18.3). Continuous centrifugation and tangential flow ultrafiltration was then used to elaborate upon this finding (Figure 18.5). Most activity in water samples from the River Murray at Corowa, the only site examined in this way, was associated with particles of sizes ranging from 0.2 to 1.0 μm and from 1 to 25 μm. There was very little activity in the <100,000 dalton fractions.

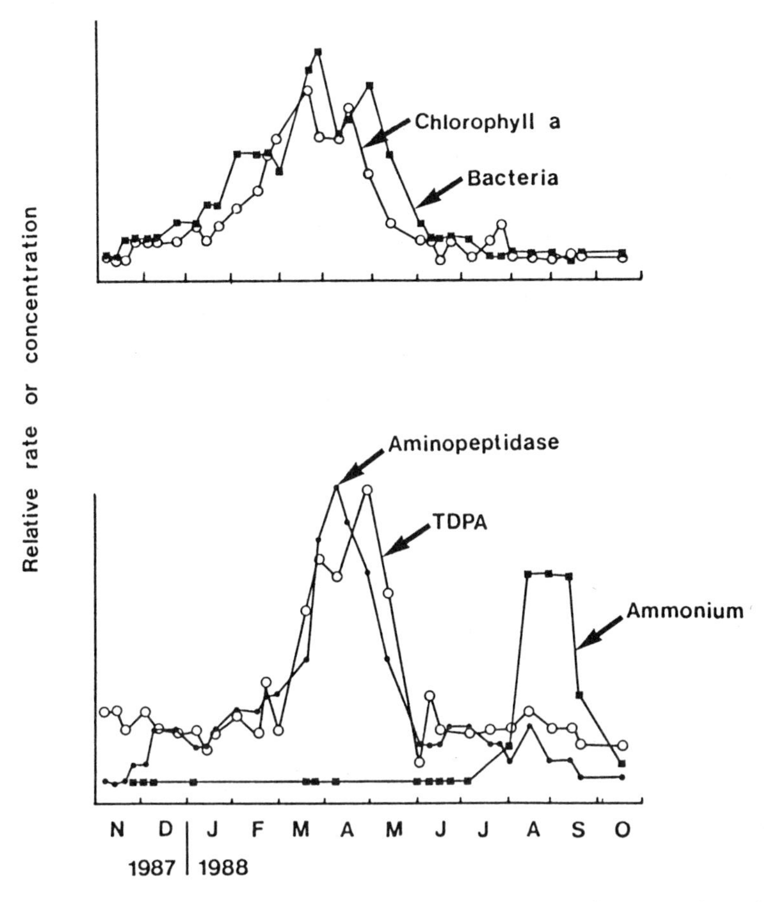

Figure 18.4 Relationship among (top) phytoplankton chlorophyll$_a$, bacterial numbers, (bottom) aminopeptidase activity, and concentrations of total dissolved primary amines (TDPA) and ammonium in the water column of Ryans 2 Billabong. Means (n = 3 to 6) are shown. Standard errors are smaller than the symbols used. Reproduced from Boon (1990) with permission from E. Schweizerbart'sche Verlagsbuchhandlung.

18.4 Discussion

Alkaline phosphatase and aminopeptidase were studied intensively because they were the most active enzymes in the billabong (Table 18.1) and were known to be important in the regeneration of phosphorus and nitrogen (Billen, 1984; Chróst et al., 1984; 1986; Degobbis et al., 1986; Boon, 1990). The data obtained in this study permit comparisons of activity not only between billabongs and rivers in Australia, but also between activity in Australian and Northern-Hemisphere freshwaters. Studies on alkaline phosphatase and aminopeptidase in Northern-Hemisphere freshwaters have shown activities to range from <0.1–

Table 18.3 Enzyme activity in filtered (0.2-μm-pore-size Nuclepore filters) and unfiltered water samples from the Kiewa River and Ryans 2 Billabong

Enzyme	Activity (μmol l^{-1} d^{-1})[1]	
	Filtered	Unfiltered
Kiewa River:		
Alkaline phosphatase	0.9 ±0.2	2.1 ±0.1
Aminopeptidase	2.6 ±0.2	3.4 ±0.1
Lipase	<0.1	0.7 ±0.1
α-D-Glucosidase	0.1 ±0.03	0.3 ±0.03
Ryans 2 Billabong:		
Alkaline phosphatase	1.4 ±0.2	42.0 ±0.5
Aminopeptidase	3.5 ±0.1	76.7 ±0.4
Lipase	0.8 ±0.1	3.9 ±0.03
α-D-Glucosidase	0.3 ±0.1	0.3 ±0.1

[1]Means ± standard errors are shown, n=3.

10 and 1–11 μmol l^{-1} d^{-1}, respectively (Chróst et al., 1984; 1986; Chróst and Overbeck, 1987; Jones, 1972). In Australian rivers, the comparable values are 1–8 and 2–32 μmol l^{-1} d^{-1} (Table 18.2). Hence, rates of enzyme activity in Australian rivers do not seem to be greatly different from those in Europe, except that the very low rates of alkaline phosphatase activity sometimes detected in the Northern-Hemisphere are not found in the antipodes. This similarity is surprising, given the many differences in water characteristics between Australian and European rivers. For instance, it might be expected that, because of the demonstrated relationships between suspended loads and enzyme activity (Table 18.3, Figure 18.5), rates should be higher in the often turbid rivers of Australia.

Two features differentiated enzyme activity in billabongs from that in rivers: the higher rate of activity, and the greater spatial and temporal variability. Aminopeptidase activity could be up to three orders of magnitude greater in Ryans 2 Billabong than in the Kiewa River or Lake Mulwala (Figure 18.2). Activity varied little along the ca. 1,100-km distance of the River Murray between Wentworth and Corowa, but varied greatly between even adjacent billabongs. For instance, Ryans 1 and 2 Billabongs were less than 50 m apart, but rates of enzyme activity in the two sites often differed by an order of magnitude. It is not surprising that activity was so variable in different billabongs, given that they are small, shallow water bodies isolated in time and space. Presumably, each billabong evolves characteristic, and perhaps specific, biological populations, influenced strongly by stochastic events. It is tempting to predict that the major factor controlling enzyme activity in different billabongs is the relative size of their phytoplankton, and hence bacterial, populations. This proposal needs to be tested further, but the strong relationships between phytoplankton and bacterial populations, proteolytic and glycosidic activity, and the regeneration of nitrogen in the water column of Ryans 2 Billabong (Figure 18.4)

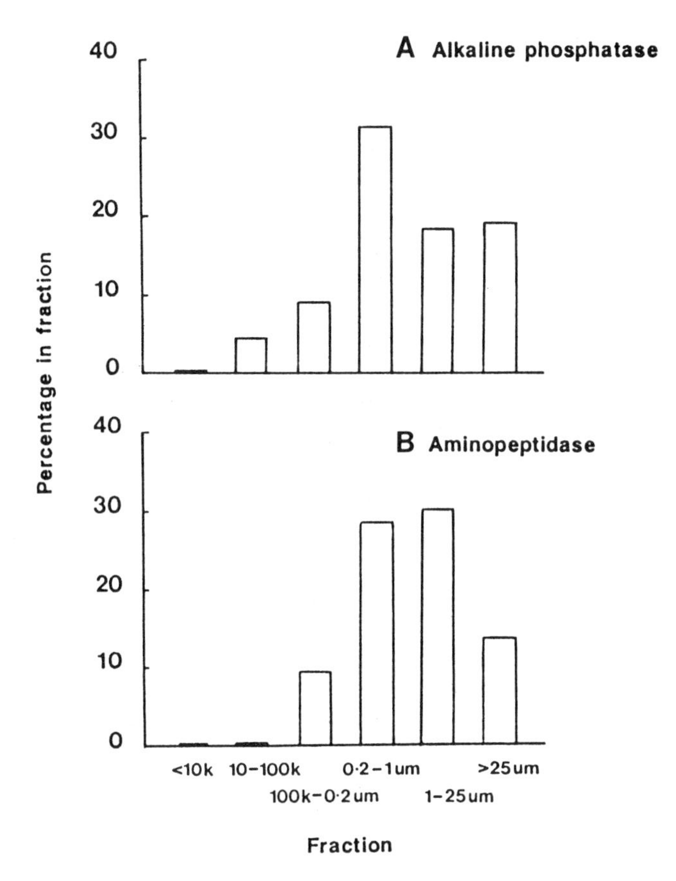

Figure 18.5 Size fractionation of enzyme activity in water samples from the River Murray at Corowa. The sizes of the six fractions are: <10,000 daltons; 10,000 to 100,000 daltons; 100,000 daltons to 0.2 μm; 0.2 to 1 μm; 1 to 25 μm; >25 μm.

gives it some credibility. These relationships can be explained in terms of the rapidly growing bacteria secreting proteolytic and glycosidic enzymes in response to the increased availability of organic substrates arising from the phytoplankton bloom. The aminopeptidases liberated free amino acids, which were eventually deaminated to yield ammonium. Bacteria, therefore, seem to have a major role in degrading organic matter and regenerating nitrogen in the water column of billabongs. Recent findings support this proposal, since 43–65% of bacteria isolated from Ryans 2 Billabong could secrete proteolytic enzymes, and 80–100% liberated ammonium from peptone (P.I. Boon, unpublished observations).

In contrast to the situation with aminopeptidase, there were few clear relationships between environmental conditions and lipase or alkaline phosphatase activity in the billabong. There was detectable lipase activity (1–12 μmol l^{-1} d^{-1}) on most occasions, and 18–80% of bacteria isolated from Ryans 2 Billabong secreted lipolytic enzymes (P.I. Boon, unpublished observations). However, lipase activity in the billabong was significantly correlated only with the load of suspended solids. The situation with phosphatase activity is more complex, because both bacteria and algae produce this enzyme (Chróst and Overbeck, 1987). The traditional view, based mainly on research with algae, is that alkaline phosphatase production is enhanced by phosphorus limitation and repressed by high concentrations of inorganic phosphorus in the external environment (Wetzel, 1983). However, more recent work by Chróst (1986) and Chróst and Overbeck (1987) suggested that the production of alkaline phosphatases by bacteria was not strongly influenced by external concentrations of orthophosphate (see also Chapter 3). Hence, a poor relationship between alkaline phosphatase activity and the concentration of inorganic phosphorus is to be expected if bacteria were the major producers of the enzyme. At least two previous papers have suggested that bacteria were responsible for much of the alkaline phosphatase activity in Northern-Hemisphere freshwaters (Stewart and Wetzel, 1982; Chróst and Overbeck, 1987). Moreover, from 30–45% of bacteria isolated from Ryans 2 Billabong could produce phosphatases (P.I. Boon, unpublished observations).

To summarize, this communication is one of the first to report on enzyme activity in Australian freshwaters. Billabongs, a characteristic aquatic environment of Australian floodplains, exhibited variable but often extremely high alkaline phosphatase and aminopeptidase activity. This was probably a function of the billabongs supporting very substantial algal and bacterial populations. Enzyme activity in rivers and billabongs was highly associated with the presence of suspended particles; preliminary results suggested the 0.2 to 25 μm fractions were the most important.

Acknowledgments This study could not have been undertaken without the excellent technical assistance of Alison Mitchell. Wendy Barron, Wendy Cowie, Kim Jenkins, and Don Omond are also thanked for their technical help. Andrew Boulton and Terry Hillman gave excellent advice on statistical design and data analysis. Grant Douglas kindly supplied the size-fractionated samples from the River Murray at Corowa.

References

APHA. 1976. *Standard Methods for the Examination of Water and Wastewater*, 14th edition. American Public Health Association, Washington DC. 1193 pp.

Billen, G. 1984. Heterotrophic utilization and regeneration of nitrogen. pp. 313–355 in Hobbie, J.E. and Williams, P.J.LeB. (editors), *Heterotrophic Activity in the Sea*. Plenum Press, New York.

Boon, P.I. 1989. Organic matter breakdown and nutrient regeneration in Australian freshwaters. I. Methods for exoenzyme assays in turbid aquatic environments. *Archiv für Hydrobiologie* 115: 339–359.

Boon, P.I. 1990. Organic matter breakdown and nutrient regeneration in Australian freshwaters. II. Spatial and temporal variation, and relation with environmental conditions. *Archiv für Hydrobiologie* 117: 405–436.

Chróst, R.J. 1986. Algal-bacterial metabolic coupling in the carbon and phosphorus cycle in lakes. pp. 360–366 in Megusar, F. and Gantar, M. (editors), *Perspectives in Microbial Ecology*. Slovene Society for Microbiology, Ljubljana.

Chróst, R.J. and J. Overbeck. 1987. Kinetics of alkaline phosphatase activity and phosphorus availability for phytoplankton and bacterioplankton in Lake Plußsee (north German eutrophic lake). *Microbial Ecology* 13: 229–248.

Chróst, R.J., Siuda, W. and Hałemejko, G.Z. 1984. Longterm studies on alkaline phosphatase activity (APA) in a lake with fish-aquaculture in relation to lake eutrophication and phosphorus cycle. *Archiv für Hydrobiologie, Supplement* 1: 1–32.

Chróst, R.J., Wciso, R. and Hałemejko, G.Z. 1986. Enzymatic decomposition of organic matter by bacteria in an eutrophic lake. *Archiv für Hydrobiologie* 107: 145–165.

Degobbis, D., Homme-Maslowska, E., Oroi, A.A., Donazzolo, R, and B. Pavoni. 1986. The role of alkaline phosphatase in the sediments of Venice Lagoon on nutrient regeneration. *Estuarine and Coastal Shelf Science* 22: 425–437.

Hart, B. T. and I.D. McKelvie. 1986. Chemical limnology in Australia. pp. 3–31 in De Deckker, P. and Williams, W.D. (editors), *Limnology in Australia*. Dr. W. Junk, Dordrecht.

Hellawell, J.M. 1978. *Biological Surveillance of Rivers*. Natural Environment Research Council (Water Research Centre), Stevenage. 332 pp.

Hillman, T.J. 1986. Billabongs. pp. 457–470 in De Deckker, P. and Williams, W.D. (editors), *Limnology in Australia*. Dr. W. Junk, Dordrecht.

Jones, J.G. 1972. Studies on freshwater microorganisms: phosphatase activity in lakes of differing degrees of eutrophication. *Journal of Ecology* 60: 777–791.

Lake, P.S., Barmuta, L.A., Boulton, A.J., Campbell, I.C. and R.M. St-Clair. 1985. Australian streams and Northern Hemisphere stream ecology: comparisons and problems. *Proceedings of the Ecological Society of Australia* 14: 61–82.

Mackay, N., Hillman, T.J. and J. Rolls. 1988. *Water Quality of the River Murray. Review of Monitoring 1978 to 1986*. Murray-Darling Basin Commission, Canberra. 62 pp.

Stewart, A.J. and R.G. Wetzel, R.G. 1982. Phytoplankton contribution to alkaline phosphatase activity. *Archiv für Hydrobiologie* 93: 265–271.

Walker, K.F. 1979. Regulated streams in Australia: the Murray-Darling River system. pp. 143–163 in Ward, J.V. and Stanford, J.A. (editors), *The Ecology of Regulated Rivers*. Plenum Press, New York.

Walker, K.F. and T.J. Hillman. 1977. *Limnological Survey of the River Murray in Relation to Albury-Wodonga, 1973–1976*. Albury-Wodonga Development Corporation and Gutteridge Haskins Davey, Albury-Wodonga. 256 pp.

Wetzel, R.G. 1983. *Limnology*. Saunders College Press, Philadelphia. 753 pp.

Wilkinson, L. 1988. *Systat: The System for Statistics*. Systat, Evanston. 785 pp.

Williams, W.D. 1988. Limnological imbalances: an antipodean viewpoint. *Freshwater Biology* 20: 407–420.

19

Hydrolytic Activities of Organisms and Biogenic Structures in Deep-Sea Sediments

Marion Köster, Preben Jensen, and Lutz-Arend Meyer-Reil

19.1 Introduction

Benthic activities such as feeding, burrow construction, and locomotion significantly affect the distribution and rate of organic matter decomposition (Aller, 1982; 1988; Aller and Aller, 1986; Aller and Yingst, 1985; Andersen and Kristensen, 1988; Kristensen, 1988; Kristensen and Blackburn, 1987); for instance, the capture of organic-rich particles by the protoplasma net of foraminiferans, the transport of freshly sedimented organic particles into deeper sediment layers by bioturbation, and the accumulation of organic matter in the digestive tract of organisms all cause organic matter degradation to be concentrated at special sites in the sediment.

Previous investigations of the impact of biogenic structures on the turnover of organic matter have concentrated on shallow-water coastal sediments. It has been shown that burrows and fecal pellets of macrofaunal organisms offer microenvironments for a broad spectrum of physiological groups of micro- and meiofauna (Meyers et al., 1987; Reichardt, 1988; Reise, 1987). Microbial decomposition rates are usually elevated in these microhabitats (Aller and Aller, 1986; Alongi, 1985; Reichardt, 1986). However, very little is known about the contribution of benthic organisms to organic matter decomposition in deep-sea sediments. Sediments of the Norwegian-Greenland Sea, located in the Jan Mayen Fracture Zone (water depth between 1,200 and 1,800 m) were colonized by mass abundances of epibenthic foraminiferans (belonging to the genera *Hyperammina* and *Reophax*), whereas sediments from the Vøring Plateau (water depth 1,240 m) were characterized by bioturbating macrofauna organisms, such as enteropneusts, echiurans, anthozoans, and sipunculans (Romero-Wetzel, 1987; 1989). These in- and epifaunal benthic organisms significantly affect organic matter decomposition rates in deep-sea sediments of the Norwegian-Greenland Sea.

The enzymatic hydrolysis of particulate organic matter is the initial and rate-limiting step of the organic carbon oxidation process in sediments. However, at the present time, the direct measurement of organic matter degradation under in situ conditions is limited by a number of methodological problems. Therefore radioactive or dye-labelled artificial substrates represent useful tools in the study of organic matter hydrolysis. Among the dye-labelled substrates, fluorescein diacetate (FDA) has been proved as a suitable model substrate, which is hydrolyzed rather unspecifically by esterases (see Chapter 5).

In this study, sediments from the Norwegian-Greenland Sea were analyzed for hydrolytic activity associated with biogenic structures (meio- and macrofauna organisms as well as their tubes and burrows) using fluorescein diacetate as a model substrate. In comparison with the hydrolytic activity measured in the bulk phase sediment, the importance of biogenic structures for organic matter decomposition in deep-sea sediments becomes obvious.

19.2 Methods

Sediment samples from the Norwegian-Greenland Sea were taken by a large Reineck box corer during two expeditions with R.S. "Meteor" in August/September 1988 (cruise M7/4-5) and in June/July 1989 (cruise M10/3). The positions of the stations are indicated and specified in Figure 19.1. Sediments were screened for dominating meio-and macrofauna organisms. Infaunal organisms were prevailing at the station on the Vøring Plateau (station 533), whereas stations located in the Jan Mayen Fracture Zone (stations 576, 625, 635, and 681) were generally characterized by a rich fauna of epibenthic foraminiferans (*Hyperammina* and *Reophax* spp.). Beside the organisms, sediment was collected from the burrows of the organisms as well as the surrounding sediment. At the time of sampling, the temperature of the sediment surface ranged between −1°C and +1°C.

In order to analyze the hydrolytic activity associated with biogenic structures in deep-sea sediments, individuals of enteropneusts (*Stereobalanus canadensis*), echiurans (not identified), and anthozoans (*Cerianthus vogtii*) were dissected. In the case of echiurans and anthozoans, only one individual tube dwelling organism was available for enzymatic analyses. All the organisms were homogenized with a teflon pestle and diluted in filter-sterilized bottom water. Organisms, that were too small to be dissected (polychaetes, ophiuroids, sipunculans, and foraminiferans) were treated as a whole. The selected dilution was dependent on the amount of homogenized material. Sediment cores were dissected at 2.5-mm intervals and diluted 1:5.

The enzymatic assay was started by adding fluorescein diacetate (Serva) at

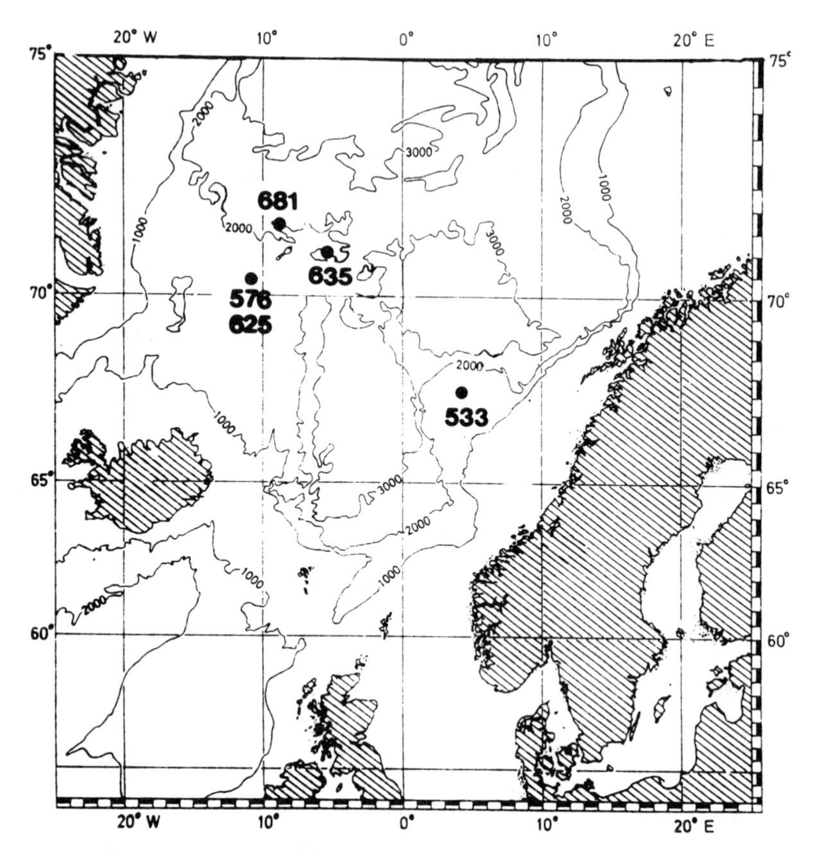

Figure 19.1 Sampling sites in the Norwegian-Greenland Sea. Station 533 is located on the Vøring Plateau; the stations 576, 625, 635, and 681 are located in the Jan Mayen Fracture Zone. Station number, sampling date, position, and water depth are as follows: no. 533 (30 August 1988) 67°44.0′ N, 05°55.6′ E: 1,240 m; no. 576 (18 September 1988) 70°20.1′ N, 10°37.8′ W: 1,745 m; no. 625 (18 June 1988) 70°20.0′ N, 10°37.5′ W: 1,710 m; no. 635 (21 June 1989) 70°57.4′ N, 05°32.7′ W: 1,751 m; no. 681 (6 July 1989) 71°37.8′ N, 08°41.0′ W: 1,168 m.

saturation level (final concentration about 100 μM) to the homogenized and diluted samples. Enzymatic assays were run in time-course experiments (usually three incubation periods) under close to in situ temperature (between 1–2°C). After the appropriate incubation, samples were centrifuged at 6,000 rpm (0°C, 10 min). The supernatant was analyzed for the release of fluorescein in a spectrofluorometer (Kontron SFM 25; excitation 470 nm, emission 510 nm). Enzymatic hydrolysis rates were extrapolated from the slope of the activity curve calculated by linear regression and standardized against a solution of fluorescein (Serva).

For sediments from burrow walls and fecal pellets of selected enteropneusts and echiurans, total bacterial numbers were determined in diluted and sonicated samples by epifluorescence microscopy as described by Meyer-Reil (1983).

19.3 Results and Discussion

Using artificial-dye-labelled substrates, the potential enzymatic activity of sediments can be estimated. The extrapolation to natural in situ enzymatic activity is difficult due to our limited knowledge of the concentrations and spectrum of naturally occurring substrates. However, the enzymatic activity rates measured reflect the pool of natural enzymes (esterases), which, in turn, is the response of previous variations in the composition and spectrum of naturally occurring substrates (Meyer-Reil, 1987). In the present study, the substrate fluorescein diacetate was used to measure the hydrolytic activity associated with organisms and their biogenic structures, as well as with bulk- phase sediment in the Norwegian-Greenland Sea.

Enzymatic activity associated with infauna Enzymatic activity associated with selected benthic macrofauna organisms (enteropneusts, echiurans, anthozoans, and other small-sized species) was analyzed in sediments from the Vøring Plateau. Data for hydrolytic activity of the fauna are presented in Tables 19.1 to 19.3. Figure 19.2 gives an overall picture of the hydrolytic activity associated with the various faunal groups.

Different segments of enteropneusts (Table 19.1) and the echiuran (proboscis, collar, gut; Table 19.2) revealed varying levels of enzymatic activity, which may reflect their physiological functions. High hydrolytic activity was measured in the gut (including tissue and sediment), due to the presence of various digestive enzymes. However, sediment sampled from the gut revealed only slightly enhanced enzymatic activity (Table 19.1, Figure 19.2). From these observations it may be concluded that digestive enzymes are preferentially active in connection with the gut tissue. The hydrolytic activity associated with the tentacles of anthozoans (Table 19.2) was even one order of magnitude higher than the enzymatic activity in the digestive tract of enteropneusts and the echiuran (Figure 19.2). The work of Tiffon (1975) suggests, that in some anthozoans, the tentacles are responsible for extracorporeal digestion, that is dependent on the presence of hydrolases in the ectoderm. The relatively high enzymatic activity measured in the body tissue of the anthozoan can be explained by the presence of rather active cells (nematocysts), which occur in all parts of the body of Ceriantharia and are continually discharged until several layers of nematocysts form the dwelling tube of the anthozoan (Hartog, 1977).

Some segments of the enteropneust, such as the proboscis and collar, collect and transport food particles by ciliary currents to the mouth. These segments revealed enzymatic activity one order of magnitude lower as compared to organs with digestive function (Table 19.1, Figure 19.2). In the mucus secreted by the animal, the potential enzymatic activity was higher than the ac-

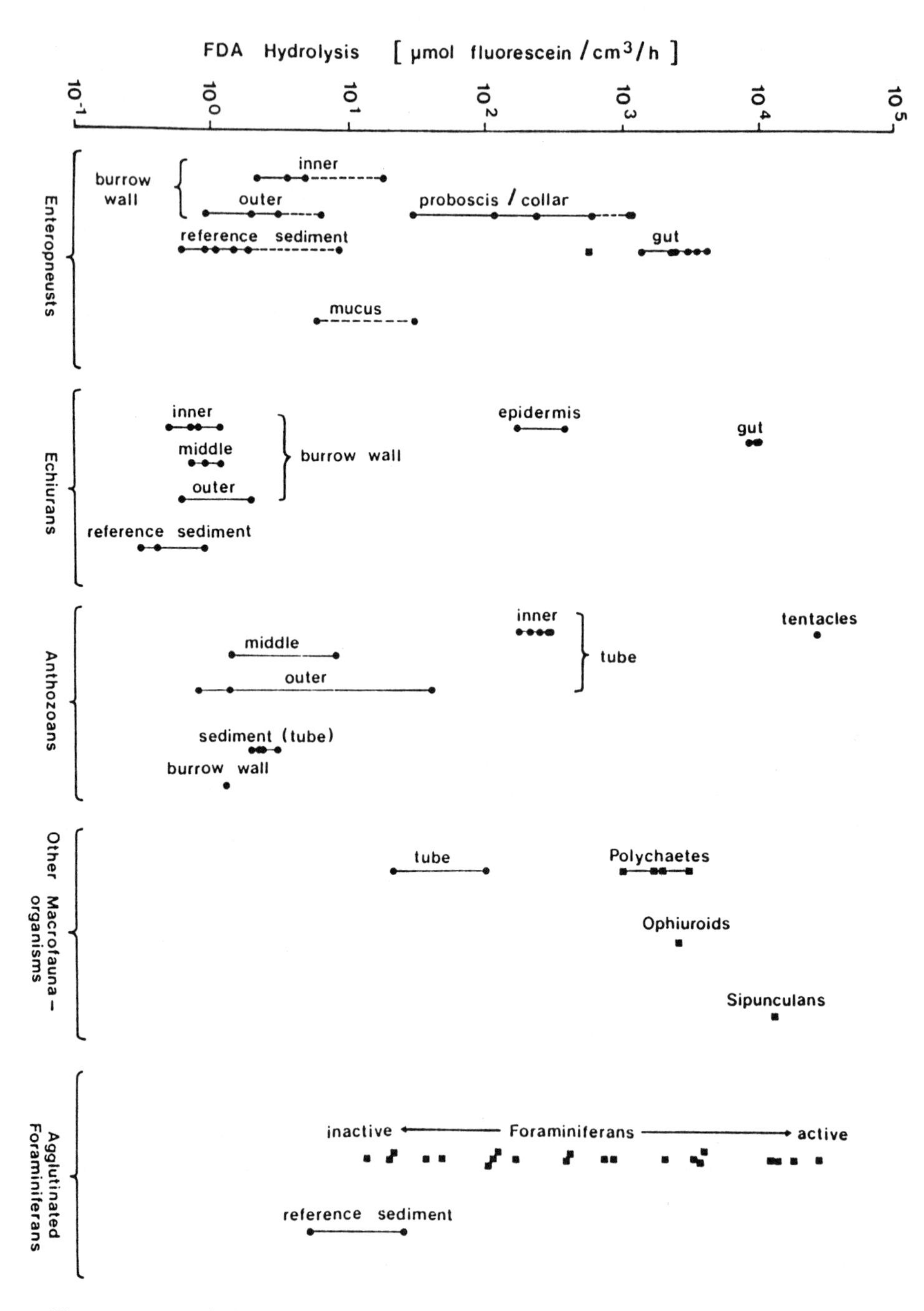

Figure 19.2 Hydrolytic activity associated with meio- and macrofauna organisms. Activity rates (fluoresceindiacetate hydrolysis) are expressed as release of fluorescein per cm^3 of material per hour. Circles indicate hydrolytic activity of segments of macrofauna organisms, burrows, and surrounding sediment, respectively. Squares indicate hydrolytic activity of individual organisms. Dashed lines indicate damaged individuals.

Table 19.1 Hydrolytic activity associated with enteropneusts, their burrow walls, and surrounding sediment

	Activity rate[1]		
Source	Range or value	Mean	Number of measurements
Proboscis	30–1,200 (1,300)	487	3 (1)
Collar	120–1,200 (1,100)	647	3 (1)
Gut:			
For-	1,400–2,900 (4,100)	2,167	3 (1)
Hind-	2,400–3,500	2,933	3
Sediment (gut)	27	27	1
Mucus	5.9 (30)	5.9	1 (1)
Total for organism	— (570)	—	— (1)
Fecal pellets:	1.9–2.7	2.3	2
Reference sediment:			
Brownish	0.9	0.9	1
Grayish	0.4	0.4	1
Burrow wall:			
Inner	2.2–5.0 (8.2–18)	3.6	3 (2)
Outer	0.9–3.0 (6.5)	2.0	3 (1)
Reference sediment	0.6–1.5 (1.9–8.8)	1.1	3 (2)

[1]Activity rates are expressed as μmoles of fluorescein released per cm^3 of homogenized material per hour (for details see text). Values in parentheses indicate values for damaged individuals.

tivity of an equivalent volume of surrounding sediment (Figure 19.2), probably due to the presence of digestive enzymes in the mucus. These enzymes may be responsible for the enhanced hydrolytic activity associated with the proboscis and collar, since both organs were covered with mucus. The levels of enzymatic activity, comparable for proboscis and collar, confirmed this hypothesis. There was a slight decrease of enzymatic activity from the inner to the outer burrow walls of enteropneusts. Compared to the surrounding sediment, hydrolytic activity associated with burrow linings was enhanced by a factor of 2 to 5 (Table 19.1, Figure 19.2). In burrow walls of echiurans, gradients of enzymatic activity were less pronounced (Table 19.2, Figure 19.2). Gradients of enzymatic activity measured between inner and outer burrow walls disappeared, if individuals were damaged by sampling (Figure 19.2—dashed line). This may be the result of the release of enzymes, thus leading to an increase of hydrolytic activity in the surrounding sediment. It is not clear where the enhanced levels of hydrolytic activity associated with the burrow walls of enteropneusts originate from, but it may be due to the metabolism of the organisms. Since bacterial numbers at the burrow linings of enteropneusts and echiurans were enhanced by a factor of about 2 over the reference sediment (Figure 19.3), bacteria may have contributed to the elevated levels of enzymatic activity in the burrow walls.

Fecal pellets of enteropneusts were found in the burrows at a sediment

Table 19.2 Hydrolytic activity associated with echiurans and anthozoans, their burrow walls, and surrounding sediment.

	Activity rate[1]		
Source	Range or value	Mean	Number of measurements
Echiurans			
Proboscis	170		1
Epidermis	3.7		1
Gut:			
For-	8,100		1
Mid-	9,500		1
Hind-	9,200		1
Epidermis	380		1
Burrow wall:			
Inner	0.5–1.2	0.8	4
Middle	0.7–1.2	0.9	3
Outer	0.6–2.0	1.3	2
Reference sediment	0.3–0.9	0.5	3
Anthozoans:			
Body tissue	1,200		1
Tentacles	25,000		1
Tube:			
Inner	180–290	244	5
Middle	1.4–7.9	4.7	2
Outer (mucus)	0.8–14	7.4	2
Sediment (tube)	2.0–2.3	2.2	3
Burrow wall (inner)	1.3		1

[1]Activity rates are expressed as μmoles of fluorescein released per cm^3 of homogenized material per hour (for details see text).

depth of 8 to 10 cm. They revealed twofold higher enzymatic activity and threefold higher bacterial density as compared to the surrounding sediment (Table 19.1, Figure 19.3). Obviously, inner burrow walls as well as fecal pellets represent microenvironments located in deeper sediment regions, where bacteria-mediated decomposing processes are favored.

Differences in hydrolytic activity linked with different regions of the anthozoan tube were recorded. The inner surface of the tube, consisting of a thin, grayish "epithel-like" layer, revealed significantly higher hydrolytic activity than the outer surface of the tube (Table 19.2, Figure 19.2), which consisted of a slimy greenish material. Since the tube of one *Cerianthus vogtii* specimen may reach a length of up to 8 m, and penetrates downwards to a depth of at least 40 cm, extended regions of enhanced enzymatic activity could be created by these organisms.

Since it was difficult to dissect smaller macrofauna organisms, individuals of polychaetes, ophiuroids, and sipunculans were homogenized as total organisms and analyzed for hydrolytic activity (Table 19.3). When interpreting

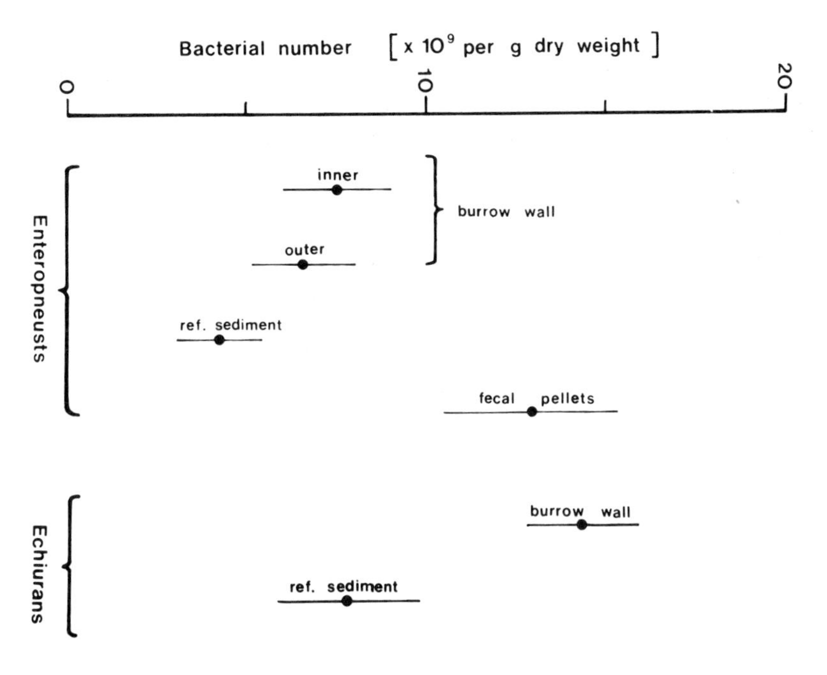

Figure 19.3 Bacterial numbers (x 10^9 per g of dry weight of sediment) associated with fecal pellets and burrow walls of enteropneusts and echiurans in comparison to the reference sediment. Error bars delineate the standard error of the mean.

Table 19.3 Hydrolytic activity associated with individual benthic organisms

Source	Volume (mm^3)	Activity rate[1] In mmol cm^{-3} h^{-1}	In μmol $ind.^{-1}$ h^{-1}
Polychaetes (free living)	212	1.9	410
	141	2.9	410
	8	1.6	13
Polychaetes (sessile)			
Body:			
For-	110	0.9	100
Hind-	110	1.0	110
Tube:			
For-	31	0.02	0.6
Hind-	31	0.1	3
Ophiuroids	14	1.2	17
Sipunculans	16	12.5	200

[1]Activity rates are expressed per cm^3 of homogenized material of the organisms (left column) and per one organism (right column), respectively.

the hydrolytic activity rates measured, one has to consider that the data reflect the activity of the total pool of enzymes present. The enzymatic activity measured for individual macrofauna organisms was similar to the hydrolytic activity associated with the digestive tract of enteropneusts and echiurans (Tables 19.1 and 19.2, Figure 19.2). However, analyzing enteropneusts as total individuals, revealed activity rates approximately one order of magnitude lower (Figure 19.2).

These suggest that the level of hydrolytic activity corresponds to the size of the organisms: the smaller the organism, the higher the enzymatic activity (e.g., sipunculans, the smallest macrofauna organisms investigated, had the highest hydrolysis rate; cf. Table 19.3, Figure 19.2). These results fit into the general picture of the dependence of metabolic activity on the size of organisms (Gerlach et al., 1985, and literature cited therein).

Enzymatic activity associated with epifauna Whereas sediments from the Vøring Plateau were dominated by infauna, sediments sampled around the Island of Jan Mayen were characterized by a rich epibenthic fauna, which consisted almost exclusively of agglutinated foraminiferans (*Hyperammina* and *Reophax* spp.). These organisms accumulate organic particles in their protoplasma nets, which extend above the sediment surface. Thus, it can be expected that the major part of the sedimented organic matter undergoes decomposition prior to reaching the sediment surface. As a result of their mode of feeding, epibenthic foraminiferans may prevent the flux of sedimented organic matter to deeper sediment layers. This lack of degradable organic matter in deeper sediment layers may be the reason for the poor macrofauna colonization observed in sediments densely colonized by epibenthic foraminiferans.

The existence of mass abundances of epibenthic foraminiferans at stations around the Island of Jan Mayen coincided with a steep gradient of hydrolytic activity rates, which became especially obvious from fine-scale investigations. Enzymatic hydrolysis was enhanced by about two orders of magnitude in the 0- to 0.5 cm horizon over horizons only a few millimeters deeper (Figure 19.4). As compared to surface sediments of the Vøring Plateau, where epifauna were less obvious, the hydrolytic activity of surface sediments colonized with foraminiferans was about 100 to 300 times higher (Figure 19.4). Analyses of hydrolytic activity of selected individual foraminiferans confirmed the hypothesis that epifauna were responsible for the magnitude and steep gradient of hydrolytic activity. Below the uppermost sediment layer, living agglutinated foraminiferans disappeared. Correspondingly, enzymatic activity rates decreased and reached levels comparable to levels of enzymatic activity measured in sediments of the Vøring Plateau (Figure 19.4).

Since the fluorometric assay is highly sensitive, even individual foraminiferans only a few millimeters in length can be analyzed for hydrolytic activity. Therefore enzymatic assays may provide an alternative approach to differentiate relatively quickly between active and inactive (dead?) organisms. Enzymatic assays may replace the traditional, more time consuming differentiation

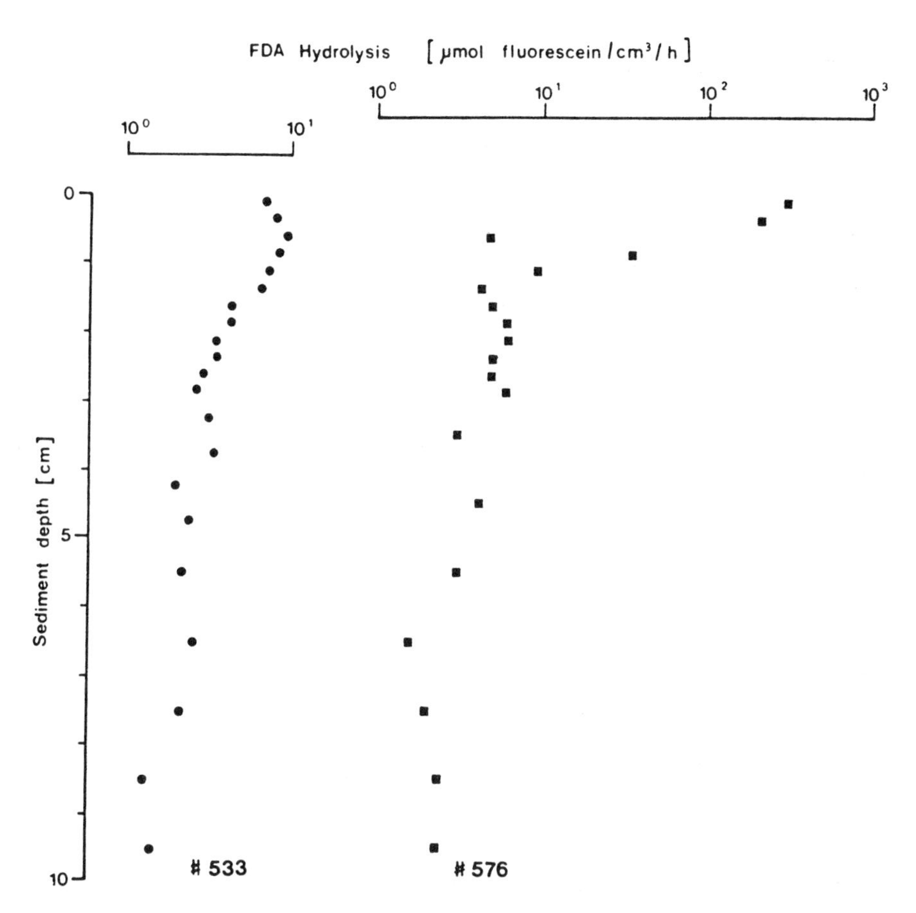

Figure 19.4 Fine-scale distribution of hydrolytic activity in sediments from the Vøring Plateau (station no. 533) inhabited by macrofauna and from sediments around the Island of Jan Mayen (station no. 576) colonized by epibenthic foraminiferans.

of the physiological state of foraminiferans by staining with Rose Bengal (Lutze, 1964; Bernhard, 1988).

Analyses of individual foraminiferans demonstrated large variations of enzymatic activity (Table 19.4, Figure 19.2), perhaps dependent on the physiological state of the organisms. Most of the foraminiferans had very low enzymatic activity rates, which were in the same range as the hydrolytic activity of subsurface sediments (without agglutinated foraminiferans). It can be concluded that these organisms were inactive with regard to organic matter decomposition. An additional indication may be derived from ATP measurements, which showed no enrichment of biomass at the sediment surface (data not shown). As described by Linke (1989), benthic foraminiferans are capable of decomposing their own protoplasma during starvation. Following organic matter supply an "awakening" reaction occurs, stimulating cell metabolism. In

Table 19.4 Hydrolytic activity associated with individual agglutinated foraminiferans

Sample number	Volume (mm^3)	Activity rate[1] In mmol cm^{-3} h^{-1}	Activity rate[1] In μmol $ind.^{-1}$ h^{-1}
1	57	0.4	23
2	57	0.1	6.3
3	64	0.02	1.3
4	49	0.03	1.7
5	49	0.02	0.9
6	127	11.0	1,400
7	57	0.1	7.5
8	31	3.2	99
9	31	0.4	12
10	41	0.05	1.9
11	41	0.9	35
12	35	2.0	69
13	38	12.6	480
14	31	0.1	3.7
15	7	0.2	1.1
16	106	0.1	10
17	31	26.5	820
18	57	3.5	200
19	41	16.3	670
20	47	3.6	170
21	106	0.7	76

[1]Activity rates are calculated per cm^3 of homogenized material of the foraminiferans (left column) and per one individual foraminiferan (right column), respectively.

addition to the dead and inactive individuals, a few organisms with extremely high enzymatic activity were found (Table 19.4).

The data indicate that the hydrolytic activity associated with an individual active foraminiferan already accounted for the total enzymatic activity recorded in one cm^3 of sediment (consisting of foraminiferans embedded in sediment) from the uppermost horizon. In comparison to deeper sediment horizons, one individual foraminiferan is about two to three orders of magnitude more active (Table 19.4, Figure 19.4), which demonstrates that agglutinated foraminiferans dominate organic matter decomposition at the sediment surface.

Conclusions Our results demonstrate that enzymatic decomposition processes in deep-sea sediments of the Norwegian-Greenland Sea are closely associated with biogenic structures. Since these extend down to a sediment depth at least of approximately 40 cm (in the case of anthozoan tubes), enhanced levels of enzymatic activity can be expected even in deeper sediment horizons. In comparison to enzymatic activity measured in the bulk-phase sediment, the importance of biogenic structures for the transformation of organic matter in these sediments becomes especially obvious. Furthermore, biogenic structures in sediments are apparently more wide-spread than was previously thought

(Romero-Wetzel, 1989). In this regard, offshore nutrient-limited sediments differ from coastal nutrient-rich sediments, where the main organic matter decomposition occurs in the bulk phase sediment (Faubel and Meyer-Reil, 1983). The dependence of hydrolytic activity on the size of organisms fits the general observation that smaller organisms are characterized by higher metabolic activity.

The close association of hydrolytic activity with in- and epifauna, respectively, and their differing modes of organic matter hydrolysis reflect different strategies for growth and survival under substrate limitation. Infauna organisms in sediments of the Vøring Plateau capture, concentrate, and degrade organic matter in their burrows. Epifauna organisms (e.g., agglutinated foraminiferans) dominating in sediments from stations around the Island of Jan Mayen capture and digest particles using their plasma nets, which extend above the sediment surface. These two modes of organic matter decomposition may play different roles in the early digenesis of organic material and its sedimentary record.

Acknowledgments We thank A. Altenbach, P. Linke, and A. Thies for sharing their samples with us. This work was supported by the Deutsche Forschungsgemeinschaft (Report No. 87 of the Sonderforschungsbereich 313 at Kiel University).

References

Aller, R.C. 1982. The effects of macrobenthos on chemical properties of marine sediment and overlying water. pp. 53–102 in McCall, P.L. and Tevesz, M.J.S. (editors), *Animal-Sediment Relations*. Plenum Publishing Co., New York.

Aller, R.C. 1988. Benthic fauna and biogeochemical processes in marine sediments. I. The role of burrow structures. pp. 301–338 in Blackburn, T.H. and Sørensen, J. (editors), *Nitrogen Cycling in Coastal Marine Environments*. John Wiley and Sons, Chichester.

Aller, J.Y. and R.C. Aller. 1986. Evidence for localized enhancement of biological activity associated with tube and burrow structures in deep-sea sediments at the HEBBLE site, western North Atlantic. *Deep-Sea Research* 33: 755–790.

Aller, R.C. and Y. Yingst 1985. Effects of the marine deposit-feeders *Heteromastus filiformis* (Polychaeta), *Macoma balthica* (Bivalvia), and *Tellina texana* (Bivalvia) on averaged sedimentary solute transport, reaction rates, and microbial distributions. *Journal of Marine Research* 43: 615–645.

Alongi, D.M. 1985. Microbes, meiofauna, and bacterial productivity on tubes constructed by the polychaete *Capitella capitata*. *Marine Ecology Progress Series* 23: 207–208.

Andersen, F.O. and E. Kristensen. 1988. The influence of macrofauna on estuarine benthic community metabolism—A microcosm study. *Marine Biology* 99: 591–603.

Bernhard, J.M. 1988. Postmortem vital staining in benthic foraminifera: Duration and importance in population studies. *Journal of Foraminifera Research* 18: 143–146.

Faubel, A. and L.-A. Meyer-Reil. 1983. Measurement of enzymatic activity of meiobenthic organisms: methodology and ecological application. *Cahiers de Biologie marine* 24: 35–49.

Gerlach, S.A., Hahn, A.E. and M. Schrage. 1985. Size spectra of benthic biomass and metabolism. *Marine Ecology Progress Series* 26: 161–173.

den Hartog, den, J.C. 1977. Descriptions of two new Ceriantharia from the Caribbean region, *Pachycerianthus curacaoensis* n.sp. and *Arachnanthus nocturnus* n.sp. with a discussion of the cnidom and of the classification of the Ceriantharia. *Zoologische Mededelingen* 51: 211–248.

Kristensen, E. 1988. Benthic fauna and biogeochemical processes in marine sediments: microbial activities and fluxes. pp. 275–299 in Blackburn, T.H. and Sørensen, J. (editors), *Nitrogen Cycling in Coastal Marine Environments*, John Wiley and Sons, Chichester.

Kristensen, E. and T.H. Blackburn. 1987. The fate of organic carbon and nitrogen in experimental marine sediment systems: influence of bioturbation and anoxia. *Journal of Marine Research* 45: 231–257.

Linke, P. 1989. Lebendbeobachtungen und Untersuchungen des Energiestoffwechels benthischer Foraminiferen aus dem Europäischen Nordmeer. Report No. 18 of the Sonderforschungsbereich 313, Kiel University.

Lutze, G.F. 1964. Zum Färben rezenter Foraminiferen. *Meyniana* 14: 43–47.

Meyer-Reil, L.-A. 1983. Benthic response to sedimentation events during autumn to spring at a shallow-water station in the Western Kiel Bight. II. Analysis of benthic bacterial populations. *Marine Biology* 77: 247–256.

Meyer-Reil, L.-A. 1987. Seasonal and spatial distribution of extracellular enzymatic activities and microbial incorporation of dissolved organic substrates in marine sediments. *Applied and Environmental Microbiology* 53: 1748–1755.

Meyers, M.B., Fossing, H. and E.N. Powell. 1987. Microdistribution of interstitial meiofauna, oxygen and sulfide gradients, and the tubes of macro-infauna. *Marine Ecology Progress Series* 35: 223–241.

Reichardt, W. 1986. Polychaete tube walls as zonated microhabitats for marine bacteria. *Ifremer Actes de Colloques* 3: 415–425.

Reichardt, W. 1988. Impact of bioturbation by *Arenicola marina* on microbiological parameters in intertidal sediments. *Marine Ecology Progress Series* 44: 149–158.

Reise, K. 1987. Spatial niches and long-term performance in meiobenthic *Plathelminthes* of an intertidal lugworm flat. *Marine Ecology Progress Series* 6: 329–333.

Romero-Wetzel, M.B. 1987. Sipunculans as inhabitants of very deep, narrow burrows in deep-sea sediments. *Marine Biology* 96: 87–91.

Romero-Wetzel, M.B. 1989. Struktur und Bioturbation des Makrobenthos auf dem Vøring-Plateau (Norwegische See). Report No. 13 of the Sonderforschungsbereich 313, Kiel University.

Tiffon, Y. 1975. Hydrolases in the ectoderm of *Cerianthus lloydii* Gosse, *Cerianthus membranaceus* Spallanzani and *Metridium senile* (L.): Demonstration of extracellular and extracorporeal digestion. *Journal of Experimental Marine Biology and Ecology* 18: 243–254.

Index